任务型语码转换式双语教学系列教材

总主编 刘玉彬 副总主编 杜元虎 总主审 段晓东

环境与生物技术

ENVIRONMENT AND BIOTECHNOLOGY

主　编 吕国忠

副主编 乌云娜 崔玉波 金 华

大连理工大学出版社

图书在版编目(CIP)数据

环境与生物技术 / 吕国忠主编. 一 大连 : 大连理工大学出版社，2014.9(2016.9 重印)
任务型语码转换式双语教学系列教材
ISBN 978-7-5611-9444-7

Ⅰ.①环… Ⅱ.①吕… Ⅲ.①环境工程一双语教学一高等学校一教材一英、汉②生物工程一双语教学一高等学校一教材一英、汉 Ⅳ.①X5②Q81

中国版本图书馆 CIP 数据核字(2014)第 180249 号

大连理工大学出版社出版
地址:大连市软件园路 80 号　邮政编码:116023
发行:0411-84708842　邮购:0411-84708943　传真:0411-84701466
E-mail:dutp@dutp.cn　URL:http://www.dutp.cn
大连理工印刷有限公司印刷　　大连理工大学出版社发行

幅面尺寸:183mm×233mm　印张:17.25　字数:563 千字
2014 年 9 月第 1 版　2016 年 9 月第 2 次印刷

责任编辑:邵　婉　责任校对:齐　跃
封面设计:波　朗

ISBN 978-7-5611-9444-7　定　价:30.00 元

总序

PREFACE

2014年的初夏，我们为广大师生奉上这套“任务型语码转换式双语教学系列教材”。

“任务型语码转换式双语教学”是双语教学内涵建设的成果，主要由两大模块构成：课上，以不影响学科授课进度为前提，根据学生实际、专业特点、学年变化及社会需求等，适时适量地渗透英语专业语汇、语句、语段或语篇，“润物细无声”般地扩大学生专业语汇量，提高学生专业英语能力；课外，可向学生提供多种选择的“用中学”平台，如英语科技文献翻译、英语实验报告、英语学术论文、英语小论文、英语课程设计报告、模拟国际研讨会、英语辩论、工作室英语讨论会等，使学生的专业英语实践及应用达到一定频度和数量，激活英语与学科知识的相互渗透，培养学生用英语学习、科研、工作的能力及适应教育国际化和经济一体化的能力。

为保证“任务型语码转换式双语教学”有计划、系统、高效、科学地持续运行，减少教学的随意性和盲目性，方便师生的教与学，我们编写了这套“任务型语码转换式双语教学系列教材”。

本套教材的全部内容均采用汉英双语编写。

教材按专业组册，涵盖所有主干专业课和专业基础课，力求较为全面地反映各学科领域的知识体系。

分册教材编写以中文版课程教材为单位，即一门课为分册教材的一章，每章内容以中文版教材章节为序，每门课以一本中文教材为蓝本，兼顾其他同类教材内容，蓝本教材绝大部分是面向21世纪的国家规划教材。

教材的词汇短语部分，注意体现学科发展的新词、新语，同时考虑课程需求及专业特点，在不同程度上灵活渗透了各章节的重要概念、定义，概述了体现章节内容主旨的语句及语段。分册教材还编写了体现各自专业特点的渗透内容，如例题及解题方法，课程的发生、发展及前沿简介，图示，实验原理，合同文本，案例分析，法条，计算机操作错误提示等。

部分教材补充了中文教材未能体现的先进理论、先进工艺、先进材料或先进方法的核心内容，弥补了某些中文教材内容相对滞后的不足；部分教材概述了各自专业常用研究方法、最新研究成果及学术发展的趋势动态；部分教材还选择性地把编者的部分科研成果转化为教材内容，以期启发学生的创新思维，开阔学生的视野，丰富学生的知识结构，从教材角度支持学生参与科研活动。

本套教材大多数分册都编写了对“用中学”任务实施具有指导性的内容，应用性内容的设计及编写比例因专业而异。与专业紧密结合的应用性内容包括英语写作介绍，如英语实验报告写作，英语论文写作，英语论文摘要写作，英语产品、作品或项目的概要介绍写作等。应用性内容的编写旨在降低学生参与各种实践应用活动的难度，提高学生参与“用中学”活动的可实现性，帮助学生提高完成“用中学”任务的质量水平。

考虑学生英语写作和汉译英的方便，多数分册教材都编写了词汇与短语索引。

“任务型语码转换式双语教学系列教材”尚属尝试性首创，是多人辛勤耐心劳作的结果。尽管在编写过程中，我们一边使用一边修改，力求教材的实用性、知识性、先进性融为一体，希望教材能对学生专业语汇积累及专业资料阅读、英语写作、英汉互译能力的提高发挥作用；尽管编者在教材编写的同时也都在实践“任务型语码转换式双语教学”，但由于我们缺乏经验，学识水平和占有资料有限，加上为使学生尽早使用教材，编写时间仓促，在教材内容编写、译文处理、分类体系等方面存在缺点、疏忽和失误，恳请各方专家和广大师生对本套教材提出批评和建议，以期再版时更加完善。

在教材的编写过程中，大量中外出版物中的内容给了我们重要启示和权威性的参考帮助，在此，我们谨向有关资料的编著者致以诚挚的谢意！

编　者

2014 年 5 月

前言
FOREWORD

本书是为了面向环境工程、环境科学及生物技术三个本科专业的学生而编写的双语教材。编写的目的是帮助学生们掌握更多的专业英文词汇，提高专业英文文献阅读能力，提升专业外语水平，同时，也有助于提高学生们的就业竞争能力和考研成功率。

为了便于学生们掌握较多的专业英文词汇，本书在相关课程的各章节中都集中列出相关专业英文词汇，并有一定数量的例句来解释概念或叙述相关原理或过程，非常容易被学生们掌握。我们相信，一旦同学们拿起这本书，利用两周或三周的时间就会掌握本专业涉及的主要专业英文词汇。读完这本书，同学们就会感受到“学好英文并不难”“提高专业英文文献阅读能力也不是一件很困难的事情”。

随着我国高校国际化办学进程的加快，我院正在积极推进该项目工作。显然，没有好的英文功底作基础，很难培养出具有国际化视野的高水平人才。另外，随着全球经济一体化进程的加快，外文(尤其英文)和专业外语作为语言工具的重要性越来越明显。在不远的将来，我们会有更多的学生赴国外留学进修，我们也会迎来更多的外国留学生与同学们同堂上课或开展实验，我们将会鼓励或培养更多的老师用英文授课，今天西方国家高校的高度国际化办学模式就是我们明天办学的方向。在我国，高校国际化办学势在必行！学好英语势在必行！

很高兴同学们选择了环境工程、环境科学、生物技术专业，这些专业不仅是国外热门专业，也是我国大力发展的朝阳专业，具有广阔的应用前景。目前，没有哪一位同学或公民不谈论环境问题和生物技术问题。随着全球经济的快速发展，环境问题会越来越突出，生物技术会变得越来越重要。那么有关环境和生物技术的专业发展也会出现更多新的专业词汇和理论，我们将不遗余力地坚持及时更新本教材，争取让更多的学生受益。

本书共涉及23门课程的双语内容，涵盖了环境工程、环境科学、生物技术三个专业的主要课程和教学内容。参加本书编写任务的教师（按姓氏笔画排序）有：王冰、王芳、乌云娜、仉春华、冉春秋、孙红杰、孙晓东、李贺、李政海、吕国忠、苏丹、张风杰、邹吉祥、金华、欧晓霞、郭鹏、崔玉波、鲍雅静、董玉瑛。本书凝聚了这些教师的心血。尽管每位参编教师反复修改，但由于时间和编写水平所限，书中一定会存在不足之处，肯请读者批评指正。另外，特别感谢李晓梅教授提出宝贵修改意见。

编 者

2014年7月

目录

CONTENTS

目录

CONTENTS

第一章　有机化学
Chapter 1　Organic Chemistry

第一节　绪论
Section 1　Introduction

碳氢化合物　hydrocarbon
电子　electron
结构　structure
原子　atom
分子　molecule
化合物　compound
有机化合物　organic compound
物质　substance
化学键　chemical bond
组成　composition
有机反应　organic reaction
化学性质　chemical property
物理性质　physical property
分子式　molecular formula
原子核　nucleus
质子　proton
中子　neutron
电荷　charge
正电荷　positive charge
负电荷　negative charge
原子序数　atomic number
原子轨道　atomic orbital
壳层　shell
亚层,亚壳层　subshell
轨道形状　shape of orbitals
垂直的　perpendicular
原子半径　atomic radius
半径(复数)　radius
两倍(的)　double
轨道方向　orientation of orbitals (direction of orbitals)
占据　occupy
(S轨道被两个电子占据)　(S orbital is occupied by two electrons.)
电子构型　electron configuration
最外层电子　outermost electron
化合价　valence
价键　valence bond
价电子　valence electron
价壳层　valence electron
价键理论　valence bond theory
键长　bond length
键能　bond energy
键角　bond angles
电负性　electron negativity
离子键　ionic bond
共价键　covalent bond
共价的　covalent
分子结构　molecular structure
相互的　mutual
相互吸引　mutual attraction
子结构式　molecular structures formula
阳离子,正离子　cation
阴离子,负离子　anion
成键电子　bonding electron
孤对电子　one-pair electron
电子对(两个电子)　electron pair
三倍(的)　triple
多个　multiple
极性的　polar
非极性的　nonpolar
诱导效应　inductive effect
杂化　hybrid
共振,共轭　resonance
共振键　resonance bond
定域的　localized
离域的　delocalized
八隅体规则　octet rule
共振结构　resonance structure
几何形状　geometry
杂化轨道　hybrid orbital
轨道杂化　orbital hybridization
共同　in common
短线式　dash formula
缩简式　condensed formula
键线式　bond-line formula

底物 substract
试剂 reagent
产物 product
反应物 reactant
Lewis 酸 Lewis acid
Lewis 碱 Lewis base
官能团,功能团 functional group
碳 carbon
氮 nitrogen
氧 oxygen
氢 hydrogen
硫 sulfur
对……熟悉 be familiar with
由……构成 be made of
被分成 be divided into
聚焦 focus on
杂化轨道理论 hybrid orbital theory
分子轨道理论 molecular orbital theory
均裂 homolytic bond cleavage
异裂 heterolytic bond cleavage
碳正离子 carbocation
碳负离子 carbanion
自由基 radical
氢键 hydrogen bond

1 Covalent bond: A chemical bond formed when electrons are shared between two atoms.
共价键:当两个原子间共用电子时形成的化学键。

2 We call the tendency for an atom to achieve a configuration where its valence shell contains eight electrons the octet rule.
我们把一个原子的价电子层形成稳定八电子构型的倾向称为八隅规则。

第二节　饱和烃:烷烃和环烷烃
Section 2　Saturated Hydrocarbons: Alkanes and Cycloalkanes

烷烃 alkane
环烷烃 cycloalkane
脂肪族的 aliphatic
甲烷 methane
乙烷 ethane
丙烷 propane
丁烷 butane
戊烷 pentane
己烷 hexane
庚烷 heptane
辛烷 octane
壬烷 nonane
癸烷 decane
碳链 carbon-chain
亚甲基 methylene
压缩的结构分子式 condensed structural formula
同分异构体 isomer
异构现象 isomerism
构造异构体(相同分子式,不同结构) constitutional isomerism
张力 strain
命名法 nomenclature
词头,前缀 prefix
词尾,后缀 suffix
取代基烷基 alkyl group substituent
甲基 methyl
乙基 ethyl
丙基 propyl
异丙基 isopropyl
丁基 butyl
异丁基 isobutyl
仲丁基 sec-butyl
叔丁基,特丁基 tert-butyl
母链 parent chain
一级的 primary
二级的 secondary
三级的 tertiary
四级的 quaternary
两个 di-
三个 tri-
支链烷烃 branched chain alkane
环型碳氢化合物 cyclic hydrocarbon
构像 conformation
构象异构体 conformer
纽缦投影图 Newman projection
重叠式构像 eclipsed conformation
角张力 angle strain
分子内的 intramolecular
分子内张力 intramolecular strain
非键相互作用力 unbonded interaction strain
环戊烷 cyclopentane
环己烷 cyclohexane

椅式构像　chair conformation
船式构像　boat conformation
垂直键　axial bond
平伏键　equatorial bond
羟基、氢氧基　hydroxyl
羟基化　hydroxylation
分子间吸引力　intermolecular force of attraction
分散力　dispersion force
卤化反应　halogenation
氧化反应　oxidation

1 Those hydrocarbons in which all of the carbon-carbon are single bonds are called alkanes.
那些分子中所有碳碳键均为单键的烃称为烷烃。

2 Cycloalkanes containing a single ring have two fewer hydrogen atoms and thus have the general formula C_nH_{2n}.
含一个环的环烷烃少两个氢原子，其通式为 C_nH_{2n}。

3 Alkanes and cycloalkanes are so similar that many of their properties can be considered side by side. Some differences remain, however.
烷烃和环烷烃非常相似，它们具有许多相似的性质，但仍存在一些不同。

4 The formal system of nomenclature used today is one proposed by the International Union of Pure and Applied Chemistry (IUPAC).
今天普遍采用的命名方法是国际纯粹与应用化学联合会(IUPAC)采纳的系统命名法。

5 Underlying the IUPAC system of nomenclature for organic compounds is a fundamental principle: each different compound should have an unambiguous name.
有机化合物的 IUPAC 系统命名法的基本原则是：每一个化合物都应该有一个明确的绝不模棱两可的名称。

6 A series of compounds, where each member differs from the next member by a constant unit, like unbranched alkanes, are called a homologous series. Members of a homologous series are called homologs.
每一个化合物与下一个化合物之间相差一个恒定单元的一系列化合物，如直链烷烃，称为同系列。同系列的各个成员称作同系物。

第三节　不饱和烃：烯烃和炔烃

Section 3　Unsaturated Hydrocarbon: Alkenes and Alkynes

不饱和烃　unsaturated hydrocarbon
烯烃　alkene
炔烃　alkyne
衍生物　derivative
偏离　deviation
未杂化的　unhybridized
相邻的　adjacent
热能　thermal energy
限制的　restricted
转换　convert
优先性　preference
乙烯　ethane
丙烯　propene
乙炔　ethylene
丙炔　propylene
乙烯基　vinyl
丙烯基　allyl
异丙烯基　isopropenyl
乙炔基　ethynyl
二烯烃　diene
三烯　triene
多烯　polyene
碳骨架　carbon skeleton
可溶解的　soluble
不溶解的　insoluble
加成反应　addition reaction
聚合物　polymer
单体　monomer
氢卤化反应　hydrohalogenation
水合作用，水化作用　hydration
溴　bromine
溴化作用　bromination
还原作用　reduction
氢化作用　hydrogenation
机理　mechanism
势能　potential energy

势能图 potential energy diagram
过渡态 transition state
动能 kinetic energy
放热的 exothermic
吸热的 endothermic
活化能 activation energy
转移 transfer
氢转移反应 proton-transfer reaction
中间体 intermediate
反应中间体 reaction intermediate
多步反应 multistep reaction
亲电子的,吸电子的 electrophilic
亲电加成反应 electrophilic addition reaction
卤化物,卤素 halide
区域选择性的 regioselective
顺式 cis
反式 trans
顺反异构现象 cis-trans isomerism
酮式-烯醇式互变异构 keto-enol tautomerism
共轭效应 conjugated effect
烯丙基正离子 allylic carbocation
超共轭 hyper conjugation
共振论 resonance theory
电环化反应 electrocyclic reaction
周环反应 pericyclic reaction
催化加氢反应 catalytic hydrogenation
异相反应 heterogeneous reaction
均相反应 homogeneous reaction
亲电试剂 electrophile
区域选择性 regioselectivity
顺式加成 syn-addition
亲核加成 nucleophilic addition

1 Alkenes are hydrocarbons whose molecules contain the carbon-carbon double bond. Hydrocarbons whose molecules contain the carbon-carbon triple bond are called alkynes.
分子中含有碳碳双键的烃类称作烯烃。分子中含有碳碳三键的烃类称作炔烃。

2 There is usually a high free energy of activation for uncatalyzed alkene hydrogenation, and uncatalyzed reaction does not take place at room temperature. However, hydrogenation will take place readily at room temperature in the presence of a catalyst because the catalyst provides a new pathway for the reaction that involves lower free energy of activation.
烯烃非催化加氢通常具有高的活化自由能,在室温下不能发生。但是,在有催化剂存在的情况下,加氢反应在室温下就能很容易进行,因为催化剂为反应提供了一条低活化自由能的新的反应途径。

3 Markovnikov's rule: in the addition of HX to an alkene, the hydrogen atom add to the carbon atom of the double bond that already has the greater number of hydrogen atoms. Reactions that illustrate Markovnikov's rule are said to be Markovnikov additions. When alkenes are treated with HBr in the presence of peroxides, an anti-Markovnikov addition occurs. This anti-Markovnikov addition occurs only when HBr is used in the presence of peroxides and does not occur significantly with HF, HCl, and HI even when peroxides are present.
马尔可夫尼可夫规则:卤化氢与烯烃加成反应中,氢原子加到含氢较多的双键碳原子上。符合马尔可夫尼可夫规则的反应称为马氏加成。在过氧化物作用下,溴化氢与烯烃反应发生反马氏加成。反马氏加成仅发生于在过氧化物存在下HBr的加成,对HF、HCl和HI则不成立,即使在过氧化物存在的情况下。

4 A reaction is stereospecific when a particular stereoisomeric of the starting material reacts in such a way that it gives a specific stereoisomeric form of the product. It does this because the reaction mechanism requires that the configurations of the atoms involved change in a characteristic way.
反应的立体选择性是在反应物具有特殊的立体构型时,反应按照一种特殊的途径进行,生成具有特殊立体构型的产物。这是因为反应机理要求相关原子的构型需要以一种特殊的方式发生改变。

第四节 芳烃:芳香性
Section 4 Aromatic Compounds

芳香族化合物 aromatic compound
苯 benzene
芳香的 aromatic
异丙苯 isopropyl-benzene
芳基 aryl
烃 arene

共振理论 theory of resonance
共振能 resonance energy
杂环的 heterocyclic
吡啶 pyridine
嘧啶 pyrimidine
呋喃 furan
吡咯 pyrrole
咪唑 imidazole
吲哚 indole
嘌呤 purine
六隅体 sextet
甲苯 toluene
苯乙烯 styrene
苯酚 phenol
苯胺 aniline
苯甲醛,安息香醛 benzaldehyde
苯甲酸 benzoic acid
苯甲醚 anisole
苯基 phenyl
苄基 benzyl
邻位 ortho-
间位 meta-
对位 para-
多环的 polynuclear
萘 naphthalene
蒽 anthracene
菲 phenanthrene
吸引,抽取 withdraw
硝基 nitro-
芳环上的亲核取代反应 electrophilic aromatic substitution
氯化 chlorination
溴化 bromination
硝化 nitration
磺化 sulfonation
烷基化 alkylation
酰化 acylation
活化基团 activating group
钝化基团 deactivating group

1 Organic compounds were classified as two broad categories: aliphatic and aromatic. To be classified as aliphatic meant then that the chemical behavior of a compound was “fatlike”. To be classified as aromatic meant then that the compound had a low hydrogen-carbon ratio and that it was “fragrant”.
有机化合物被分为两大类:脂肪烃和芳香烃。划分为脂肪烃意味着该化合物的化学行为像“脂肪”;划分为芳香烃意味着化合物的氢碳比很低,有“芳香味”。

2 Hückel’s rule is concerned with compounds containing one planar ring in which each atom has a p orbital as in benzene. Planar monocyclic rings with $4n+2\pi$ electrons, where $n=0,1,2,3,\ldots$, delocalized electrons should be aromatic.
休克尔规则讨论的是像苯分子一样的、含一个平面环的且平面环中每一个原子都有一个 p 轨道的化合物。有 $4n+2\pi$ 个电子的平面单环($n=0,1,2,3,\cdots$),其离域的电子应该具有芳香性。

3 C60, a molecule shaped like a soccer ball and called buckminsterfullerene, is a member of an exciting new group of aromatic compounds called fullerenes.
C60,分子形状像足球,被称作巴克明斯特富勒烯,是被称作富勒烯的一组新的有趣的芳香族化合物的成员之一。

4 Nanotubes are a relatively new class of carbon-based material related to buckminsterfullerenes. Nanotubes are very tough — about 100 times as strong as steel. They also have shown to act as electrical conductors or semiconductors and many other applications.
纳米管是一族相对新的与巴克明斯特富勒烯相关的碳材料。纳米管非常硬 —— 大约是钢的 100 倍。它们也应用在电子导体或半导体以及其他方面。

第五节 卤代烃
Section 5 Alkyl Halides

卤代烷 haloalkane
伯卤代烷 primary alkyl halide
仲卤代烷 secondary alkyl halide
叔卤代烷 tertiary alkyl halide
烯丙基溴 allyl bromide
卞基氯 benzyl chloride
亲核取代反应 nucleophilic substitution reaction
水解反应 hydrolysis

双分子亲核取代反应机理 substitution nucleophilic bimolecular
单分子亲核取代反应机理 substitution nucleophilic unimolecular
双分子消降反应机理 elimination bimolecular
单分子消降反应机理 unimolecular elimination
卤化物 halide
氯代 -chloro-
溴代 -bromo-
卤仿 haloform
氯仿 chloroform
溴仿 bromoform
碘仿 iodoform
亲核的 nucleophilic
亲核加成 nucleophilic addition
亲核试剂 nucleophile
协同的,同时的 concerted
双分子的 bimolecular
单分子的 unimolecular
翻转 inversion
构型 configuration
离去集团 leaving group
氢氧负离子 hydroxide
甲醇盐 methoxide
氨化物 amide
质子化溶剂 protic solvent
疏质子溶剂 aprotic solvent
消除 elimination
消除反应 elimination reaction
指定,分配 assign
脱卤化氢作用 dehydrohalogenation
支配,起主导作用 predominate
物种,成分 species

1 Nucleophilic substitution reaction is a substitution reaction which is initiated by a nucleophile. S_N2 reaction, means substitution, nucleophilic, bimolecular.

亲核取代反应是被亲核试剂引发的取代反应。S_N2 反应,其含义是取代、亲核、双分子。

2 This reaction is said to be second order overall. Because bond formation and bond breaking occur simultaneously in a single transition state, the S_N2 reaction is an example of what is called a concerted reaction.

这一类反应是二级反应。因为键的形成和键的断裂是通过一个过渡态同时发生的。S_N2 反应是协同反应的一个例子。

3 S_N2 reaction always occur with inversion of configuration. S_N1 reaction (substitution, nucleophilic, unimolecular) is said to be unimolecular (first order) in the rate-determining step. If a given alkyl halide and nucleophile react rapidly by an S_N2 mechanism but slowly by an S_N1 mechanism under a given set of conditions, then an S_N2 pathway will be followed by most of the molecules, vice versa.

S_N2 反应总是伴随构型的翻转。S_N1 反应(取代、亲核、单分子)认为速控步是单分子反应(一级)。如果在指定条件下,指定烷基卤和亲核试剂以 S_N2 反应机理反应快于 S_N1 反应机理,则 S_N2 机理是主要的,反之亦然。

4 Experiments have shown that a number of factors affect the relative rates of S_N1 and S_N2 reactions. The most important factors are:

a. the structure of the substrate;
b. the concentration and reactivity of the nucleophile (for bimolecular reactions only);
c. the effect of the solvent;
d. the nature of the leaving group.

实验表明,很多因素会影响 S_N1 反应和 S_N2 反应的相对速率。最重要的一些因素如下:

a. 底物的结构;
b. 亲核试剂的浓度和反应活性(仅对双分子反应而言);
c. 溶剂的影响;
d. 离去基团的性质。

5 Elimination reactions occur by a variety of mechanisms. One mechanism, called the E2 reaction, is bimolecular in the rate-determining step; the other mechanism is the E1 reaction, which is unimolecular in the rate-determining step.

消除反应可以以不同的反应机理进行。其中一种称作 E2 消除，速控步是双分子反应；另一种是 E1 反应，速控步是单分子反应。

6 Functional group inter-conversions of methyl, primary, and secondary alkyl halides using S_N2 reactions:

采用 S_N2 反应进行的甲基、一级烷基和二级烷基卤之间的官能团相互转换图：

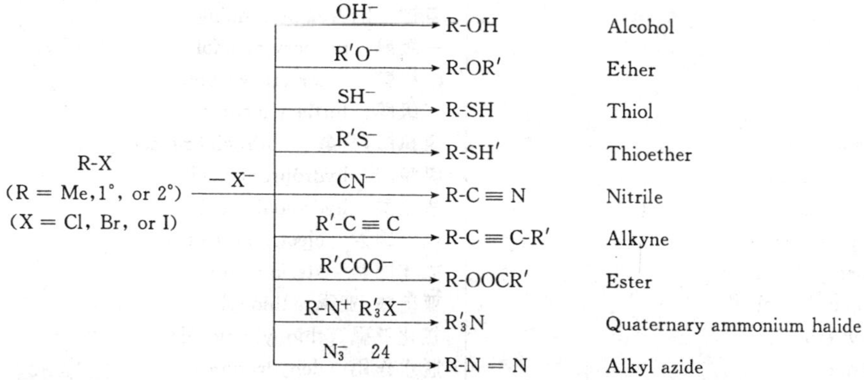

7 Adding a phase-transfer catalyst solves this problem by transferring the ionic reactant into the organic phase. And, again, because the reaction medium is aprotic, an S_N2 reaction occurs rapidly.

相转移催化剂的加入通过把离子性的反应物转移到有机相中，从而解决了这一问题。因为反应媒介是非质子性的，所以 S_N2 反应可以很快进行。

8 Compounds called crown ethers are also phase-transfer catalysts and are able to transport ionic compounds into an organic phase.

冠醚化合物也是一种相转移催化剂，能把离子性的化合物转移到有机相中。

第六节　立体化学

Section 6　Stereochemistry

异构现象　isomerism
立体异构　stereoisomerization
构象异构体　conformational isomer
手性分子　chiral molecule
非手性分子　achiral molecule
手性　chirality
对映异构体　enantiomer
旋光性　optical activity
旋光仪　polarimeter
手性，立体化学　chirality
立体的　spatial
反射，反应，倒影　reflection
镜像　mirror image
立体异构体　stereoisomer
立体异构现象　stereoisomerism
立体阻碍　steric hindrance
非对映体　diastereomer
非手性的　achiral
对称性　symmetry
对称面　plane of symmetry
乳酸　lactic acid
非对映异构体　diastereomers
旋光纯度　optical purity
阻转异构现象　atropisomerism
立体专一性　stereospecific
立体选择性　stereoselectic
不能重叠放置的　nonsuperposable
叠置，重叠放置　superpose
立体中心　stereocenter
非环形分子　acyclic-molecule
偏振光　polarized light
外消旋的　racemic
外消旋混合物　racemic mixture
分离　separation
拆分　resolution
用……的方法　by means of ...
拆分试剂　resolving reagent

第七节 醇、醚和硫醇

Section 7 Alcohols, Ethers and Thiols

醇 alcohol
醚 ether
硫醇 thiol
甲醇 methanol
乙醇 ethanol
丙醇 propanol
丁醇 butanol
环己醇 cyclohexanol
烯丙醇 allyl alcohol
苯甲醇 menzel alcohol
异丙醇 isopropyl alcohol
异丁醇 isobutyl alcohol
乙醚 diethyl ether
醇的 alcoholic
相互转换 interconversion
羟基，氢氧基 hydroxyl
氢硫基，巯基 sulfhydryl
二醇 diol
三醇 triol
邻二醇 glycol
词尾、后缀 suffix
穿入，插入，中缀 infix
环氧化物 epoxide
烷氧基 alkoxy group
环氧乙烷，氧化乙烯 ethylene oxide
四氢呋喃 tetrahydrofuran
四氢吡喃 tetrahydropyran
1,4-二氧六环 1,4-dioxane
类似物 analog
硫醇 mercaptan
乙硫醇 ethanethiol
硫化物，硫醚 sulfide
二甲硫醚 dimethyl sulfide
氢键 hydrogen bonding
一级醇 primary alcohol
二级醇 secondary alcohol
三级醇 tertiary alcohol
氢氯酸，盐酸 hydrochloric acid
氢溴酸 hydrobromic acid
氢碘酸 hydroiodic acid
电子因素 electronic factor
立体因素 steric factor
亚硫酰，亚砜 thionyl
氯化亚砜 thionyl chloride
脱水作用 dehydration
同时发生的 simultaneous
乙醛 aldehyde
羧酸 carboxylic acid
变化，重排作用 transformation
铬酸 chromic acid
氧化铬 chromium oxide
重铬酸钾 potassium dichromate
吡啶 pyridine
沉淀 precipitate
环氧化物 epoxide
高氯酸 perchloric acid
质子化的 protonated
二甲醚 dimethyl ether
二苯醚 diphenol ether
甲乙醚 ethyl methyl ether
甲基叔丁醚 tert-butyl methyether
环氧丙烷 propylene oxide

1 Alcohols are compounds whose molecules have a hydroxyl group attached to a saturated carbon atom.
醇是指分子中有羟基连接在饱和碳原子上的化合物。

2 Compounds that have a hydroxyl group attached directly to a benzene ring are called phenols.
羟基直接连在苯环上的化合物称作苯酚。

3 Ethers differ from alcohols in that the oxygen atom of an ether is bonded to two carbon atoms.
醚不同于醇，醚中的氧原子与两个碳原子成键。

4 With phase-transfer catalysis, we usually use two immiscible solutions that are in contact—often an aqueous phase containing an ionic reactant and an organic phase containing the organic substrate. Normally the reaction of two substances in separate phases like this is inhibited because of the inability of the reagents to come together. 相转移催化中，我们通常采用两种相互接触但互不混溶的溶剂——水相中含离子性反应物，有机相中含有机底物。通常情况下，像这种各自在两种独立相中的两种底物之间的反应是被禁阻的，因为反应物不能彼此接近。

5 Crown ethers are cyclic polymers of ethylene glycol. The relationship between the crown ether and the ion that it transports is called a host-guest relationship.
冠醚是一种环状的乙二醇聚合物。冠醚和其输送的离子之间的关系称作主－客体关系。

6 The crown ether acts as the host, and the coordinated cation is the guest.
冠醚作为主体，配位的阳离子作为客体。

7 Some synthetic connections of alkynes, alcohols, alkyl halides, and ethers:
炔烃、醇、烷基卤和醚之间的合成联系：

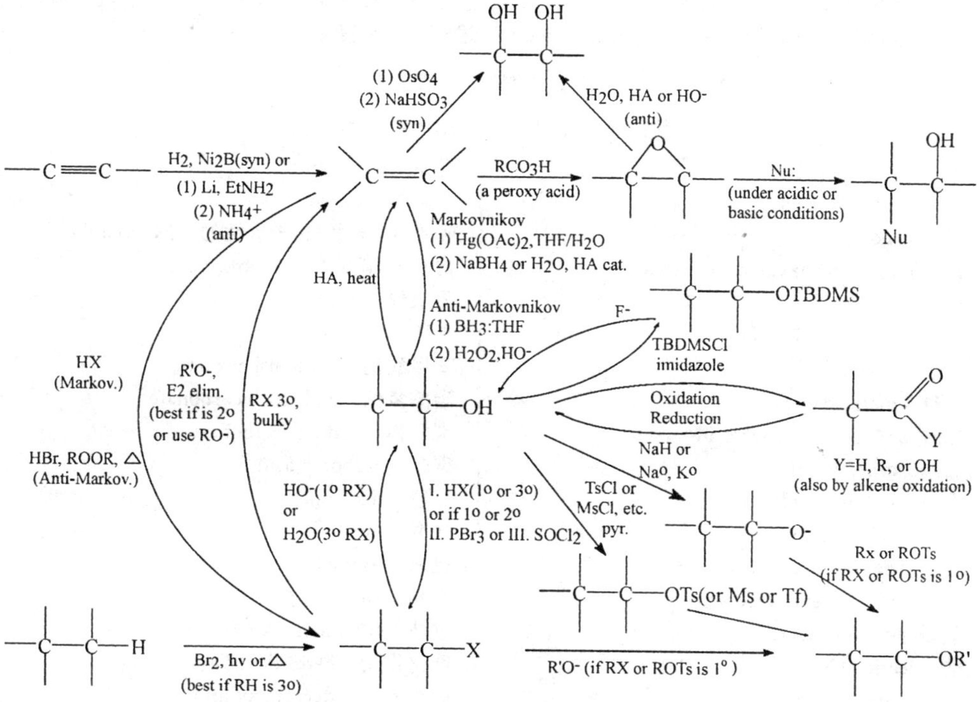

第八节　醛、酮和醌

Section 8　Aldehydes, Ketones and Quinone

醛　aldehyde
酮　ketone
甲醛　formaldehyde groups in organic chemistry.
乙醛，蚁醛　acetaldehyde
丙酮　acetone
羰基　carbonyl
有机镁　organomagnesium
格式试剂　Grignard reagent
负碳离子　carbanion
氨基化，胺化　amination
金属有机化合物　organometallic compound
半缩醛　hemiacetal
缩醛　acetal
酮式　keto
环已酮　cyclohexanone
烯醇　enol
互变体　tautomer
硼氢化钠（$NaBH_4$）sodium borohydride
半缩酮　hemiketal
氢化铝锂　lithium aluminum hydride
同碳二醇　geminal
羟醛缩合　aldol condensation

1 All aldehydes have a carbonyl group, bonded on one side to a carbon and on the other side to a hydrogen. In ketones, the carbonyl group is situated between two carbon atoms.
所有醛都含有一个羰基基团,一边与碳原子相连,一边与氢原子相连。酮的羰基则是位于两个碳原子之间。

2 Grignard reagents, organomagnesium halides, discovered by the French chemist Victor Grignard in 1900, have great use in organic synthesis. The actual structures of Grignard reagents are more complex than the general formula RMgX indicates.
格林雅试剂,即卤化有机金属镁,法国化学家维克多·格林雅发现于1900年,在有机合成中具有很大用处。其实际结构比通常给出的分子式 RMgX 表述的要复杂得多。

第九节 羧酸

Section 9 Carboxylic acids

酰基 acyl group
羧酸 carboxylic acid
羧基 carboxyl
羧酸衍生物 carboxylic acid derivatives
二酸 dicarboxylic acid
草酸,乙二酸 oxalic acid
己二酸 adipic acid
丙二酸 malonic acid
丁二酸,琥珀酸 succinic acid
戊二酸 glutaric acid
苯甲酸 benzoic acid
水杨酸 salicylic acid
邻苯二甲酸 phthalic acid
对苯二甲酸 terephthalic acid
甲酸,蚁酸 formic acid
乙酸 acetic acid
丙酸 propanoic acid
丁酸 butanoic acid
戊酸 pentanoic acid
己酸 hexanoic acid
乙酰基 aceto
联系 associate
二聚体 dimmer
吸水的,亲水的,亲水性的 hydrophilic
憎水的,疏水的 hydrophobic
无限地 infinitely
常数 constant
苯甲酸钠 sodium benzoate
碳酸氢钠 sodium bicarbonate
碳酸钠 sodium carbonate
碳酸 carbonic acid
分解 decompose
演变 evolve
提取 extract
水相 aqueous phase
过滤 filtration
酯化作用 esterification
脱羧基化 decarboxylation
重新分布 redistribution
羧化作用 carbonation
脱羧反应 decarboxylation

1 The carboxyl group, —COOH, is one of the most widely occurring functional groups in chemistry and biochemistry. Not only are carboxylic acids themselves important, but the carboxyl group is the parent group of a large family of related compounds called acyl compounds or carboxylic acid derivatives.
在化学和生物化学中,羧基是最广泛存在的官能团之一。羧酸不仅自身十分重要,而且羧基基团也是大家族酰基化合物和羧酸衍生物的母体官能团。

2 Reactions of carboxylic acids:
羧酸的反应:
(1) As acid 作为酸:

$$RCO_2H + NaOH \longrightarrow RCO_2^- Na^+ + H_2O$$

$$RCO_2H + NaHCO_3 \longrightarrow RCO_2^- Na^+ + H_2O + CO_2$$

(2) Reduction 还原：

$$RCO_2H + LiAlH_4 \xrightarrow[(2)H_2O]{(1)Et_2O} RCH_2OH$$

(3) Conversion to acyl chlorides 转化成酰氯：

$$RCO_2H \xrightarrow{SOCl_2 \text{ or } PCl_5} RCOCl$$

(4) Conversion to esters (Fisher esterification) or lactones 转化成酯（费希尔酯化）或内酯：

(5) Conversion to amide 转化成胺：

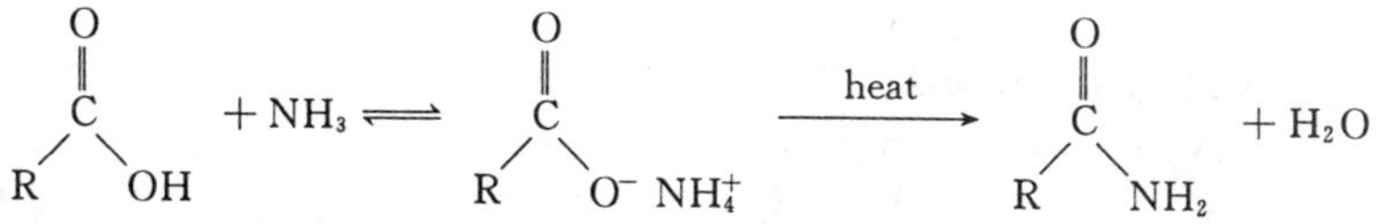

(6) Decarboxylation 去酰羰基反应：

$$R\overset{O}{\overset{\|}{C}}CH_2\overset{O}{\overset{\|}{C}}OH \xrightarrow{heat} R\overset{O}{\overset{\|}{C}}CH_3 + CO_2$$

$$HO\overset{O}{\overset{\|}{C}}CH_2\overset{O}{\overset{\|}{C}}OH \xrightarrow{heat} CH_3\overset{O}{\overset{\|}{C}}OH + CO_2$$

第十节　羧酸衍生物

Section 10　Functional Derivatives of Carboxylic Acids

衍生　derive
羧基　carboxyl
酸酐　carboxylic acid anhydride
酰卤　acid halide
晴　nitrogen
酯　ester
酰胺　amide
酰基　acyl group
丁二酸　succinic acid
邻苯二甲酸　phthalic acid
内酯　lactone
内酰胺　lactam
亚酰胺　imide
典型的反应，特征反应　characteristic reaction
亲核试剂　nucleophile
酰基的亲核取代反应　nucleophilic acyl substitution
回流　reflux
含水的　aqueous
相反　reverse
可逆的　reversible
耗费，消耗，消费　consume
皂化，皂化作用　saponification
激烈的条件　vigorous conditions
乙酸酐　acetic anhydride
中和　neutralize
副产物　by-product
烯醇负离子　enolate anion
可比较的，相似的电负性　comparable electronegativity
醇醛，羟醛　aldol
醇醛缩合　aldol condensation
引起，发生　bring about
稀酸　dilute acid
共轭　conjugate
挤出，排挤出　expel
交叉醇醛缩合反应　crossed aldol reaction
分子内醇醛缩合反应　intramolecular aldol reaction
迪克曼缩合反应　Dieckmann condensation
弹出　eject

再生，重生　regenerate

交叉克莱森缩合反应　crossed Claisen condensation

丙酰氯　propanone chloride

乙酸乙酯　ethyl acetate

苯甲酸乙酯　ethyl benzoate

甲乙酸酐　acetic formatic anhydride

乙酰胺　acetamide

N,N-二甲基甲酰胺　N,N-dimethyl formamide

皂化反应　saponification

1 Reactions of acyl chlorides:

酰卤的反应：

(1) Conversion to acids 转化成酸：

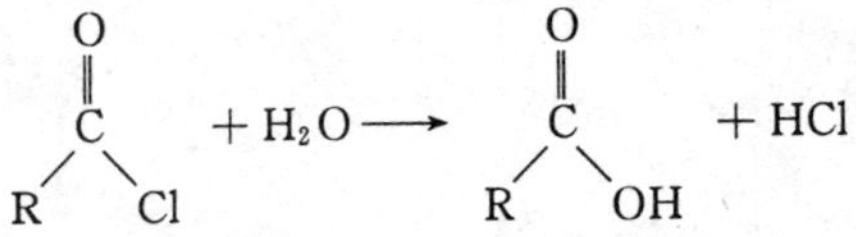

(2) Conversion to anhydrides 转化成酸酐：

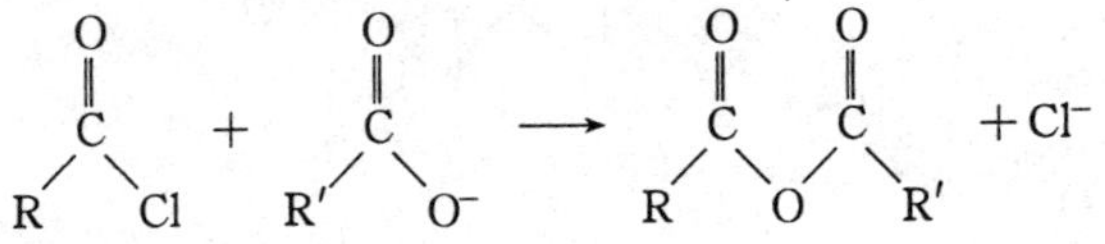

(3) Conversion to esters 转化成酯：

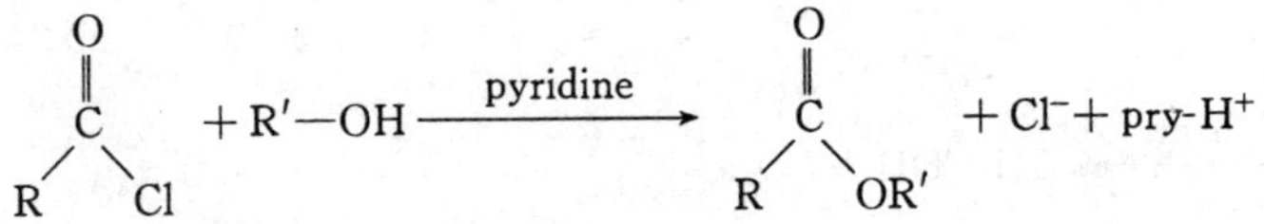

(4) Conversion to amides 转化成胺：

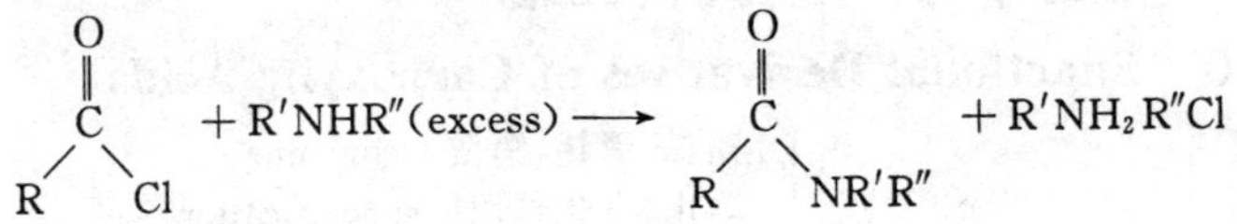

R′ and/or R″ may be H.

(5) Conversion to ketones 转化成酮：

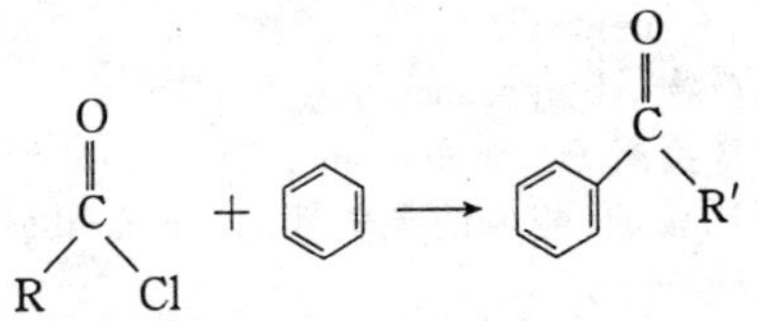

Friedel-crafts acylation
付氏酰基化反应

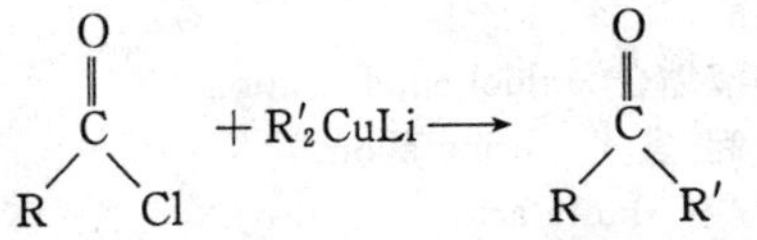

Reactions of dialkylcuprates
二烷基铜锂化反应

(6) Conversion to aldehydes 转化成醛：

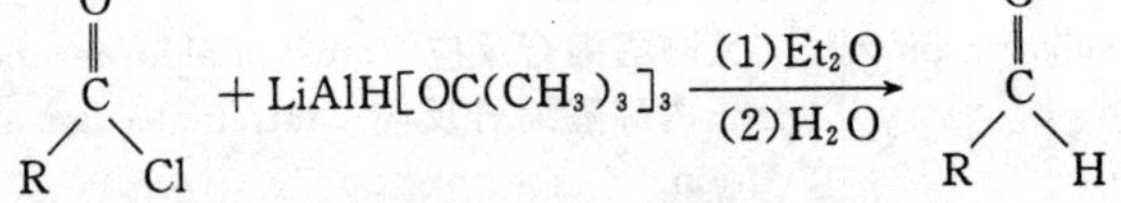

2 Reactions of acid anhydrides：

酸酐的反应：

(1) Conversion to acids 转化成酸：

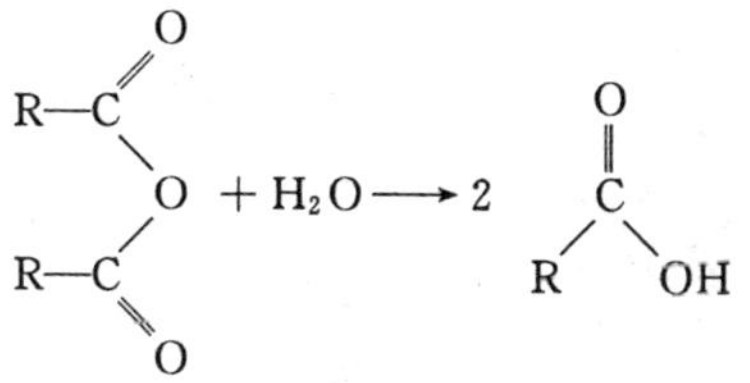

(2) Conversion to esters 转化成酯：

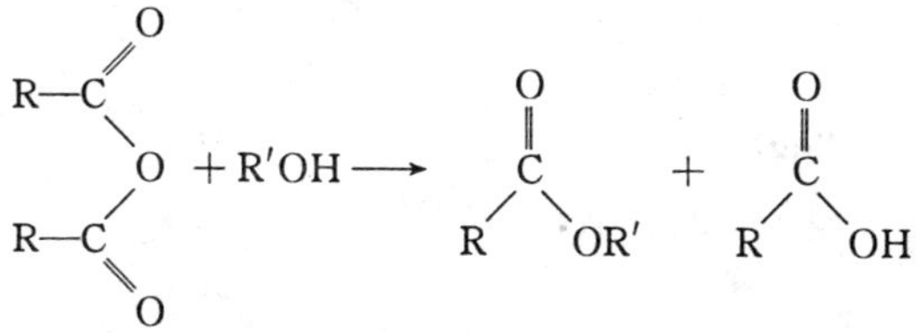

(3) Conversion to amides 转化成胺：

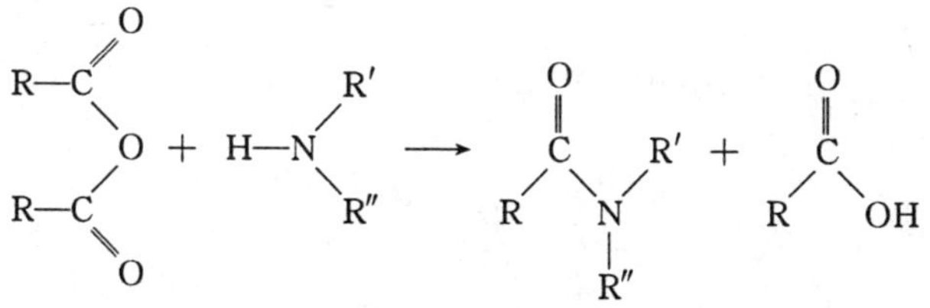

R′ and/or R″ may be H.

(4) Conversion to aryl ketone 转化成芳基酮：

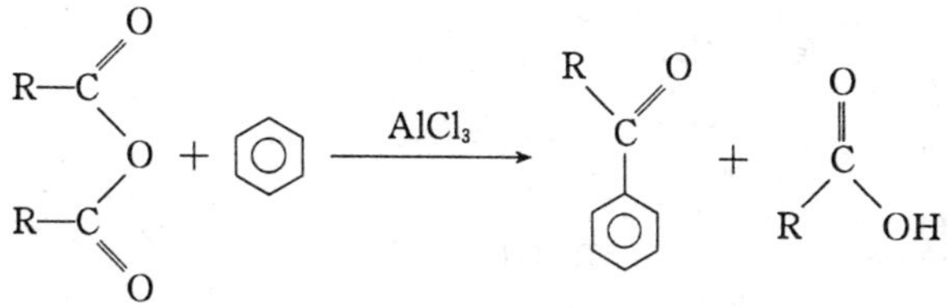

3 Reactions of esters：

酯的反应：

(1) Hydrolysis 水解：

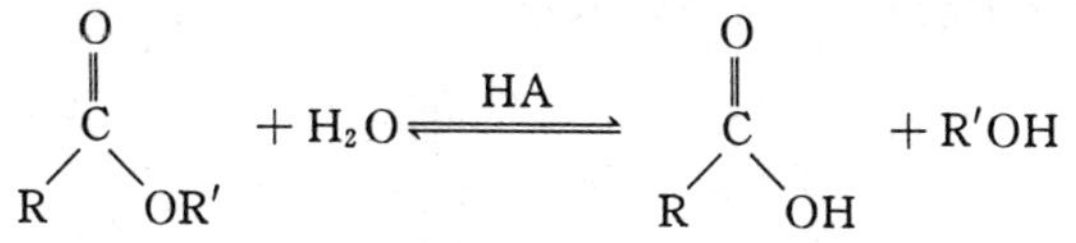

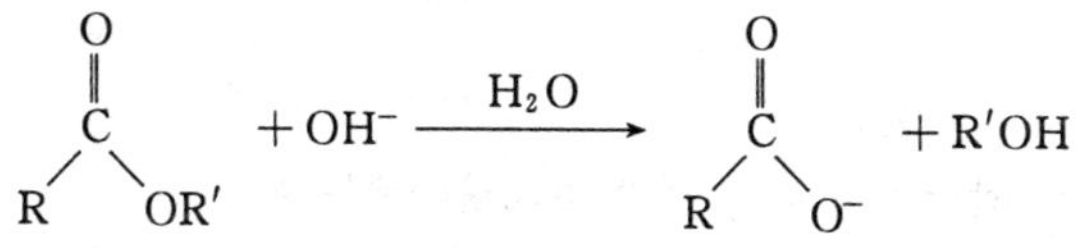

(2) Conversion to other esters－transesterification 转化成其他酯——酯化转移反应：

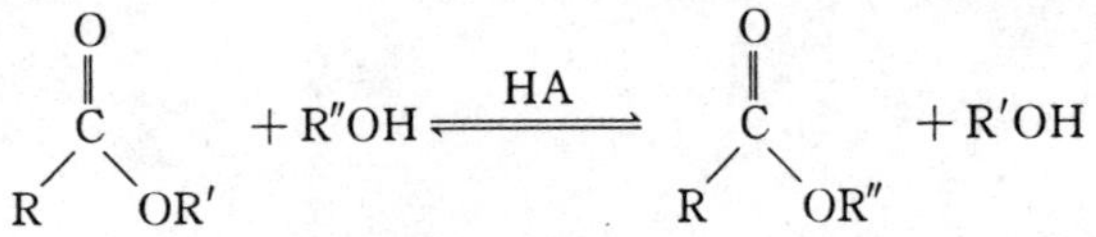

(3) Conversion to amides 转化成胺：

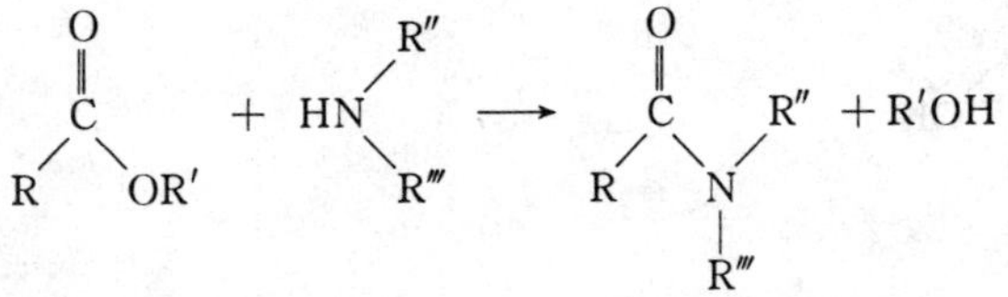

R′ and/or R″ may be H.

(4) Reaction with Grignard reagents 与格氏试剂反应：

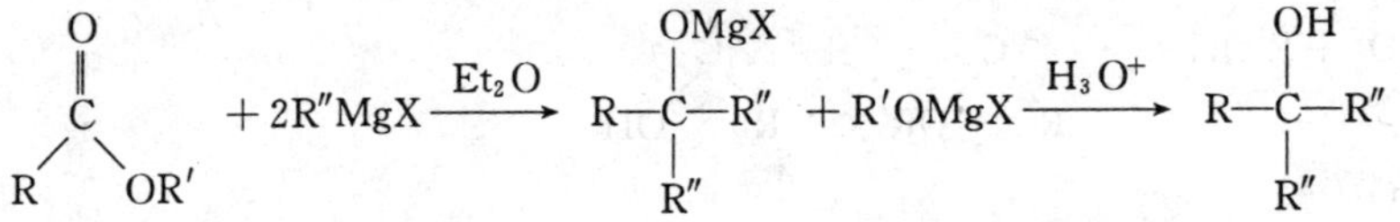

(5) Reduction 还原：

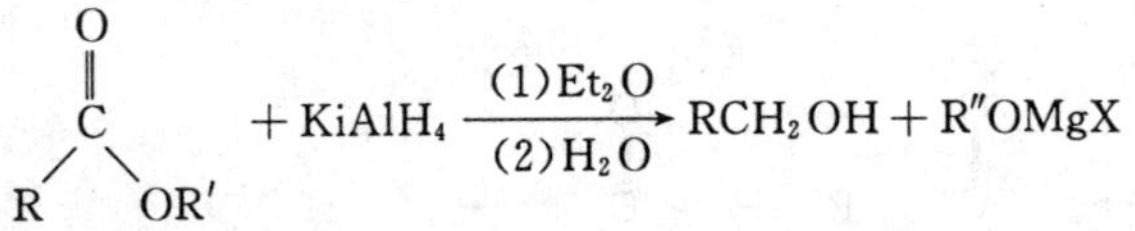

4 Reactions of amides：

胺的反应：

(1) Hydrolysis 水解：

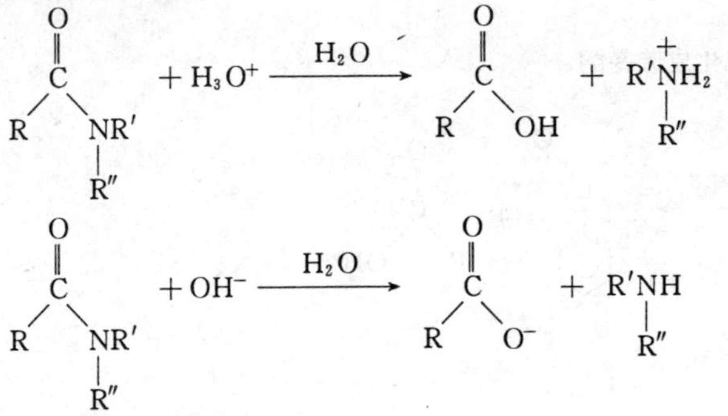

R′ and/or R″ may be H.

(2) Conversion to nitriles－dehydration 转化成腈——脱水反应：

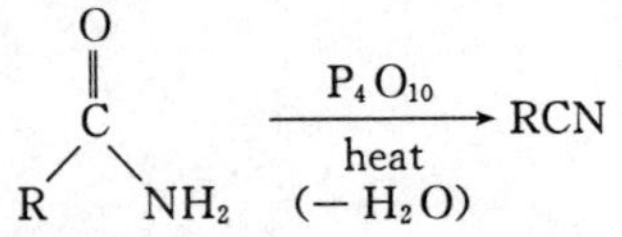

5 Reactions of Nitriles：

腈的反应：

(1) Hydrolysis to a carboxylic acid or carboxylate anion 水解生成羧酸或羧酸盐：

$$R—C\equiv N \xrightarrow{H_3O^+,\text{heat}} RCO_2H$$

$$R\text{—}C\equiv N \xrightarrow{OH^-, H_2O, heat} RCO_2^-$$

(2) Reduction to an aldehyde with $(i\text{-}Bu)_2AlH$(DIBAL-H) 用$(i\text{-}Bu)_2AlH$(DIBAL-H) 还原成醛：

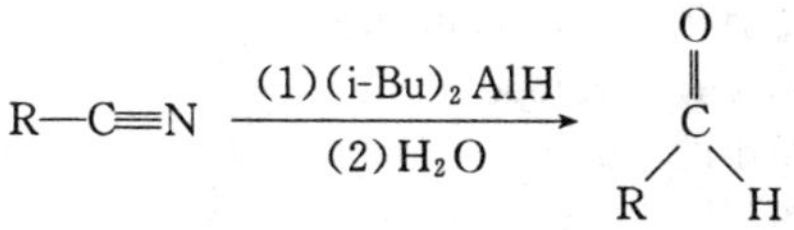

(3) Conversion to a ketone by a Grignard or organolithium reagent 通过格氏试剂或有机金属锂转化成酮：

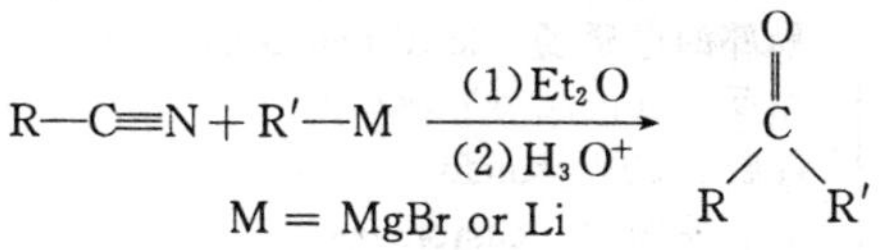

第十一节　有机含氮化合物

Section 11　Organic Nitrogen Compound

胺　amine
叔丁胺　t-butylamine
苄胺　benzylamine
苯胺　aniline
氨气　ammonia
芳香族硝基化合物　aromatic nitro-compound
硝基　nitro-group
碱性　basicity
亲核性　nucleophilicity
亚硝酸　nitrous acid
喹啉　quinoline
异喹啉　isoquinoline
季铵盐　quaternary ammonium salt
季铵碱　quaternary ammonium hydroxide
重氮化反应　diazotization
乙腈　methyl cyanide
丁腈　butanenitrile
苯甲晴　benzonitrile

1 Amides with no substituent on the nitrogen react with solutions of bromine or chlorine in sodium hydroxide to yield amines through a reaction known as the Hofmann rearrangement or Hofmann degradation.
氮原子上没有取代基的酰胺在氢氧化钠的作用下与溴或氯气反应，通过霍夫曼重排反应或霍夫曼降解反应得到胺。

2 The Curtius rearrangement is a rearrangement that occurs with acyl azides.
克尔提斯重排是指酰基叠氮化合物的重排。

3 When a quaternary ammonium hydroxide is heated, E_2-type elimination takes place. The products are an alkene, water and a tertiary amine. This reaction is called Hofmann elimination.
加热条件下氢氧化铵发生 E_2 消除，产物是烯烃、水和叔胺。这一反应称作霍夫曼消除反应。

4 Although most elimination involving neutral substrates tend to follow the Zaitsev rule, elimination with charged substrates tend to follow what is called the Hofmann rule and yield mainly the least substituted alkene. Tertiary amine oxides undergo the elimination of a dialkylhydroxylamine when they are heated. This reaction is called the Cope Elimination.
尽管多数包含中性底物的消除反应倾向于遵从扎依采夫规则，但带电荷的底物倾向于遵从霍夫曼规则，主要生成取代基最少的烯烃。在加热条件下，叔胺氧化物消去一个二烷基羟胺，这个反应称为科普消除反应。

第十二节　核磁共振光谱

Section 12　Nuclear Magnetic Resonance Spectroscopy

原子核的　nuclear
磁性的　magnetic

共振 resonance
光谱学,波谱学 spectroscopy
核磁共振波谱学 nuclear magnetic resonance spectroscopy
中心主题 central theme
红外的,红外线 infrared
红外光谱 infrared spectroscopy
电磁辐射 electromagnetic radiation
辐射能 radiant energy
射线 ray
g-射线 Gamma ray
X-射线 X-ray
紫外线 ultraviolet light
可见光 visible light
红外辐射 infrared radiation
微波 microwave
无线电 radio wave
光谱 spectrum
电磁波频谱,电磁波谱 electromagnetic spectrum
电磁辐射 electromagnetic radiation
波 wave
波长 wavelength
频率 frequency
区域 region
赫兹 hertz
速度 velocity
热辐射 heat radiation
流 stream
粒子 particle
粒子流 stream of particle
光子 photon
千卡 kilocalorie
毫米 millimeter
微米 micrometer
纳米 nanometer
埃 angstrom
能伸缩的、可弯曲的结构 flexible structure
拉长 stretch
弯曲 bend
可伸缩、弯曲的弹簧 flexible spring
原子核 nuclei
自旋 spin
量子化的 quantized
辐射,照射 irradiate
吸收 absorb
放射 emit
起源 origin
普通化学 general chemistry
同位素 isotope
微小的 tiny
棒 bar
跃迁 flip
信号 signal
局部的电磁场 local magnetic field
相反 oppose
屏蔽,保护 shield
依照惯例的 customary
共振频率 resonance frequency
参照 reference
基准物 reference compound
四甲基硅烷 tetramethylsilan
漂移,迁移,移动 shift
化学位移 chemical shift
光谱仪 spectrometer
氘 deuterium
氘代氯仿 deuterochloroform
氧化氘,氘代水,重水 deuterium oxide
样品 sample
池 cell
样品池 sample cell
命名法 terminology
低磁场 downfield
高磁场 upfield
等价的氢原子 equivalent hydrogen
区分 distinguish
光谱 spectra
积分,累积 integration
成正比的 proportional
信号裂分 signal splitting
峰 peak
双重峰 doublet
三重峰 triplet
四重峰 quartet
多重峰 multiplet
去屏蔽 deshield
鉴别性的,有显著特征的 distinctive
检索 index
不足 deficiency

1 The nuclei of certain elements and isotopes behave as though they were magnets spinning about an axis.

某些元素和同位素的核绕轴自旋，像磁铁一样。

2 The nuclei of ordinary hydrogen (^{1}H) and carbon-13 (^{13}C) nuclei have this property.
普通氢1和碳13的核都有这种性质。

3 When one places a compound containing ^{1}H or ^{13}C atoms in a very strong magnetic field and simultaneously irradiates it with electromagnetic energy, the nuclei of the compound may absorb energy through a process called magnetic resonance.
当含^{1}H和^{13}C原子的化合物置于强磁场中且同时给予适当的电磁照射时，这些化合物的核可以通过磁共振过程吸收能量。

4 This absorption of energy is quantized and produces a characteristic spectrum for the compound.
这一能量的吸收过程是量子化的，并为该化合物产生特定的光谱。

5 Absorption of energy does not occur unless the strength of the magnetic field and the frequency of electromagnetic radiation are at specific values.
能量的吸收仅在磁场强度和电磁辐射频率均处于某特定值时才能发生。

6 We are already familiar with the concept of electron spin and with the fact that the spins of electrons confer on them the spin quantum states of $+1/2$ or $-1/2$.
我们已经熟悉电子自旋的概念，电子自旋的自旋量子态是$+1/2$或$-1/2$。

7 The nuclei of certain isotopes also spin, and therefore these nuclei possess spin quantum numbers I. The nucleus of ordinary hydrogen, ^{1}H (i.e., a proton), is like the electron; its spin quantum number I is 1/2 and it can assume either of two spin states: $+1/2$ or $-1/2$.
特定同位素的核也发生自旋，因此，这些核也有这些量子数I。普通氢的核^{1}H，像电子；其自旋量子数为1/2，具有两种自旋量子态：$+1/2$或$-1/2$。

8 These correspond to the magnetic moments allowed for $I = 1/2$, $m = +1/2$ or $m = -1/2$. Other nuclei with spin quantum numbers $I = 1/2$ are ^{13}C, ^{19}F, and ^{31}P.
与自旋量子数$I = 1/2$相关的磁量子数m可以为$+1/2$或$-1/2$。其他自旋量子数为1/2的核还有^{13}C、^{19}F和^{31}P。

9 Some nuclei, such as ^{12}C, ^{16}O, and ^{32}S, have no spin ($I = 0$), and these nuclei do not give an NMR spectrum. Other nuclei have spin quantum numbers greater than 1/2.
像^{12}C、^{16}O和^{32}S这样的核，自旋量子数为0，所以这些核不具有NMR光谱。有些核自旋量子数大于1/2。

第十三节 红外光谱
Section 13 Infrared Spectroscopy

红外光谱 infrared spectroscopy	整齐的，整洁的 neat
红外线 infrared radiation	透明的 transparent
振动红外光谱 vibrational infrared	分子振动 molecular vibration
波数 wave number	指纹区 fingerprint region
倒数的 reciprocal	相关 correlation
量程 scale	累积的 cumulative
透过率 transmittance	强度 intensity

1 Infrared (IR) spectroscopy provide a simple and rapid instrumental technique that can give evidence for the presence of various functional groups.
红外光谱提供了一种简单快速的仪器测试手段，可以给出不同官能团的信息。

2 As all forms of spectroscopy, it depends on the interaction of molecules or atoms with electromagnetic radiation. Infrared radiation causes atoms and groups of atoms of organic compounds to vibrate with increased amplitude about the covalent bonds that connect them.
像所有形式的光谱一样，红外光谱依靠电磁辐射下分子或原子之间的相互作用。红外辐射引起有机化合物原子或原子团围绕其共价键的振幅增大。

3 Since the functional groups of organic molecules include specific arrangements of bonded atoms, absorption of IR energy by an organic molecule will occur in a manner characteristic of the types of bonds and atoms present in the specific functional groups of that molecule.

既然有机分子的官能团包含特定的原子键合排列，其红外吸收也会与分子中特殊官能团键和原子的类型特征相关。

4 These vibrations are quantized, and as they occur, the compounds absorb IR energy in particular regions of the IR portion of the spectrum.

这些振动是量子化的，振动发生时，化合物在光谱特定的红外区产生红外吸收。

5 Not all molecular vibrations result in the absorption of IR energy. In order for a vibration to occur with the absorption of IR energy, the dipole moment of the molecule must change as the vibration occurs.

并不是所有分子振动都会产生红外吸收。能吸收红外能量的振动，在振动时分子的偶极矩必须发生改变。

6 Methane does not absorb IR energy. Symmetrical vibrations of the carbon-carbon double and triple bonds of ethane and ethyne do not result in the absorption of IR radiation, either.

甲烷不产生红外吸收。乙烯和乙炔碳碳双键和三键对称的振动也不产生红外吸收。

第二章　化工原理
Chapter 2　Unit Operation of Chemical Engineering

第一节　流体流动
Section 1　Fluid Flow

单元操作　unit operation
动量传递　momentum transfer
热量传递　heat transfer
质量传递　mass transfer
过程传递速率　rate of transfer process
过程推动力　driving force
过程阻力　resistance
连续介质　continuum
流体　fluid
流体动力学　fluid dynamics
流体静力学　fluid statics
可压缩流体　compressible fluid
不可压缩流体　incompressible fluid
黏度　viscosity
黏性力　viscous force
动力黏度　dynamics viscosity
运动黏度　kinematic viscosity
质量力　body force
表面力　surface force
牛顿型流体　Newtonian fluid
非牛顿型流体　non-Newtonian fluid
塑性流体　plastic fluid
假塑性流体　pseudo-plastics fluid
非稳态　unstable state
稳态　stable state
定态　steady state
非定态　unsteady state
连续过程　continuous process
间歇过程　batch process
半连续过程　semi-continuous process
质量流量　mass flow rate
质量流速　mass velocity
平均流速　average velocity
体积流量　volumetric flow rate
连续性方程　continuous equation
机械能方程　mechanical energy equation
柏努利方程　Bernoulli's equation
雷诺实验　Reynolds experiment
雷诺准数　Reynolds number
位能　potential energy
静压能　static energy
流型　flow pattern
层流(滞流)　streamline flow/laminar flow
湍流(紊流)　turbulent flow
边界层　boundary layer
层流底层　laminar sublayer
摩擦损失　friction loss
摩擦系数　friction coefficient
摩擦因数　friction factor
动压头　kinetic head
静压头　static head
速度分布　velocity distribution
粗糙度　roughness
绝对粗糙度　absolute roughness
相对粗糙度　relative roughness
压降　pressure drop
阻力损失　drag loss
突然缩小　sudden contraction
突然扩大　sudden expansion
缩脉(流颈)　vena contract
管件　pipe fitting
弯头　elbow
三通　T-piece
截止阀(球心阀)　globe value
闸阀　gate value
哈根-泊谡叶方程　Hagen-poiseuille equation
当量直径　equivalent diameter
当量长度　equivalent length
水力平均直径　hydraulic mean diameter
水力半径　hydraulic radius
润湿周边　wetted perimeter
液柱压力计　manometer
皮托管　Pitot tube
文丘里管　Venturi tube
文丘里流量计　Venturi meter
孔板流量计　orifice meter

转子流量计　rotameter

流量系数(孔流系数)　flow coefficient

1 In a Newtonian fluid, the shear stress is proportional to the shear rate, and the proportionality constant is called the viscosity.
在牛顿型流体中,剪应力与剪切速率成正比,比例常数称为黏度。

2 The most common types of full-bore meters are Venturi and orifice meters and variable-area meters such as rotameters.
最常见的流量计有文丘里流量计和孔板流量计,还有变截面的转子流量计。

第二节　流体输送机械
Section 2　Fluid-moving Machinery

离心泵　centrifugal pump
容积式泵　positive displacement pump
活塞泵　piston pump
往复泵　reciprocating pump
轴流泵　axial flow pump
混流泵　mixed flow pump
涡轮泵　turbine pump
屏蔽泵　canned-motor pump
齿轮泵　gear pump
回转泵　rotary pump
隔膜泵　diaphragm pump
计量泵　metering pump
叶轮　impeller
蜗壳　volute
气缚　air binding
特性曲线　characteristic curve
设计点　design point
扬程　height
容量　capacity
汽蚀　cavitation
汽蚀余量　net positive suction head,NPSH
必需汽蚀余量　required NPSH
临界汽蚀余量　critical NPSH
有效汽蚀余量　available NPSH
通风机　fan
罗茨鼓风机　Roots blower
回转鼓风机　rotary blower
涡轮鼓风机　turbo blower
叶片式鼓风机　vane type blower
真空泵　vacuum pump
压缩机　compressor
往复式活塞压缩机　reciprocating piston compressor
多级离心泵　centrifugal pump of multistage type
立式离心泵　centrifugal pump with vertical axis

1 The centrifugal pump consists essentially of one or more impellers rotating in the center of a casting.
离心泵基本上由一个或多个叶轮组成,叶轮可在壳体中旋转。

2 The cavitational equation, cavitational condition and control equations are analyzed on the basis of the centrifugal pump.
以旋离心泵为研究对象,对其汽蚀方程、汽蚀条件和控制方程进行了分析。

第三节　过滤
Section 3　Filtration

均相混合物　homogeneous system
非均相混合物　non-homogeneous system
球形度　sphericity
颗形群　granular group
真密度　actual density
堆积密度　accumulative density
重力沉降　gravity settling
自由沉降　free settling
沉降速度　settling velocity
终端速度　terminal velocity
滤饼过滤　cake filtration
深层过滤　deep bed filtration
动态过滤　dynamic filtration
复合过滤　complex filtration
滤饼　filter cake
滤液　filtrate
过滤介质　filtration medium
固液分离　solid-liquid separation
机械分离　mechanical separation
重力分离　gravity separation

惯性分离 inertial separation
离心分离 centrifugal separation
助滤剂 filter aid
混凝剂 coagulant
絮凝剂 flocculant
离心沉降 centrifugal settling
悬浮体 suspensoid
真空过滤机 vacuum filter
板框压滤机 plate-and-frame filter press
过滤离心机 filtration centrifuge
沉降槽 settling bath
旋风分离器 cyclone separator
加压叶滤机 pressure leaf filter
超滤机 ultrafilter
转筒真空过滤机 rotary vacuum drum filter
降尘室 dust-settling chamber

1 Mathematic model of filtration resistance and filtration velocity of deep bed filtration under constant pressure is established.
建立了恒压操作下深层滤层的过滤阻力和过滤速度数学模型。

第四节 传热
Section 4 Heat Transfer

热传导 heat conduction
热对流 heat convection
自然对流 natural convection
强制对流 forced convection
辐射 radiation
传热速率 heat transfer rate
热流量 heat flow
热通量 heat flux
加热剂 heat agent
冷却剂 coolant
对流传热 convective heat transfer
总传热系数 overall heat transfer coefficient
传热膜系数 film heat transfer coefficient
温度梯度 temperature gradient
导热系数 thermal conductivity
非定态传热 unsteady state heat transfer
温度场 temperature field
温度分布 temperature distribution
温度[分布]剖面图 temperature profile
等温面 isothermal surface
温度梯度 temperature gradient
热导率 thermal conductivity
平壁热传导 heat transfer through plane wall
圆筒壁热传导 heat transfer through cylindrical wall
努塞尔特准数 Nusselt number
雷诺准数 Reynolds number
普兰特准数 Prandtl number
格拉斯霍夫准数 Grashof number
强制对流 forced convection
自然对流 free convection
冷凝 condensation
滴状冷凝 dropwise condensation
膜状冷凝 filmwise condensation
核状沸腾 nucleate boiling
膜状沸腾 film boiling
液体沸腾 liquid boiling
池沸腾(大容积沸腾) pool boiling
温差 temperature difference
对数平均温差 logarithmic mean temperature difference
算术平均温差 arithmetic mean temperature difference
并流 co-current flow
逆流 countercurrent flow
错流 cross flow
热阻 thermal resistance
污垢 fouling
传质单元数 number of transfer units (NTU)
总传热系数 overall heat transfer coefficient
换热 heat exchange
换热器 heat exchanger
间壁式换热器 dividing wall type heat exchanger
夹套式换热器 jacketed heat exchanger
管壳换热器 shell-and-tube heat exchanger
套管换热器 double-pipe heat exchanger
板式换热器 plate heat exchanger
螺旋板换热器 spiral plate heat exchanger
板翅换热器 plate-fin heat exchanger
固定管板换热器 fixed tube-sheet heat exchanger
沉浸式蛇管换热器 submerged coil heat exchanger
浮头式换热器 floating head heat exchanger
喷淋式冷却器 water drop cooler

U形管换热器　U-tube heat exchanger
列管式换热器　shell and tube heat exchanger
管束　tube bundle
盘管　coil
蓄热器　recuperator/heat accumulator
冷却器　cooler/chiller
明火加热炉　fired heater
空气冷却器　air-cooled heat exchanger/air cooler
预热器　preheater
管板　tube sheet
挡板　baffle
翅片管　finned tube
直列管排　in-line tube arrangement
错列管排　staggered tube arrangement
紧凑型换热器　compact heat exchanger
管程　tube pass
壳程　shell pass
夹套　jacket
汽化器　vaporizer
热管　heat-pipe
热管换热器　heat-pipe exchanger
载热体　heating medium/heat transfer medium/heat carrier
隔热　thermal insulation
隔热层　lagging

1 The basic relation for heat flow by conduction is the proportionality between heat flux and the temperature gradient. It is known as Fourier's law. The proportionality constant k is a physical property of the substance called the thermal conductivity. It, like the Newtonian viscosity, is one of the so-called transport properties of the material.
传导的热流量基本关系是和热通量、温度梯度之间的比例关系，称为傅立叶定律。其中比例常数 k 是物质的一个物理性质，称为导热系数，它像牛顿黏度一样，是物质的传输性质之一。

第五节　吸收
Section 5　Absorption

物理吸收　physical absorption
化学吸收　chemical absorption
吸收等温线　absorption isotherm
吸收速率　absorption rate
溶解度　solubility
溶液　solution
溶质　solute
溶剂　solvent
惰性气体　inert gas
解吸，脱附　desorption
吸附　adsorption
变压吸附　pressure swing adsorption (PSA)
非等温吸收　non-isothermal absorption
吸收因子　absorption factor
分子扩散　molecular diffusion
对流传质　mass transfer by convection
扩散通量　rate of diffusion
气液传质设备　gas-liquid mass transfer equipment
等摩尔反向扩散　equimolar counter diffusion
总体流动　bulk flow
填料塔　packed column
板式塔　tray column
气液相平衡　gas liquid equilibrium
双膜理论　two-film theory
气膜控制　gas-film control
液膜控制　liquid-film control
拉西环　Raschig ring
鲍尔环　Pall ring
阶梯环　cascade ring
螺旋环　spiral ring
液泛　flooding
泛点　flooding point
载液　carrier liquid
载点　loading point
泛点速度　flooding velocity
润湿表面积　wetted surface area
传质单元高度　height of transfer unit
润湿率　irrigation rate
喷淋密度　specific liquid rate/spray density
空隙率　voidage
持液量　liquid holdup
持气率　gas holdup
填料支承板　supporting plate
分布板　distribution plate
泡罩板　bubble cap tray
筛板　sieve tray
浮阀板　floating valve tray

穿流塔板 dual-flow tray
穿流栅板 turbo-grid tray
垂直筛板 vertical sieve tray (VST)
多降液管塔板 multidowncomer tray (MD tray)
塔板间距 tray spacing
降液管 downcomer
降液管液柱高度 downcomer backup
溢流堰 overflow weir
堰高 weir height
堰上溢流液头 height of crest over weir
雾沫夹带 entrainment
漏液 weeping
塔板效率 plate efficiency/tray efficiency
级效率 stage efficiency
点效率 point efficiency
默弗里效率 Murphree efficiency
操作弹性 flexibility
负荷性能图 capacity graph
鼓泡塔 bubble column

1 A common apparatus in gas absorption and certain other operations is the packed column. The device consists of a cylindrical column, or tower, equipped with a gas inlet and distributing space at the bottom; a liquid inlet and distributor at the top; gas and liquid outlets at the top and bottom, respectively.
气体吸收和一些其他操作中常见的设备是填料塔,填料塔设备由圆柱形或塔组成,塔底安装有一个气体进口和气体分布空间,塔顶部安装有液体入口和液体分布器;气体和液体出口分别安装在塔体的顶部和底部。

第六节 蒸馏
Section 6 Distillation

连续蒸馏 continuous distillation
平衡蒸馏 equilibrium distillation
间歇蒸馏 batch distillation
精馏 rectification
馏出液 distillate
残液 residue
再沸器 reboiler
冷凝器 condenser
道尔顿定律 Dalton's law
挥发度 volatility
相对挥发度 relative volatility
简单蒸馏 simple distillation
提馏 stripping
塔板 plate/tray
理论级 theoretical stage
理论板 theoretical plate
实际板 actual plate
全凝器 total condenser
分凝器 partial condenser
操作线 operating line
回流比 reflux ratio
全回流 total reflux
最小回流比 minimum reflux ratio
适宜回流比 optimum reflux ratio
芬斯克方程 Fenske's equation
理想溶液 ideal solution
非理想溶液 non-ideal solution
二元混合物 binary mixture
多元混合物 multicomponent mixture
物料衡算 mass balance
精馏段 rectification section
操作线方程 operating line equation
提馏段 stripping section
回收率 recovery
返混 backmixing
液沫夹带 liquid entrainment
共沸蒸馏 azeotropic distillation
萃取蒸馏 extractive distillation
水蒸气蒸馏 steam distillation
分子蒸馏 molecular distillation
真空蒸馏 vacuum distillation
反应蒸馏 distillation with chemical reaction

1 The function of distillation is to separate a liquid mixture of miscible and volatile substance into individual components.
蒸馏的作用是把具有挥发性的液体混合物分离成单个组分。

2 Distillation may be carried out by either of two methods, without reflux (distillation) and with reflux (rectification). Either of these methods may be conducted as a continuous process or a batch process.

蒸馏可通过两种方法进行，没有回流的蒸馏和有回流的精馏，这两种方法可以是连续过程或是间歇过程。

第七节 干燥
Section 7 Drying

干燥 drying
湿物料 moist material
湿含量 moisture content
临界湿度含量 critical moisture content
游离水分 free moisture
结合水分 bound moisture
干燥速率 drying rate
干燥曲线 drying curve
恒速干燥阶段 constant rate drying period
降速干燥阶段 falling rate drying period
厢式干燥器 tray dryer
隧道干燥器 tunnel dryer
带式干燥器 belt dryer
回转干燥器 rotary dryer
圆盘干燥器 disk dryer
喷流干燥器 jet dryer
滚筒干燥器 rotating drum dryer
气流干燥器 pneumatic dryer
旋转闪蒸干燥器 spin flash dryer
螺旋干燥器 spiral dryer
离心干燥器 centrifugal dryer
流化床干燥器 fluidized bed dryer
喷雾干燥器 spray dryer
喷动床干燥器 spouted bed dryer
红外干燥 infrared drying
冷冻干燥 freeze drying
真空干燥 vacuum drying

1 This section is a review of methods for the dehydration of foods, including some novel drying technology, such as microwave-freeze drying, ball drying, ultrasonic drying, and centrifugal fluidized bed drying. The advantages and disadvantages of each method were compared with each other, and the applications of each method were also discussed.

本节综述了食品干燥技术，如微波冷冻干燥、热球干燥、超声波干燥、离心式流化床干燥，并对每种方法的优点与缺点做了对比，对其应用也做了介绍。

第三章　物理化学
Chapter 3　Physical Chemistry

第一节　化学热力学基础
Section 1　Fundamentals of Chemical Thermodynamics

系统　system
环境　surroundings
敞开系统　open system
封闭系统　closed system
隔离系统　isolated system
宏观性质　macroscopic property
强度性质　intensive property
广度性质　extensive property
相　phase
均相系统　homogeneous system
多相系统　heterogeneous system
状态　state
状态函数　state function
热力学平衡　thermodynamic equilibrium
热平衡　heat equilibrium
力平衡　force equilibrium
相平衡　phase equilibrium
化学平衡　chemical equilibrium
过程　process
定温过程　isothermal process
定压过程　isobaric process
定容过程　isochoric process
绝热过程　adiabatic process
循环过程　cyclic process
自由膨胀过程　free expansion process
凝聚相　condensed phase
流体相　fluid phase
饱和蒸汽压　saturated vapor pressure
标准沸点　standard boiling point
临界状态　critical state
临界参量　critical parameters
超临界流体　excess critical fluid
热　heat
化学计量数　chemical stoichiometric number
反应进度　extent of reaction
功　work
体积功　volume work
非体积功　non-volume work
准静态过程　quasi-static process
可逆过程　reversible process
不可逆过程　irreversible process
热力学能　thermodynamic energy
第一类永动机　first kind of perpetual motion machine
理想气体　perfect gas
热容　heat capacity
恒容摩尔热容　molar heat capacity at constant volume
恒压摩尔热容　molar heat capacity at constant pressure
吉布斯自由能　Gibbs free energy
焓　enthalpy
熵　entropy
相变焓　enthalpy of phase transition
生成焓　enthalpy of formation
燃烧焓　enthalpy of combustion
热化学方程式　thermochemical equation
卡诺定理　Carnot theorem
自发过程　spontaneous process
熵增原理　the principle of the increase of entropy
亥姆霍茨自由能　Helmholtz free energy
亥姆霍茨函数判据　Helmholtz function criterion
吉布斯函数判据　Gibbs function criterion
热力学基本方程　master equations of thermodynamics
麦克斯韦关系式　Maxwell's relations

1 Heat is the transfer of energy that makes use of chaotic molecular motion. The chaotic motion of molecules is called thermal motion. When a system heats its surroundings, molecules of the system stimulate the thermal motion of the molecules in the surroundings.
热是利用分子无序运动而造成的能量传递。分子的无序运动称为热运动。当一个系统向环境放热时，系统的分子激发环境分子的热运动。

2 Work is the transfer of energy that makes use of organized motion. When a system does work, it stimulates orderly motion in the surroundings. The most common forms of work are pressure-volume work and electrical work.
功是利用分子有序运动而造成的能量传递。当系统做功时，它激发环境中的有序运动。功最常见的形式是压力 —— 体积功和电功。

3 The physical properties of a perfect gas are completely described by the amount of substance, temperature, pressure and volume. These four parameters are not independent, and the relations between them are expressed by the equation of gas state.
理想气体的物理性质由其物质的量、温度、压力和体积决定。这四个参数不是相互独立的，它们的关系可由气体状态方程描述。

4 Gibbs and Helmholtz free energies are state functions which do not have measurable absolute values. The free energy change represents the maximum amount of work, other than volume expansion work.
吉布斯自由能和亥姆霍茨自由能是状态函数，它们的绝对值不能测量。自由能的变化代表着功的最大值，而不是体积膨胀功。

5 The enthalpy of the solid state reaction to prepare the precursor of $ZnFe_2O_4$ was calculated to be 406.21 kJ/mol, indicating that the reaction is endothermic.
计算了固相反应制备 $ZnFe_2O_4$ 前驱物的焓变：$\Delta Hr = 406.21$kJ/mol，证明了该反应是吸热反应。

第二节　相平衡热力学
Section 2　Thermodynamics of Phase Equilibrium

相平衡　phase equilibrium
克拉佩龙方程　Clapeyron equation
特鲁顿规则　Trouton rule
多组分系统　multicomponent system
溶液　solution
溶剂　solvent
溶质　solute
质量浓度　mass concentration
摩尔分数　mole fraction
质量摩尔浓度　molality of solute
偏摩尔量　partial molar quantity
化学势　chemical potential
物质平衡判据　equilibrium criterion of substance
拉乌尔定律　Raoult's law
亨利定律　Henry's law
理想液态混合物　mixture of ideal liquid
理想稀溶液　ideal dilute solution
分配定律　distribution law
依数性质　colligative properties
渗透压　osmotic pressure
正偏差　positive deviation
负偏差　negative deviation
活度　activity
活度因子　activity factor
凝固点下降系数　freezing point lowering coefficients
沸点升高系数　boiling point elevation coefficients
物种数　number of substance
组分数　number of components
自由度数　number of degrees of freedom

1 A phase can be a solid, liquid, vapour (gas) or aqueous solution which is uniform in both chemical constitution and physical state.
相态可以是固体、液体、蒸汽(气) 或者是化学成分和物理状态均匀的水溶液。

2 Osmosis is the movement of a solvent from a solution of lower solute concentration (higher solvent concentration) to one of higher solute concentration (lower solvent concentration). In osmotic pressure measurements, a semi-permeable membrane, which is permeable only to solvent separating two liquids. This means that this technique is only used for relatively large solute molecules, often polymers or biological macromolecules.
渗透是溶剂从溶质浓度低(溶剂浓度高) 的溶液向溶质浓度高(溶剂浓度低) 的溶液的流动。在渗透压的测量中，采用一种只能透过溶剂分子的半透膜来隔离两种溶液。这意味着该技术只能用来测量较大分子，例如聚合物或生物大分子。

3 Raoult's Law states that the vapour pressure of a component in a mixture at a given temperature is directly proportional to its mole fraction in the liquid mixture, and is equal to the product of its mole fraction in the liquid mixture and the vapour pressure of the pure component at that temperature.
拉乌尔定律规定同一温度下液态混合物中某一组分的蒸汽分压与它的摩尔分数成正比，等于同一温度下该纯组分的摩尔分数与其饱和蒸汽压的乘积。

4 An ideal solution is a mixture of two species, A and B, which show similar molecular interactions between molecules of A, molecules of B and molecules of B and A.
理想溶液是指由 A、B 两物质组成的溶液中，A-A、B-B、A-B 相互作用力相同的液态混合物。

第三节 化学平衡热力学
Section 3 Thermodynamics of Chemical Equilibrium

限度 limit
标准平衡常数 standard equilibrium constant
标准亲和势 standard affinity
生成吉布斯函数 Gibbs function of formation
分压比 ratio of partial pressure
逸度 fugacity
逸度因子 fugacity factor
分逸度比 ratio of partial fugacity
主反应物 principal reactant
分解压 dissociation pressure
同时反应 simultaneous reaction
独立反应 independent reaction
耦合反应 coupling reaction
定温方程 isothermal equation
平衡移动 equilibrium shift
惰性气体 inert gas
解离 dissociate
缔合 associate
方向 direction
转化率 conversion rate
标准摩尔生成吉布斯函数 standard molar Gibbs function of formation

1 Law of chemical equilibrium: at a given temperature, a chemical system may reach a state in which a particular ratio of reactant and product concentrations is a constant value.
化学平衡定律：在一定温度下，达到化学平衡状态时，反应物浓度与产物浓度之比为常数。

2 An equilibrium is established when the rate of the forward and backword reactions are equal. At equilibrium, the Gibbs free energy changes for both the forward and backword reactions are zero.
当一个反应向两边进行的速率相同时，该反应就达到了化学平衡。在平衡点，吉布斯自由能的变化为零。

3 Le Chatelier's principle: If a stress is applied to a system at equilibrium, the system shifts in the direction that relieves the stress.
勒夏特勒原理：当外界对平衡体系产生压力时，平衡体系总是向着减少这种压力的方向移动。

4 When an inert gas is added in a given volume of the container, there is no effect on the equilibrium position, which can be contributed to that the addition of an inert gas increases the total pressure but has no effect on the concentration or partial pressures of the reactants or products.
在体积不变的容器中加入惰性气体，对平衡无影响，因为惰性气体的加入使总压增加，但对反应物、产物的浓度和分压没有影响。

第四节 化学动力学基础
Section 4 Fundamentals of Chemical Kinetics

动力学 kinetics
化学动力学 chemical kinetics
化学反应速率 chemical reaction rate
反应机理 mechanism of reaction
宏观反应动力学 macroscopic reaction kinetics
速率方程 rate equation
微观反应动力学 microscopic reaction kinetics
转化速率 rate of conversion

反应速率 rate of reaction
消耗速率 dissipate rate
增长速率 increase rate
微分(的) differential
积分(的) integral
反应级数 order of reaction
反应总级数 overall order of reaction
一级反应 first order reaction
间歇式反应器 batch reactor
连续式反应器 continuous reactor
反应半衰期 half life of reaction
静态法 stop state method
流动态法 flow state method
微分反应器 differential reaction
积分反应器 integral reaction
总反应器 overall reaction
元反应 elementary reaction
指前参量 pre-exponential parameter
活化能 activation energy
活化分子 activated molecule
复合反应 combined reaction
对行反应 opposing reaction
可逆反应 reversible reaction
连串反应 consecutive reaction
静态 steady state
平衡态 equilibrium state
表现速率常数 apparent rate coefficient
表观活化能 apparent activation energy
链引发 chain initiation
活性的 active
活性中心 active center
催化剂寿命 life of catalyst
中毒 poison
载体 carrier
链传递 chain transfer/propagation
链终止 chain termination
直链反应 straight chain reaction
支链反应 side chain reaction
热爆炸 heat explosion
链爆炸 chain explosion
催化剂 catalyst
催化作用 catalysis
阻化剂 inhibitor
自催化作用 autocatalysis
均相催化 homogeneous catalysis
多相催化 heterogeneous catalysis
广义的酸碱 generalized acid-base
络合催化 complex catalysis
酶催化 enzyme catalysis
主催化剂 principal catalyst
助催化剂 promoter
再生 reactivation

1 For any reversible reaction, when reaching equilibrium state to a significant extent, the molecule of reactants and resultants must be stable, and the entropy is defined as a precise value. The symbol of reaction equilibrium is that the concentrations of each substance in the system stay constant.

对于任一可逆的化学反应,在一定条件下达到化学平衡状态时,体系中各反应物和生成物的物质的量不再发生变化,其活度熵为一定值。化学反应平衡的标志是化学反应体系内的各物质的浓度不再随时间的变化而变化。

2 The rates of reactions:

Suppose the chemical reaction is of the form

$$A + B \longrightarrow Y + Z$$

and that concentrations of the species A, B, Y and Z are [A], [B], [Y] and [Z] respectively. The rate of the reaction can be expressed as the rate of change of the concentration of any of the species. Thus the rate of formation of product is $d[Y]/dt$, and the rate of destruction of A is $-d[A]/dt$, the formula of $-d[A]/dt = -d[B]/dt = d[Y]/dt = d[Z]/dt$ is obtained.

反应速率:

假设 A 与 B 反应生成 Y 和 Z,并分别用[A]、[B]、[Y]和[Z]来表示 A、B、Y 和 Z 的浓度。反应速率可用这几种物质浓度的变化率来表示。Y 的生成速率为 $d[Y]/dt$,A 的消耗速率为 $-d[A]/dt$,并且有 $-d[A]/dt = -d[B]/dt = d[Y]/dt = d[Z]/dt$ 成立。

3 Though the molecules of substance in the chemical equation seem easy, in micromechanics, a chemical reaction is accomplished through steps generally.

虽然化学方程式中各物质的计量比看似简单,但微观上,一个化学反应通常是经过几步完成的。

第五节　电解质溶液
Section 5　Electrolytic Solution

电解质　electrolyte
电解质溶液　electrolyte solution
强电解质　strong electrolyte
弱电解质　weak electrolyte
真正电解质　real electrolyte
潜在电解质　potential electrolyte
解离　dissociate
电离　ionize
电解　electrolyze
电离度　ionization degree
活度　activity
活度因子　activity factor
离子平均活度　ionic mean activity
离子平均活度因子　ionic mean activity factor
电解质活度　activity of electrolyte
正离子活度　activity of positive ion
负离子活度　activity of negative ion
电解质质量摩尔浓度　molality of electrolyte
离子强度　ionic strength
离子氛　ionic atmosphere
导体　conductor
电子导体　electronic conductor
离子导体　ionic conductor
电导　conductance
电导池　conductance cell
电导率　conductivity
摩尔电导率　molar conductivity
漂移速率　drift rate
无限稀薄摩尔电导率　limiting molar conductivity
电迁移率　electric mobility
离子独立运动定律　law of the independent migration of ion

1 The simplest way of studying the motion of ions in solution is through their conductivity, their ability to conduct electricity, which can be determined by measuring their electrical resistance.
研究溶液中离子的运动最简单的方法是通过电导率，电导率就是一种物质导电的能力。要确定溶液的电导率，首先要测定其电阻。

2 An ion in solution is subject to variety of forces. If there are concentration gradients, the ions will spread from the regions of high concentration into those of low. The ions also respond to an electrical force. This motion can be interpreted as the result of an electrical force operating on the ions, the magnitude of the force being proportional to the potential gradient.
溶液中的离子会受到很多力的作用，如果溶液存在浓度梯度，离子要从高浓度区域向低浓度区域扩散。在电流的作用下，离子也要运动，因其受到了电场力的作用，该力的大小与电势梯度成正比。

第六节　电化学系统的热力学及动力学
Section 6　Thermodynamics and Kinetics of Electrochemical System

电化学系统　electrochemical system
相间电势　potential between two phases
接触电势　contact potential
液接电势　liquid-junction potential
扩散电势　diffusion potential
盐桥　salt bridge
电池图式　cell diagram
电化电池　electrochemical cell
原电池　primitive cell
电解池　electrolytic cell
可逆电池　reversible cell
不可逆电池　irreversible cell
干电池　dry battery/dry cell
蓄电池　storage cell/secondary cell
燃料电池　fuel cell
电极　electrode
阳极　anode
阴极　cathode
正极　positive electrode
负极　negative electrode
氢电极　hydrogen electrode
标准氢电极　standard hydrogen electrode
甘汞电极　calomel electrode
氧化　oxidation
还原　reduction
电动势　electromotive force

能斯特方程　Nernst equation	分解电压　decomposition voltage
电极电势　electrode potential	腐蚀　erosion
标准电极电势　standard electrode potential	防护　protection
温度系数　temperature coefficient	电化学传感器　electrochemical sensor
极化　polarization	氧化还原反应　redox reaction
超电势　overpotential	生物电化学　bioelectrochemistry
电催化　electrocatalysis	量子电化学　quantum electrochemistry

1 An electrochemical cell consists of two electrodes dipping into an electrolyte solution. If the two electrodes use a common electrolyte solution, the solution potential is common to both electrodes. Such an arrangement is called a cell without a liquid junction. When two electrodes are immersed in different electrolytes, the potentials of the solutions differ.

两个电极浸在一种电解质中便构成了一个电化学电池，如果这两个电极共用一种电解质，溶液电势对两个电极来说是相同的，这样的电池也叫无液接电池。当两个电极浸入不同的电解质时，溶液电势不同。

2 The potential of a single electrode can't be measured, but one electrode may arbitrarily be assigned the value zero and all others referred to it. The electrode chosen to have zero potential is the standard hydrogen electrode. The standard potential is reached when the hydrogen ion activity is unity, the hydrogen gas is at unit fugacity and the temperature is 298.15K.

我们无法测量单个电极的电势，但可以假定某一个电极电势为零，其他电极都以它为参照。标准氢电极就是我们假定电势为零的电极，其标准的工作条件为氢离子活度为1，氢气的逸度为1，温度为298.15K。

3 The main topic of electrochemistry is investigation of the properties of ionic conductors and of electric circuits containing ionic conductors, and investigation of phenomena occurring during passage of an electric current through such circuits.

电化学的主要议题是研究离子导体和含离子导体电路的性能，研究通过这类电路过程中发生的电流现象。

4 In the general sense of the term all electrodes involve reduction (electronation) and oxidation (de-electronation), but the term oxidation-reduction electrode, or redox electrode, is normally confined to the case where a species exists in the solution in two oxidation states.

从广义角度讲，所有的电极或者为还原电极（得电子），或者为氧化（失电子）电极，但这里所说的氧化-还原电极只局限于某物质在溶液中以两种氧化态的形式存在的情形。

第四章　环境化学
Chapter 4　Environmental Chemistry

第一节　基本概念
Section 1　Basic Conception

太阳系　solar system
地核　earth's core
地幔　earth's mantle
地壳　earth's crust
圈层　segment
岩石圈　lithosphere
土壤圈　pedosphere
水圈　hydrosphere
大气圈　atmosphere
生物圈　biosphere
化学圈　chemosphere
板块构造　plate tectonics
火山活动　volcanic activity
大陆漂移　continental drift
地质年代　geological age
生态系统　ecosystem
开放生态系统　open ecosystem
封闭生态系统　closed ecosystem
生物群落　biotic community
生产者　producer
自养生物　autotroph
消费者　consumer
分解者　decomposer
非生物因子　abiotic factor
能量食物　energy giving food
生态金字塔　ecological pyramid
生态危机　ecological crisis
生态演替　ecological succession
环境　environment
地球环境　earth environment
自然环境　natural environment
社会环境　social environment
人类环境　human environment
生境　habitat
内环境　internal environment
室内环境　indoor environment
环境因素　environmental factor
环境容量　environmental capacity
环境破坏　environmental disruption
环境问题　environmental problem
环境污染　environmental pollution
公害　public nuisance
环境保护　environmental protection
环境学　environmentology
环境化学　environmental chemistry
来源　source
排放　emission
排放标准　amount of emission
迁移　migration
分布　distribution
相互作用　interaction
转化　transformation
形态　morphology
效应　effect
归宿　fate
化学平衡　chemical equilibrium
平衡常数　equilibrium constant
化学变化　chemical change
相　phase
相变　phase transition
相平衡　phase equilibrium
热力学第一定律　first law of thermodynamics
内能　internal energy
热力学第二定律　second law of thermodynamics
不可逆过程　irreversible proccss
可逆过程　reversible process
状态函数　function of state
焓　enthalpy
反应热　reaction heat
标准状态　normal state
燃烧热　heat of combustion
生成热　heat of formation
熵　entropy
吉布斯自由能　Gibbs free energy
生成自由能　free energy of formation
化学势　chemical potential

活度 activity
逸度 fugacity
化学动力学 chemical kinetics
反应方程式 reaction equation
反应速度 reaction rate
质量作用定律 law of mass action
反应速率常数 reaction rate constant
反应级数 reaction order
零级反应 zero-order reaction
一级反应 first-order reaction
半衰期 half life
准一级反应 pseudo first-order reaction
二级反应 second-order reaction
基元反应 elementary reaction
总反应 overall reaction
速度控制步骤 rate-determining step
反应机理 reaction mechanism
可逆反应 reversible reaction
平行反应 parallel reaction
连串反应 consecutive reaction
链式反应 chain reaction
催化反应 catalytic reaction
均相反应 homogeneous reaction
范特霍夫定律 van't Hoff's law
阿仑尼乌斯方程 Arrhenius equation
活化能 activation energy

1 Two human activities that are strongly connected with environmental chemistry in general and pollution in particular are energy production and waste disposal.
通常来说，有两种人类活动尤其和污染密切相关，这两种人类活动是能源生产和废弃物处理。

2 Clearly, understanding of a disposal problem involves knowledgeof physical process such as mixing, chemical processes involved in degradation reactions, chemical reactions of the waste and its products with various aspects of the environment, and biological processes, meteorological processes, geology, oceanography, and so on.
很显然，对一种废弃物处置问题的理解需要很多交叉学科知识，物理工艺中的混合理论，化学工艺中涉及的降解反应、废弃物的化学反应过程和处理后的产品，还涉及环境学、生物学、气象学、地理学、海洋学等知识。

第二节 能量流动和物质循环
Section 2 Energy Flow and Matter Cycle

资源 resource
能量资源 energy resource
化石能源 fossil fuel
核燃料 nuclear fuel
清洁能 clean energy
生物能 bioenergy
石油 petroleum
煤 coal
天然气 natural gas
初级能源 primary energy
可再生性能源 renewable energy
太阳辐射 solar radiation
发射光谱 emission spectrum
辐射能 radiation energy
电磁辐射 electromagnetic radiation
太阳常数 solar constant
白云岩 dolomite
页岩 shale
腐殖质 humic substance
总无机碳 total inorganic carbon
总有机碳 total organic carbon
二氧化碳 carbon dioxide
一氧化碳 carbon monoxide
甲烷 methane
非甲烷烃 non-methane hydrocarbon
碳水化合物 carbohydrate
单糖 monosaccharide
二糖 disaccharide
多糖 polysaccharide
淀粉 starch
纤维素 cellulose
半维生素 hemicellulose
糖原 glycogen
脂 lipid
蛋白质 protein
氨基酸 amino acid
肽键 peptide linkage
辅酶 coenzyme
光合作用 photosynthesis
光反应 light reaction

暗反应　dark reaction
环式光合磷酸化作用　cyclic photophosphorylation
呼吸作用　respiration
好氧呼吸　aerobic respiration
酵解　glycolysis
三羧酸循环　tricarboxylic acid cycle (TCA)
乙酰辅酶　acetyl coenzyme
厌氧呼吸　anaerobic respiration
酒精发酵　alcoholic fermentation
发酵　fermentation
甲烷发酵　methane fermentation
酸性发酵　acidic fermentation
氧循环　oxygen cycle
氮循环　nitrogen cycle
氮氧化物　nitrogen oxide
有机氮　organic nitrogen
氨氮　ammonia-nitrogen
硝酸盐氮　nitrate-nitrogen
总氮　total nitrogen(TN)
固氮作用　nitrogen fixation
氮同化作用　nitrogen assimilation
硝化作用　nitrification
反硝化作用　denitrification
氧化作用　oxidation
硫循环　sulfur cycle
二氧化硫　sulfur dioxide
硫化氢　hydrogen sulfide
磷循环　phosphorus cycle

1 Ozone is a meta-stable molecule, its equilibrium concentration is extremely low, but isolated molecules decompose very slowly. They can, however, be destroyed rapidly by catalytic chain reaction. In such reactions, the ozone is converted to dioxygen by a chain carrier, X, which is itself restored in the process. The general reactions are $X + O_3 = XO + O_2$; $XO + O = X + O_2$. Adding these two reaction together produces reaction: $O_3 + O = 2O_2$.
臭氧是一个相对稳定的分子，它的平衡浓度很低，但是孤立的分子分解很慢。然而，在催化剂的作用下，通过链式反应很快就能分解。在这类反应中，通过链传递者 X，臭氧转化为分子氧，而 X 在反应中保持不变。反应的通式如下：$X + O_3 = XO + O_2$；$XO + O = X + O_2$，上述两反应相加得：$O_3 + O = 2O_2$。

2 The catalytic converter has evolved from a device that simply catalyzes the conversion of carbon monoxide and hydrocarbon to carbon dioxide, to a single unit that uses the reducing capacity of the gases emitted from the engine to first reduce NO_X to N_2 and then later to oxidize the reduced gases to carbon dioxide upon addition of air.
接触式反应器经逐步改进，已经由单一的把一氧化碳和烃类催化转化为二氧化碳改造成能够降低发动机向大气中排放氮氧化物和氮气，同时能够在添加空气的条件下进一步把剩余气体转化为二氧化碳。

第三节　化学污染物
Section 3　Chemical Pollutant

化学品　chemical
化学污染物　chemical pollutant
优先污染物　priority pollutant
毒物　poison
毒剂　toxicant
环境激素　environmental hormone
废弃物　waste
黏度　viscosity
粒度　granularity
熔点　melting point
沸点　boiling point
闪点　flash point
蒸汽压　vapour pressure
挥发度　volatility
蒸发　evaporation
热容　heat capacity
表面张力　surface tension
偶极矩　dipole moment
折射率　refractive index
耗氧性　oxygen consumptivity
富营养作用　eutrophication
腐蚀性　corrosivity
可燃性　combustibility
温室效应　greenhouse effect
臭氧耗竭　ozone depletion
酸性沉积　acid deposition

化学烟雾 chemical smog
毒性 toxicity
急性毒性 acute toxicity
半致死剂量 median lethal dose(LD50)
半致死浓度 median lethal concentration(LC50)
慢性毒性 chronic toxicity
亚急性毒性 subacute toxicity
致癌性 carcinogenicity
致畸性 teratogenicity
致突变性 mutagenicity
扩散 diffusion
凝结 condensation
挥发 volatilization
凝固 solidification
升华 sublimation
浸出 leaching
降尘 dust-falling
沉降 settling
淀积 sedimentation
斯托克斯定律 Stoke's law
雨洗 rainout
洗脱 washout
湿沉积 wet deposition
干沉积 dry deposition
散落 fallout
溶解 dissolution
溶解度 solubility
分配系数 partition coefficient
亨利定律 Henry's law
亨利系数 Henry's coefficient
克劳修斯-克莱佰珑方程 Clausius-Clapeyron equation
同离子效应 common-ion effect
盐效应 salt effect
溶度积 solubility product
脂溶性 fat-solubility
吸附 adsorption
物理吸附 physical adsorption
化学吸附 chemical adsorption
交换吸附 exchange adsorption
吸附等温线 adsorption isotherm
弗里德里方程 Freundlich equation
郎格谬尔方程 Langmuir equation
吸附系数 adsorption coefficient
胶体 colloid
溶胶 sol
胶粒 colloidal particle
等电点 isoelectric point
聚集 aggregation
凝集 coagulation
絮凝 flocculation
水解 hydrolysis
酸离解常数 acid dissociation coefficient
碱离解常数 base dissociation coefficient
水离子积 ion product of water
缓冲作用 buffer action
碳酸体系 carbon acid system
络合反应 complex reaction
配位化合物 complex compound
稳定系数 stability constant
螯合物 chelate compound
氧化值 oxidation number
氧化还原电对 redox pair
电极电位 electrode potential
电子活度 electron activity
生物化学氧化还原反应 biochemical redox reaction
光化学氧化还原反应 photochemical redox reaction
接种 inoculation
驯化 acclimatization
稳定期 stationary phase
衰亡期 death phase
对数期 logarithmic phase
残留 residue
生物蓄积 bioaccumulation
生物浓集 bioconcentration
风化 weathering
净化 scavenging
生物放大 biological amplification
生物转化 biotransformation
生物降解 biodegradation
底物 substrate

1 Determinations of dissolved oxygen serve as the basis of the BOD test, thus, they are the foundation of the most important determination used to evaluate the pollutional strength of domestic and industrial wastes.

因为溶解氧浓度是BOD测定依据,所以溶解氧浓度一直是评价生活污水和工业废水的重要指标。

2 Primary pollutants are released directly from sources.
直接由污染源排放的污染物称为一次污染物。

3 Secondary pollutants are formed through chemical reactions of the primary pollutants and the constituents of the unpolluted atmosphere in the air.
与一次污染物和未受污染的空气的组分发生反应生成的污染物是二次污染物。

第四节　水环境化学
Section 4　Environmental Chemistry of Water

天然水　natural water
地表水　surface water
地下水　ground water
原水　raw water
工业污水　industrial water
农业用水　agricultural water
生活污水　domestic water
饮用水　drinking water
废水　waste water
城市污水　municipal sewage
生活污水　domestic sewage
水质　water quality
水质标准　water quality standard
臭味　odour
臭阈值　threshold odour number (TON)
浊度　turbidity
透明度　transparence
电导率　electric conductivity
氧化还原电位　oxidation reduction potential (ORP)
溶解氧　dissolved oxygen(DO)
硬度　hardness
碱度　alkalinity
酸度　acidity
铬　chromium
汞　mercury
镉　cadmium
铅　lead
镓　gallium
砷　arsenic
镍　nickel
氰化物　cyanide
亚硝酸盐　nitrite
磷酸盐　phosphate
绿色化学　green chemistry
锂　lithium
膜，隔膜　membrane
己糖　hexose
葡萄糖　glucose
生化需氧量　biochemical oxygen demand(BOD)
化学需氧量　chemical oxygen demand(COD)
总需氧量　total oxygen demand(TOD)
总有机碳　total organic carbon (TOC)
水中有机物染物　waterborne organic pollutant
单环芳烃　monocyclic aromatic
多环芳烃　polycyclic aromatic hydrocarbon (PAH)
酚　phenol
氯代化合物　chlorinated compound
三卤甲烷　trihalomethane(THM)
邻苯二甲酸酯　phthalate
有机氮化合物　nitrogenous organic compound
烷基汞　alkyl mercury
有毒化合物　toxic chemical
有毒金属　toxic metal
非靶标生物毒性　toxicity to non-target insect
维生素　vitamin
表面活性物质　surface active substance
理论需氧量　theoretical oxygen demand(ThOD)
硫胺素(维生素 B1)　thiamine
解毒　detoxification
二恶英　dioxin
汞的生物甲基化　biomethylation of mercury
维生素 H，生物素　biotin
抗氧化剂　antioxidant
抗坏血酸盐　ascorbate
阿特拉津，莠去津　atrazine
苯并蒽　benzanthracene
乙醛　acet aldehyde
链烷，烷烃　alkane
烯烃，链烯　alkene
炔　alkyne
氨基化合物　amide
胺　amine
病菌，病原体　pathogen
缩氨酸　peptide

浮游植物　phytoplankton
高原　plateau
点源污染　point sources of pollution
(使)聚合　polymerize
回流污泥　return sludge
持久,稳固的　persistent
十氯酮　kepone
酮　ketone
维生素 E　tocopherol
硝化细菌　nitrifying bacteria
腈　nitrile
氮和食物生产　nitrogen and food production
阴极　cathode
绿色革命　green revolution
氧垂曲线　oxygen sag curve
生物效应　biological effect
蓝绿藻类　blue-green algae
绿藻类　green algae
硅藻类　diatom
过滤　filtration
超过滤　ultrafiltration
微孔过滤　micropore filtration
反渗透　reverse osmosis(RO)
电渗析　electrodialysis
浮渣　scum
硅酸盐　silicate
污泥沉降和干燥　sludge thickening or drying
中和　neutralization
固氮酶　nitrogenase
非持久的,非持续的　nonpersistent
非点源污染　non-point source of pollution
石灰中和　limestone neutralization
灌溉,冲洗　irrigation
高岭石　kaolinite
健康危害　health hazard
亲水的,吸水的　hydrophilic
亲水的离子表面　hydrophilic ionic surface
功能的　functional
全球酸化　global acidification
超营养作用　eutrophication
脂肪的,含脂肪的,脂肪状的　fatty
脂肪酸　fatty acid
酸雨的生态效应　ecosystem effect of acid rain
阳离子　cation
阳离子交换　cation exchange
木炭　charcoal
氯化物　chloride
慢性的,延续很长的　chronic
甘油,丙三醇　glycerol
石灰石　limestone
重力沉降　gravity settling
凝集沉降　coagulative sedimentation
上浮　flotation
气提　stripping
溶剂萃取　solvent extraction
湿式氧化　wet oxidation
一级处理　primary treatment
二级处理　secondary treatment
三级处理　tertiary treatment
深度处理　advanced treatment
脱色　decoloration
预处理　pretreatment
脱磷　dephosphorylation
日用品　domestic production
生物量　biomass
生物脱氮　biological denitrification
病源性物质　disease-causing agent
耗氧性物质　oxygen-consuming agent
农药　pesticide
溶解性固体　dissolved solid
放射性物质　radioactive substance
悬浮物和沉积物　suspended solids and sediment
化学致癌物　chemical carcinogen

1 In the water environment, chromium exists primarily in the form of chromate [Cr(VI)]. Trivalent forms [Cr(III)] are hydrolyzed completely in natural water, and the chromium precipitates as the hydroxide, leaving minor amounts in solution. Furthermore, there is no evidence to indicate that the trivalent form is detrimental to human health.

铬在水环境中主要以六价铬的形式存在。在天然水中三价铬完全水解产生铬的氢氧化物沉淀,只有少量的溶解形态。同时,没有证据表明三价铬对人体健康有害。

2 Chromium is used extensively in industry to make alloys, refractories, catalysts, chromic oxide, and chromate salts.

工业上,铬被广泛应用于生产合金、耐火材料、催化剂、氧化铬以及铬酸盐。

3 Nickel is used in electroplating, and the rinse waters from these operations constitute the major avenue by which salts of these metals gain access to the aquatic environment. It appears to be of low toxicity in humans, but there is some evidence that inhaled nickel may be carcinogenic.
镍主要来源于电镀行业,电镀厂的漂洗废水是构成水体镍污染的主要途径。很显然,镍对人体的毒性是很低的。但是有证据表明,人体吸入镍是可以致癌的。

4 The redox potential of a system can be indicated by p*E*. When p*E* is large, the electron activity is low and the system tends to be an oxidizing one: i. e. , half reactions tend to be driven to the left. When p*E* is small, the system is reducing and reactions tend to be driven to the right.
氧化还原电位(p*E*)可以用来衡量体系的氧化还原能力。从 p*E* 概念可知,p*E* 越大,电子浓度越低,体系接受电子的倾向越强,半反应应趋于向左移动。相反,p*E* 越小,电子浓度越高,体系提供电子的倾向越强,半反应应趋于向右移动。

5 Hydrolysis: The process by which a substrate is split by the intervention of water to form two end products.
水解作用:一种物质与水反应引起自身分解并形成两种新化合物的过程。

6 Reduction: A reaction in which atoms or molecules either lose oxygen or gain hydrogen or electrons.
还原作用:还原作用就是被还原物质(即氧化剂)的原子得到电子,还原剂失去电子的过程。

7 Oxidation: A reaction in which atoms or molecules gain oxygen or lose hydrogen or electrons.
氧化作用:氧化作用就是被氧化物质的原子丢掉电子,氧化剂获得电子的过程。

第五节 大气环境化学
Section 5 Environmental Chemistry of Atmosphere

大气 atmosphere
成层作用 stratification
对流层 troposphere
平流层 stratosphere
臭氧层 ozone layer
中间层 mesosphere
热层 thermosphere
电离层 ionosphere
外逸层 exosphere
大气组成 atmospheric composition
大气水 atmospheric water
自由基 free radical
气溶胶 aerosol
悬浮颗粒物 suspended particulate
尘粒 grit
飘尘 airborne dust
降尘 dustfall
飞灰 fly ash
淀粉酶 amylase
类似的,相似的,可比拟的 analogous
空燃比 air/fuel ratio
烟炱 smut
平均自由程 mean free path
标准大气 standard atmosphere
大气污染 atmospheric pollution
工艺装置 process
焚烧 incineration
本底 background
污染域 pollution scale
环境空气质量 ambient air quality
总悬浮颗粒物 total suspended particulate (TSP)
可吸入颗粒物 inhalable particulate(IP)
臭氧 ozone
苯并芘 benzopyrene
氟化物 fluoride
空气污染指数 air pollution index(API)
大气污染物 atmospheric pollutant
生存期 lifetime
消除 removal
物理消除 physical removal
光电离 photoionization
n-型半导体 n-type semiconductor
海洋能 ocean energy
戊糖 pentose
光激发 photoexcitation
失活 deactivation
光离解 photodissociation
光分解 photolysis
原油 crude oil

光氧化 photooxidation
温室效应 greenhouse effect
地球变暖 global warming
臭氧层破坏 ozone layer destruction
光化学反应 photochemical reaction
氯氟烃 chlorofluorocarbon(CFC)
氟利昂 freon
臭氧损耗潜势值 ozone depletion potential (ODP)
酸雨 acid rain
酸性降水 acid precipitation
光化学烟雾 photochemical smog
光敏剂 photosensitizer
光化学氧化剂 photochemical oxidant
过氧乙酰硝酸酯 peroxyacetyl nitrate(PAN)
能源消费 energy consumption
汽车的能源效率 energy efficiency in automobile
能源利用 energy utilization
居室空气污染 indoor air pollution
油漆 oil varnish
黏合剂 adhesive
化妆品 cosmetic
涂料 paint
烟碱 nicotine
废气治理 waste gas treatment
旋风分离器 cyclone separator
焚烧炉 combustor
静电除尘器 electrostatic precipitator
催化氧化 catalytic oxidation
催化还原 catalytic reduction
烟气脱硫 flue gas desulfurization
烟气脱氮 flue gas denitrogenation
汽车污染控制 automotive pollution control
化石能源 fuel energy
无铅汽油 unleaded gasoline
伦敦烟雾 London smog
洛杉矶烟雾 Los Angeles smog

1 The greenhouse effect is the trapping of heat from below by the atmosphere. Earth's atmosphere admits the visible rays from the sun, but traps the infrared rays emanating from Earth's surface. Some of Earth's heat is carried from the surface by air currents or by the evaporation of water, but most of it is radiated to the atmosphere, which traps it and radiates much of it back, releasing the rest to space.
温室效应是指来自下部的热量被大气截留的现象。地球大气层接受来自太阳的可见射线，但截留由地球表面放射的红外线。地球热量的一部分由表面气流传播，或由水分蒸发而传播，但其多数被辐射到大气层中，大气层将其截留，并将其多数放射回地面，其余部分被释放到太空。

2 Stratospheric ozone protects lives on Earth from harmful effects of UV radiation.
平流层中的臭氧主要保护地球表面生物免受紫外线辐射的伤害。

3 Ozone is a strong oxidant, reacts with molecules containing C = C double bonds, forming epoxides.
臭氧具有强氧化性，能够与含有碳碳双键的化合物发生反应，生成环氧化合物。

4 Ozone region: The atmospheric layer at 10-50 km altitude in which ozone (O_3) is concentrated at 1-10 parts per million. The ozone layer limits the amount of ultraviolet radiation reaching the ground surface.
臭氧层：臭氧层是指大气层的中 10～50 km 中臭氧浓度相对较高的部分，臭氧的浓度在 1～10 ppm，臭氧的作用是防止紫外线到达地面。

第六节 土壤环境化学
Section 6 Environmental Chemistry of Soil

土壤 soil
土壤剖面 soil profile
土层 soil horizon
有机质层 organic matter horizon
腐殖质层 humus layer
生物活性 biological activity
渗透 penetration
淋溶层 eluvial horizon
沉淀层 illuvial horizon
母质层 parent matter horizon
母岩层 bedrock horizon
土壤粒级 soil particle classification
砾石 gravel
土壤组成 soil composition

土壤矿物　soil mineral
硅酸盐　silicate
氢氧化物　hydroxide
原生矿物　primary mineral
石英　quartz
长石　feldspar
云母　mica
副矿物　accessory mineral
辉石　pyroxene
闪石　amphibole
橄榄石　olivine
次生矿物　secondary mineral
碳酸盐　carbonate
重碳酸盐 bicarbonate
卤化物　halide
黏土矿物　clay mineral
铝硅酸盐的结构　structure of aluminosilicate
晶体结构　crystal structure
水合氧化铝　hydrated alumina
土壤有机质　soil organic matter
土壤水　soil water
土壤溶液　soil solution
土壤胶体　soil colloid
土壤空气　soil air
土壤群落　edaphon
土壤微生物　soil microorganism
土壤污染源　soil pollution source
土壤污染物　soil pollutant
金属污染物　metallic pollutant
有机氯的　organochlorine
有机氯杀虫剂　organochlorine insecticide
有机磷酸酯　organophosphate
有机磷杀虫剂　organophosphate insecticide
对硫磷　parathion
持久性杀虫剂　persistent insecticide
持久性有机物染物　persistent organic pollutant
杀虫剂　insecticide
过硫碳酸盐 perthane
杀螨剂　miticide
杀菌剂　bactericide
除草剂　herbicide
熏蒸剂　fumigant
除虫菊酯　pyrethrin
乐果　dimethoate
西维因　sevin
林丹　lindane
艾氏剂　aldrin
滴滴涕　DDT
马拉硫磷　malathion
赛力散　ceresin
浮游生物　plankton
靶器官　target organ
多氯联苯　polychlorinated biphenyl
多氯二苯并二恶英　polychlorinated dibenzodioxins (PCDDs)
多氯二苯并呋喃　polychlorinated dibenzofurans (PCDFs)
吡哆醇　pyridoxol
五氯酚钠　sodium pentachlorophenate
肥料　fertilizer
天然肥料　natural fertilizer
化学肥料　chemical fertilizer
绿肥　green manure
氮肥　nitrogen fertilizer
磷肥　phosphatic fertilizer
钾肥　potassium fertilizer
土壤反应　soil reaction
酸性土壤　acid soil
土壤酸度　soil acidity
中性土壤　neutral soil
碱性土壤　alkaline soil
土壤碱度　soil alkalinity
盐碱土　saline-alkali
盐基饱和度　base saturation percentage
永久电荷　permanent charge
可变电荷　variable charge
羟基化作用　hydroxylation
质子化作用　protonation
单晶硅　crystalline silicon
持留　retention
非专性吸附　nonspecific adsorption
专性吸附　specific adsorption
分子吸附　molecular adsorption
土壤退化　soil degradation
土壤保护　soil conservation
土壤自净作用　soil self-purification
水俣病　minamata

例 Soil also contains organic matter, the humus, consisting of the remains of decomposed vegetation.

The remains that are most resistant to degradation, and therefore most abundants in soils are complex polymeric materials, which have a high content of aromatic groups. Partial oxidation in the soil introduces many carboxylate and phenolic OH groups.
土壤包括有机质,即腐殖质,它是由腐烂的植物残体组成的。这些残体大部分不易分解,因此,土壤中大部分成分是复杂的聚合物,它们富含芳香族化合物。土壤中部分氧化作用产生羧酸盐和带有羟基(OH)化合物。

2 Collectively these are called humic substances, and are sometimes divided into humin, humic acid and fulvic acid.
总体上,它们统称为腐殖质,有时它们又被划分为腐黑物、腐殖酸和富里酸。

3 Active acidity: The capability of hydrogen ion in the aqueous phase of a soil. It is measured and expressed as a pH value.
活性酸度:是指土壤溶液中给出氢离子的能力,常用 pH 值表示。

4 Reserve acidity: Reserve acidity is neutralized by lime or other alkaline materials.
潜性酸度:被石灰或其他碱性物质中和的酸度

第七节 环境放射性
Section 7 Environmental Radioactivity

核素 nuclide
同位素 isotope
放射性衰变 radioactive decay
人为来源 man-made source
衰变系数 decay constant
半衰期 half life
生物蓄积 bioaccumulation
放射毒性 radiotoxicity
白血病 leukemia
原子能的危害 hazard of nuclear power
中子 neutron
辐射接触 radiation exposure
放射能 radioactivity
放射性同位素 radioisotope
可再生能源 renewable energy
回收 reprocessing
半导体 semiconductor
贫营养的 oligotrophic

1 Because it is hydrophobic, DDT readily penetrates the waxy outer coating of insects and, once inside, quickly paralyzes the insect. DDT acts by binding to the nerve cells of insects in a way that holds open the molecular channels that admit sodium ions, which in turn lead to uncontrolled firing of the nerve.
因为 DDT 的疏水性,DDT 很容易透过昆虫蜡制的体表,一旦进入体内,很快使昆虫麻痹。DDT 主要通过作用于昆虫的神经细胞,形成开放的分子空隙,并允许钠离子进入,最终导致神经系统失去控制。

第五章　环境工程微生物学
Chapter 5　Microbiology of Environmental Engineering

第一节　绪论
Section 1　Introduction

环境的　environmental
工程(学)　engineering
微生物学　microbiology
可持续的　sustainable
发展　development
可持续发展　sustainable development
微生物　microorganism
原理　principle
应用、运用　application

1 Environmental biotechnology utilizes microorganisms to improve environmental quality. These improvements include preventing the discharge of pollutants to the environment, cleaning up contaminated environments, and generating valuable resources for human society. Environmental biotechnology is essential to society and truly unique as a technical discipline.

环境生物技术利用微生物来改善环境质量,包括阻止污染物向环境释放,清洁受污环境和为人类产生有价值的资源等改善措施。环境生物技术是一门必要的、具有独特意义的学科。

第二节　病毒
Section 2　Virus

病毒　virus
实体　entity
新陈代谢　metabolism
复制　replicate
亚微观的　submicroscopic
遗传的　genetic
宿主　host
种类　species
噬菌体　phage
传染,感染　infection
疱疹　herpe
天花　smallpox
流行性感冒　influenza
生物体,有机体　organism
水上的　waterborne
内涵物　inclusion
埃希氏菌属　*Escherichia*
核的　nucleic

1 In general, viruses are submicroscopic genetic elements consisting of nuclei acid (DNA or RNA) surrounded by protein and, occasionally, other components. When the viral DNA or RNA is inserted into a proper host cell, it can cause the host cell to redirect its metabolic machinery into the production of duplicate viral particles. When the viral number reaches a sufficient size, the host cell dies and breaks open, and the new virus particles are released for infection of new host cells.

一般情况下,病毒是由核酸(脱氧核糖核酸或核糖核酸)外包蛋白质构成的亚显微遗传粒子,有时还有其他组分。当病毒 DNA 或 RNA 插入到适当的宿主细胞时,能够引起宿主细胞指导自身代谢机制生产复制病毒粒子。一旦病毒数量达到一定程度,宿主细胞死亡、破裂,新的病毒粒子就被释放出来,感染新的宿主细胞。

第三节　原核生物
Section 3　Prokaryotes

原核生物　prokaryote
细菌　bacteria

细胞 cell
细胞壁 cell wall
细胞膜 cell membrane
膜、隔膜 membrane
原生质 protoplasm
颗粒 granule
内含物 inclusion
聚β-羟基丁酸(PHB) poly-β-hydroxybutyric acid
核糖体 ribosome
硫 sulfur
糖原 glycogen
气泡 gas vacuole
拟核 nucleoid
荚膜 capsule
黏液 slime
层 layer
衣鞘 sheath
鞭毛 flagella
胞囊 cyst
繁殖 reproduction
菌丝 filament
生长、发育 growth
分枝杆菌 mycobacterium
诺卡氏菌属，土壤丝菌属 *Nocardia*
革兰(姓氏) Gram
革兰氏染色法 Gram stain procedure
纯的，纯粹的 pure
培养 culture
嗜酸的 acidophilic
极端的，极度的 extreme
极端温度 extreme temperature
蓝细菌 Cyanobacteria
[植]殖段体 hormogonium
多醣，聚糖，多聚糖 polysaccharide
肽聚糖 peptidoglycan
产烷生物 methanogen
脂质，油脂 lipid
甘油，丙三醇 glycerol
脂肪的，含脂肪的 fatty
进化的 evolutionary
单细胞 single cell
空泡，液泡 vacuole

1 The morphology of bacteria includes their shape, size, structure and their spatial relationship to one another. Bacteria have three general shapes as illustrated in Figure 1.2. Those with a spherical shape are cocci, with a cylindrical shape are rods or bacilli, and with a helical shape are spirilla.
细菌形态学研究细菌的形态、大小、结构以及彼此的空间联系。细菌有三种基本形状，如图 1.2 所示。球形的是球菌，圆柱状的为杆菌，螺旋状的为螺旋菌。

第四节 真核生物
Section 4 Eukarya

菌丝 filament
真菌类 fungus
镰孢霉 *Fusarium*
霉菌，(植物的)霉病 mildew
藻类，海藻 algae
海藻的 algal
原生动物 protozoa
轮虫 rotifer
线虫类 nematode
浮游动物 zooplankton
分解体 decomposer
全动物营养性的 holozoic
自养的 holophytic
腐生的 saprophytic
形态学，形态论 morphology
酵母，发酵粉 yeast
菌丝体 mycelium
霉菌 mold
游动的孢子囊 zoosporangium
粉孢子梗 oidiophore
厚垣孢子 chlamydospore
子囊 ascus
担子 basidium
孢囊孢子 sporangiospore
生境 habitat
浮游植物 phytoplankton
光 light
甲壳纲动物，甲壳纲的 crustacea
氧的，含氧的 oxygenic
光合作用 photosynthesis
讨厌的人或东西 nuisance
依靠氧气的，需氧的 aerobic
碱度 alkalinity
毒素 toxin

球的，球形的　spherical
球形的　spherical-shaped
杆，棒　rod
棒状的　rod-shaped
纺锤　spindle
纺锤形的　spindle-shaped
小球藻　chlorella
叶绿素 chlorophyll
眼虫属　*Euglena*
腰鞭毛虫　dinoflagellate
硅藻属　*Diatom*
阿米巴，变形虫　amoeba
纤毛虫　ciliate
草履虫　paramecium
钟形虫　vorticella
甲壳类的，甲壳类　crustacean
多细胞的　multicellular
毛霉属　*Mucor*
根霉属，根霉　*Rhizopus*
曲霉菌　aspergillus
蘑菇　mushroom
马勃(菌)　puffball

1 The fungi, together with Bacteria and Archaea, are the primary decomposers in the world. The decomposers are responsible for oxidation of dead organic material, which returns the resulting inorganic elements back into the environment to be recycled again by other living forms. Fungi do not have the wide-ranging metabolic characteristics of bacteria. Essentially all are organolithotrophic, and none are phototrophic. Most fungi tend to favor terrestrial habitats, although some prefer aquatic systems. They are important in the decomposition of leaves, dead plants and trees, and other lignocellulosic organic debris, which accumulates in soil.

真菌、细菌和古菌都是初级分解者。分解者负责氧化死体有机物质，使之转变成无机物归还给环境，以便其他生物再循环利用。真菌不像细菌有多种代谢途径。本质上都属于有机营养，没有光合营养类型。大多数真菌喜欢陆地栖息，少数喜欢水生系统。它们对于分解如下物质具有重要作用，树叶、植物死体、其他土壤中聚集的含木质纤维素的有机碎片等。

第五节　微生物生理学
Section 5　Microbial Physiology

酶　enzyme
催化剂　catalyst
反应速率　the rate of the reaction
蛋白质，蛋白质的　protein
分子　molecule
初级的　primary
一级结构　primary structure
次要的，二级的　secondary
二级结构　secondary structure
三级结构　tertiary structure
氢　hydrogen
连接，结合　bonding
特异性、特征　specificity
培养基、酶作用底物　substrate
辅酶　coenzyme
米-门动力学　michaelis-mentenkinetics
反应，反动　reactivity
调节　regulating
酶活力调节　regulating the activity of enzyme
需求、要求　requirement
培养基　culture media
生长需求　requirement for growth
基础培养基　chemically defined media
复合的　complex
选择的，选择性的　selective
复合培养基　complex media
加富培养　enrichment culture
选择培养基　selective media
富集　enrichment
质壁分离　plasmolyze
渗透的，渗透性的　osmotic
渗透压　osmotic pressure
碳(元素符号 C)　carbon
氮　nitrogen
磷　phosphorus
自养生物　autotroph
异化作用　catabolism
氨基的　amino
合成代谢　anabolism
乙酰辅酶 A　acetyl-CoA
磷酸化(作用)　phosphorylation
光(合)磷酸化　photophosphorylation

醋酸盐 acetate
柠檬酸盐 citrate
氧化磷酸化 oxidativephosphorylation
柠檬的 citric
水解 hydrolysis
氧化 oxidation
电子 electron
供体 donor
电子供体 electron donor
受体 acceptor
动力(学) dynamical
电子受体 electron acceptor
呼吸,呼吸作用 respiration
好氧呼吸 aerobic respiration
厌氧性的 anaerobic
乙酰基,醋酸基 acetyl
厌氧呼吸 anaerobic respiration
二氧化碳 carbon dioxide
发酵 fermentation
甲烷生成 methanogenesis
能量 energy
具有氧化特性的 oxidative
氧化/还原对 oxidation/reduction couple
酒精,酒 alcohol
链烷,烷烃 alkane
醛,乙醛 aldehyde
碳水化合物,糖类 carbohydrate
纤维素 cellulose
己糖 hexose
草酰乙酸盐(或酯) oxalacetate
脂肪酸β氧化 β-oxidation of fatty acids
淀粉 starch
延胡索酸盐(或酯) fumarate
琥珀酸盐(或酯) succinate
琥珀酰 succinyl
柠檬酸循环 citric acid cycle
酮戊二酸盐(或酯) ketoglutarate
苹果酸盐(或酯) malate
硝化细菌 nitrobacteria
卡尔文 Calvin
卡尔文循环 The Calvin cycle
营养,营养学 nutrition

1 Enzymes are organic catalysts produced by microorganisms and used by them to speed the rate of the thousands of individual energy-yielding and cell-building reactions that occur within the cell. Enzymes comprise the largest and most specialized group of protein molecules within the cell. Proteins are polymers of a mixture of the 20-plus different amino acid subunits. Enzymes are large macromolecules, with molecular weights generally between 10,000 and a million. Enzymes have a primary structure that is dictated by the sequence of the chain of attached amino acids. Enzymes also have a secondary structure that forms when the amino acid chain twists itself into a three-dimensional configuration held together by hydrogen bonding between nearby amino acids and sulfur to sulfur (disulfide) linkages that are common to some of the amino acids. Many proteins also have a tertiary structure, in which the protein is folded back upon itself and again held together by hydrogen bonding. This tertiary structure is generally broken easily by heat or chemicals, causing enzyme denaturation or inactivation.

酶是微生物体内合成的,用来催化细胞内成千上万的能量产生和细胞构建等化学反应的生物催化剂。酶包含细胞内最大且具有特殊作用的蛋白质分子。蛋白质是由 20 多种氨基酸亚单位构成的聚合体。酶是高分子,分子量一般在一万到百万之间。酶蛋白的一级机构是指多肽链本身的氨基酸排列顺序。二级结构是由多肽链盘绕形成的三维空间构型,由氢键和二硫键维持其稳定性。三级结构是在二级结构的基础上,多肽链进一步弯曲盘绕形成的更复杂的构型,由氢键等维持其稳定性。三级结构容易被高温或化学制品等破坏,引起酶变性或灭活。

第六节 生长、繁殖和生存因子
Section 6 Growth, Propagate, Survival Factors

伸长的 elongate
细菌的 bacterial
二元的 binary
二分裂 binary fission
培养 cultivate
批量培养 batch growth
一批 batch
连续的,持续的 continuous

连续培养 continuous cultivate
产生，一代 generation
世代时间 generation time
分开 division
分裂，裂变 fission
落后，迟延 lag
平板 plate
平板计数 plate count
接种 inoculate
群体 colony
过滤，筛选 filtration
混浊，混乱 turbidity
死，死亡 death
新陈代谢的 metabolic
杀菌 sterilization
消毒 disinfection
防腐，消毒 antisepsis
化学药品 chemical
杀菌剂 germicide
细菌抑制 bacteriostasis
化合物 compound
无菌，灭菌法 asepsis
卫生设施 sanitation
热 heat
潮湿的，潮湿 moist
巴斯德杀菌法 pasteurization
干燥 desiccation
铵 ammonium
放射，放射线(物) radiation
卤素 halogen
酚，石碳酸 phenol
酚醛塑料 phenolics
双胍 biguanide
四，四进制的 quaternary
抗生素 antibiotic
对抗(状态)，对抗性 antagonism
(使)氧化 oxidize
竞争，竞赛 competition
捕食 predation
防腐剂 preservative
寄生状态 parasitism
共生，合作关系 symbiosis
静止的 stationary

例1 All organisms grow at the expense of potential energy stored in inorganic or organic molecules or present in radiant energy, such as sunlight. In order to obtain the energy, organisms transform the chemicals through oxidation and reduction reactions. They invest some of that energy into reproducing themselves. In Environmental Biotechnology, the oxidation/reduction reactions and the production of new biomass carried out by microorganisms remove materials that we consider pollutants.

生物体需要消耗储存在无机或有机分子中的能量或辐射能(如太阳能)生长。为了获得能量，有机体依靠化学物质的氧化还原反应。生物利用其中部分能量进行繁殖。在环境生物技术领域，氧化还原反应和生物量的生成都是通过微生物去除我们所谓的污染物来实现的。

第七节 微生物遗传和变异
Section 7 Microbial Heredity and Aberrance

遗传，形质遗传 heredity
变异，畸变 aberrance
核苷 nucleotide
脱氧核糖核酸 deoxyribonucleic acid
基因 gene
聚合酶 polymerase
复制 replication
信史 Messenger
核糖核酸 Ribonucleic Acid (RNA)
信息 information
嘌呤、咖啡碱 purine
嘧啶 pyrimidine
去氧核糖 deoxyribose
碱基 base
碱基对 base pair
腺嘌呤 adenine
鸟嘌呤 guanine
胞嘧啶 cytosine
胸腺嘧啶 thymine
糖苷的 glycosidic
糖苷键 glycosidic bond
编码 encode
染色体 chromosome
质粒 plasmid

转移 transfer
后滞的 lagging
前导的 leading
前导链 leading strand
后随链 lagging strand
DNA 复制 DNA replication
DNA 双链 DNA double strand
复制叉 replication fork
限制性片段长度多态 restriction fragment length polymorphism (RFLP)
聚合酶链式反应 polymerase chain reaction (PCR)
DNA 聚合酶 DNA polymerase enzyme
核酸外切酶 exonuclease
链 strand
启动子 promoter
转导 transduction
转化 transformation
结合 conjugation
寡核苷酸 oligonucleotide
核酸探针 oligonucleotide probe
探针 probe
多态现象 polymorphism
指纹,手印 fingerprint
放大,增强 amplify
凝胶体,成胶 gel
电泳 electrophoresis
限制,约束 Restriction
片段 fragment
长度 length
由分子组成的 molecular
嵌套的 nested
核糖体的 ribosomal
翻译 translation

■ Like all living organisms, microorganisms employ a complex network of information flow to control who they are and what they do. Information flow is synonymous with genetics, because the repository of information is in the gene. However, the gene is only the starting point for genetics and information flow. Several other kinds of molecules are essential.

如同所有生物有机体一样,微生物利用复杂的信息流来控制自己是谁和该做什么。信息流是遗传学的同义词,因为信息库存在于基因中。然而,基因只是遗传学和信息流的起点。许多其他类型的分子对于信息流动也很必要。

第八节 微生物生态学
Section 8 Microbial Ecology

生物地球化学的 biogeochemical
生态系统 ecosystem
(生态)生物群系 biome
(生物)群落 community
群落结构 community structure
微生物,细菌 microbe
土壤微生物 soil microbes
水生微生物 water microbes
交互作用,交感 interaction
选择 selection
聚集,集合 aggregate
絮状物 floc
生物膜 biofilm
多功能性 versatility
假单胞菌 pseudomonas
适应 adaptation
生物净化 biopurification
生物修复 bioremediation
紫色的,紫色 purple
海滨的,沿海地 littoral
指示 indicator
指示生物 indicator organisms

■ Because of their complex nature, microbial ecosystems are able to respond in a very dynamic manner to changes in their environments, particularly changes that stress the community. Examples of stresses include changes in temperature, pH, or salinity: exposure to a toxic material; exposure to xenobiotic organic molecules; and large changes in the availability of substrates.

由于微生物生态系统的复杂性,它能够对环境变化作出适时反应,尤其是对种群产生胁迫的变化。如:温度、pH 和盐度的改变;暴露于毒性物质中;接触异型生物质(药物、杀虫剂、致癌物等);底物可用性的巨大改变。

第九节 微生物在物质循环中的作用
Section 9 The Role of Microorganisms in Substance Cycle

周期，循环 cycle
碳循环 carbon cycle
食物链 food chain
球形的，全球的 global
地球升温 global warming
碳酸盐 carbonate
石灰石 limestone
碳酸根 carbonate ions
氮 nitrogen
氮循环 nitrogen cycle
磷循环 phosphorus cycle
固定 fixation
氨化(作用) ammonification
铵 ammonium
去氨基 deamination
反硝化作用 denitrification
氮化合，硝化作用 nitrification
亚硝酸盐 nitrite
硝酸盐、硝酸钾 nitrate
含氮的 nitrous
固氮 nitrogen fixation
异形细胞 heterocyst
铁 iron
锰(Mn) manganese

1 Denitrification is the dissimilatory reduction of NO_3^- or NO_2^- to (mainly) N_2 gas. In order words, NO_3^- or NO_2^- is the electron acceptor used in energy generation. Denitrification is widespread among heterotrophic and autotrophic bacteria, many of which can shift between oxygen respiration and nitrogen respiration.

反硝化作用即异化还原硝酸根和亚硝酸根为(主要)氮气的过程。换句话说，硝酸根和亚硝酸根为能量产生的电子受体。反硝化作用广泛存在于异养和自养细菌中，其中的许多种可进行氧呼吸和氮呼吸的转换。

第十节 生态工程废水处理中的微生物原理
Section 10 Microbial Principles in Waste Water Treatment of Ecological Engineering

微生物的 microbial
微生物 microorganism
下水道、污水 sewage
处理 treatment
一级污水处理 primary sewage treatment
二级污水处理 secondary sewage treatment
三级污水处理 tertiary sewage treatment
生化需氧量 biochemical oxygen demand (BOD)
生物絮凝 microbial aggregation
反应器 reactor
悬浮生长反应器 suspended-growth reactors
生物塔 biological towers
滴滤系统 trickling filter system
流态化的 fluidized
流化床 fluidized-bed
流化床膜反应器 fluidized-bed biofilm reactors
有活性的 activated
活性污泥 activated sludge
固体，固体的 solid
保持力 retention
构造，配置 configuration
更改，修改 modification
氧供给调节 oxygen supply modifications
负载 loading
混合液悬浮固体 mixed-liquor suspended solids (MLSS)
消化力 digestion
污泥消化 sludge digestion
淤泥，污泥 sludge
生物转盘 rotating biological contactor
通风 aeration
布满泡沫的 foaming
将浮渣去除掉 scum
沉淀物 settling
沉淀问题 settling problem
负载标准 loading criteria

离心的 centrifugal
离心分离 centrifugal separations
膜，隔膜 membrane
膜分离 membrane separations
生物学的 biological
厌氧性的 anaeróbic
厌氧处理 anaerobic treatment
光合的 photosynthetic
光合细菌 photosynthetic bacteria
缺氧的 anoxic
缺氧池 anoxic tank
沉淀池 settling tank
流出物，污水 effluent
向上流 upflow
上流式厌氧活性污泥床 upflow anaerobic sludge blanket reactor (UASB)

1 Two crucial characteristics define the kinds of microorganisms in activated sludge. First, the activated sludge contains a wide variety of microorganisms. Prokaryotes (bacteria) and eukaryotes (protozoa, crustacea, nematodes and rotifers) generally are present, and bacteriophage, which are bacterial viruses, probably reside in the sludge, too. Fungi seldom are important members of the community. Second, most of them are held together within flocs by naturally produced organic polymers and electrostatic forces.

活性污泥中的微生物有两个关键特点。第一，活性污泥中含有大量多种微生物，包括原核生物（细菌）和真核生物（原生动物、甲壳类、线虫类、轮虫），噬菌体（细菌病毒）也存在其中。真菌类不是该群落中的重要分子。第二，绝大多数微生物聚集在由有机聚合物和静电力形成的絮凝物中。

第十一节 废水深度处理和饮用水预处理中的微生物原理
Section 11 Microbial Principles in Waste Water Deep Treatment and Pretreatment of Drinking Water

除磷，去磷 dephosphorization
营养盐 nutrient
去除 removal
营养盐去除 nutrient removal
先后顺序 sequencing
序批式反应器 sequencing batch reactor (SBR)
水力的，水压的 hydraulic
水力停留时间 hydraulic residence times (HRT)
居住，住处 residence
体积 volume
污泥容积指数 sludge volume index (SVI)
索引，指数，指标 index
工艺，技术 technology
机制 mechanism
紫外的 ultraviolet
紫外辐射 ultraviolet radiation
氯胺 chloramine
氯 chlorine
臭氧 ozone
过氧化氢酶 catalase
电解 electrolyse
来源，水源 source
厌氧工艺 anaerobic/aerobic technology (A/O)
厌氧/缺氧/好氧工艺 anaerobic/anoxic/aerobic technology (A_2O)
Bardenpho 工艺 Bardenpho technology
UCT 工艺 University of Cape Town Process
污染，玷污 pollution
器官的，有机的 organic
氨氮 ammonia－nitrogen
酚，羟基苯 hydroxybenzene
氰化物 cyanide
重金属 heavy metal
溶解 dissolve
溶解氧 dissolved oxygen (DO)
杀虫剂 pesticide

1 Biological treatment to produce safe and pleasing drinking water is beginning to play a key role in North America, and it already is common throughout much of Europe. Biological treatment has its most widespread use as a means to remove biodegradable electron donors include biodegradable organic matter (BOM), ammonium, nitrite, ferrous iron, manganese(Ⅱ), and sulfides. The presence of even small concentrations of these instability components can foster the growth of bacteria

in the distribution system. Although the bacteria grown through oxidation of instability components seldom are directly harmful to humans, they create several undesirable water-quality changes as a result of their presence and metabolic activity.

在北美洲，生物处理技术正承担着安全生产、洁净饮用水的重要任务，它在欧洲也很普遍。生物处理技术具有普遍应用性，是去除生物可降解电子供体的手段之一。这些电子供体包括：生物可降解有机物、铵、亚硝酸盐、铁离子、二价锰、硫化物等。即便只存在极微量这些不稳定成分，也可以培养分布区系的细菌使其生长。尽管细菌通过氧化这些不稳组分生长的结果很少会给人类带来直接伤害，但是细菌的存在及其代谢活动也会影响水质。

第十二节 微生物群在处理固体废弃物和废气中的应用
Section 12 Environmental Applications of Microbial Population in Treatment of Solid Wastes and Waste Gases

废物，废弃的 waste
固体废弃物 solid waste
塑胶 plastic
建筑，建筑物 construction
建筑材料 construction materials
市政的 municipal
工业的 industrial
市政和工业固体废弃物 municipal and industrial solid waste
混合肥料，堆肥 compost
需氧气的 aerobic
好氧堆肥 aerobic compost
挥发性的，挥发物 volatile
挥发性有机物 volatile organic compounds (VOCs)
传统的 conventional
燃烧 combustion
吸附 adsorption
二次污染 secondary pollution
生物降解 biodegradation
生物过滤器 biofilter
增湿器，湿润器 humidifier
支撑，支持 support
支撑材料 support material
飞沫，喷射，喷溅 spray
喷射室 spray chamber
氮氧化物 nitrogen oxides
硫氧化物 sulphur dioxide
清洁的，干净的 clean
清洁空气 clean air
硫 sulphur

1 Some gases such as carbon dioxide, water vapor, methane, chlorofluorocarbons, oxides of nitrogen, and ozone absorb thermal radiation near the earth's surface and are called "greenhouse gases". The concentrations of these gases are increasing in the earth's atmosphere due to human activities. Since they do trap heat trying to escape from the earth, they have the potential for warming the earth, a process sometimes called global warming.

二氧化碳、水汽、甲烷、碳氟化物、氮氧化物、臭氧等气体被称为"温室气体"，可以吸收地表的热辐射。由于人类的活动，大气中这些气体的浓度正在逐渐增加。温室气体可以捕获逃离地球的热，具有使地球升温的潜能，这一过程又叫全球变暖。

第十三节 环境生物技术
Section 13 Environmental Biotechnology

固定化酶 immobilized enzyme
固定化微生物 immobilized organisms
胞外多聚物 extracellular polymers
生物乳化剂 bioemulsifier
生物表面活性剂 biosurfactan
生物利用度 bioavailability
污染物 pollutant
生物可降解污染物 bioavailability pollutants
原位生物修复 in situ bioremediation
生物活性的 bioactive
注水井 injection well
生物传感器 biosensor

中空纤维 hollow fibre | 二氧化碳固定 CO_2 fixation
细胞外的 extracellular | 基因工程菌 genetic engineering bacteria

1 Biotechnology is a very important means in eliminating toxic chemical compounds. That microbial biochemistry and genetics have been studied comprehensively is helping to construct the so-called "superbugs" according to expected objective. In this paragraph, a brief introduction was given on the metabolic pathways and the role of plasmids in the biodegradation of chlorinated aromatic compounds. The construction of new strains with novel biodegradative pathways was also summarized.

生物技术对消除有毒化学品已显示出广阔的应用前景，深入研究有毒化学品微生物降解的生物化学和遗传学，可按照预期目的构建具有更大降解能力的遗传工程菌。段落简要介绍了氯代芳香化合物生物降解过程中的代谢途径、质粒作用。另外，还总结了具有新型生物降解途径的新菌株的构建过程。

第六章　水污染控制工程
Chapter 6　Water Pollution Control Engineering

绪论
Introduction

废物　waste
废水　wastewater
固体　solid
悬浮的　suspended
处理　treatment
样品　sample
色度　color
浊度　turbidity
嗅，气味　odor
生物固体　bio-solid
废水特性　wastewater characteristics
组成　composition
消毒　disinfection
出水　effluent
营养物　nutrient
总磷　total phosphorus
化学需氧量　chemical oxygen demand (COD)
总有机碳　total organic carbon (TOC)
总固体　total solids (TS)
总悬浮物　total suspended solids (TSS)
五日生化需氧量　five-day biochemical oxygen demand (BOD_5)
废水处理　wastewater treatment
工业废水　industrial wastewater
生活污水　domestic sewage
水质指标　water quality index
水质标准　water quality standard
预处理　pretreatment
一级处理　primary treatment
二级处理　secondary treatment
三级处理　tertiary treatment
总挥发性固体　total volatile solids
总附着性固体　total fixed solids
挥发性悬浮固体　volatile suspended solids
附着性悬浮固体　fixed suspended solids
总溶解固体　total dissolved solids
挥发性溶解固体　volatile dissolved solids
总附着性溶解固体　total fixed dissolved solids
可沉淀固体　settleable solid
颗粒分布　particle size distribution
电导率　conductivity
游离氨　free ammonia
有机氮　organic nitrogen
总凯氏氮　total kjeldahl nitrogen
亚硝酸盐　nitrite
硝酸盐　nitrate
无机磷　inorganic phosphorus
有机磷　organic phosphorus
碱度　alkalinity
氯化物　chloride
硫酸盐　sulfate
金属　metal
特定的无机元素和化合物　specific inorganic elements and compounds
各种气体　various gases
大肠杆菌　coliform organism
毒性　toxicity
苯　benzene
三溴甲烷　bromoform
氯苯　chlorobenzene
氯甲烷，氯代甲烷　chloromethane
甲苯　toluene
四氯乙烷　tetrachloroethane
三氯乙烯　trichloroethene
氯乙烯　vinyl chloride
甲烷　methane
氯仿，三氯甲烷　chloroform
醛　aldehyde
卤乙酸　haloacetic acid
1-3 二氯苯　1-3 dichlorobenzene
萘　naphthalene
芴　fluorene
七氯　heptachlor
蒽　anthracene
除草剂　herbicide
消毒副产物　disinfection by-product

氯气 chlorine
紫外线 ultraviolet (uv)
臭氧 ozone
氧化 oxidation
初始混合 initial mixing

1 Pretreatment: Removal of wastewater constituents such as rags, sticks, floatable, grit, and grease that may cause maintenance or operational problems with the treatment operations, processes, and ancillary systems.

预处理:从废水中去除能引起处理系统或工艺以及辅助系统故障的成分,如碎屑、枯枝草根、漂浮物、砂砾、油脂。

2 Primary treatment: Removal of a portion of the suspended solids and organic matter from the wastewater.

一级处理:从废水中去除部分悬浮固体和有机物。

3 Secondary treatment: Removal of biodegradable organic matter (in solution or suspension) and suspended solids. Disinfection is also typically included in the definition of conventional secondary treatment.

二级处理:去除可生物降解的有机物(溶解性或悬浮的)和悬浮固体。消毒也是典型的传统二级处理的一部分。

4 Tertiary treatment: Removal of residual suspended solids (after secondary treatment), usually by granular medium filtration or micro-screens. Disinfection is also typically a part of tertiary treatment. Nutrient removal is often included in this definition.

三级处理:通常使用中等粒径滤料过滤或中格栅去除二级处理后剩余的悬浮固体。消毒也是三级处理的一部分,此外也包括营养物质的去除。

5 Advanced treatment: Removal of dissolved and suspended materials remaining after normal biological treatment when required for various water reuse applications.

深度处理:为满足各种废水回用,去除普通生物处理后未能去除的溶解性和悬浮性物质。

6 Biosolids: The materials that remain after sludge and septage are stabilized biologically or chemically.

生物固体:污泥或化粪池垃圾经生物或化学稳定后的物质。

7 Wastewater characteristics: The general classes of wastewater constituents such as physical, chemical, biological, and biochemical.

废水特性:废水成分的一般分类,如物理、化学、生物和生化。

8 Wastewater composition: The makeup of wastewater, including the physical, chemical and biological constituents.

废水构成:废水的组成,包括物理、化学和生物成分。

9 Constituents: The individual components, elements, or biological entities such as suspended solids or ammonia nitrogen.

成分:个体化合物、元素或生物体,例如悬浮固体或者氨氮。

10 Decentralized wastewater management: The collection, treatment, and reuse of wastewater at or near its source of generation.

分散式污水管理:在污染源头或附近进行废水收集、处理和再利用。

11 Effluent: The liquid discharged from a processing step.

出水:从一个处理过程排出的液体。

12 Parameter: A measurable factor such as temperature.

参数:可测量的参数,如温度。

13 Pollutants/Contaminants/Impurities: The constituents added to the water supply through use.

污染物质:水经使用后进入到水里的外来成分。

14 Reclamation: Treatment of wastewater for subsequent reuse application.

回收再利用:废水经处理后重新利用。

15 Recycled water: The water suitable for reuse, replaces reclaimed water.
循环水:适合再利用的水,取代再生水。

16 Repurification: Treatment of wastewater so that it can be used for a variety of applications including indirect potable reuse.
再纯化:废水经处理达到各种使用目的,包括再用于间接饮用水源。

17 Reuse: Beneficial use of reclaimed or repurified wastewater.
再用:再生水或再纯化水的有益使用。

18 Septage: The semiliquid material that is pumped out of septic (or interceptor) tanks, consisting of liquid, scum and sludge.
化粪池垃圾:化粪池中清掏出的半流态物质,由液体、浮渣和污泥组成。

19 Sludge: The material that settles out of wastewater in clarifiers (primary and secondary) lagoons, and aquatic and land treatment systems.
污泥:沉淀池(初沉池和二沉池)以及水生或土地处理系统从废水中沉淀出来的物质。

第一部分 废水的物理处理

Part 1 Physical Disposal of Wastewater

筛滤 screening
过滤 filtration
范围 range
粗格栅 coarse screen
中格栅 micro screen
细格栅 fine screen
人工清洗 hand cleaned
机械清洗 mechanically cleane

1 Screening: The first unit operation generally encountered in wastewater-treatment plants is screening. A screen is a device with openings, generally of uniform size, that is used to retain solids found in the influent wastewater to the treatment plant or in combined wastewater-collection systems.
格栅:在废水处理设施中,第一个操作单元通常是格栅。格栅是一种有孔眼的装置,其孔径通常是均匀的,用于截留废水处理设施或废水集水系统中的固体。

2 Classification of Screens: Two general types of screens, coarse screens and fine screens, are used in preliminary treatment of wastewater. Coarse screens have clear openings ranging from 6 to 150mm; fine screens have clear openings less than 6mm. Micro-screens, which generally have screen openings less than 50μm, are used principally in removing fine solids from treated effluents.
格栅的分类:在废水的初级处理中格栅通常有两种类型,即粗格栅和细格栅。粗格栅的有效孔径为6-150mm;细格栅的有效孔径小于6mm。中格栅的孔径通常小于50μm,主要用于去除流出水中的细小粒子。

第一节 沉淀的基本理论

Section 1 Basic Principle of Settling

分离 separation
原理 theory
微粒 particle
絮状物 floccule
沉淀物 settling
自由沉淀 discrete particle settling
絮凝沉淀 flocculent settling
区域沉淀 hindered settling
压缩沉淀 compression settling
理想沉淀 ideal discrete particle settling
重力沉降 gravity settling
流入区 inlet zone
污泥区 sludge zone
沉淀区 settling zone
流出区 outlet zone

1 Discrete particle settling: Refers to the setting of particles in a suspension of low solids concentration by gravity in a constant acceleration field. Particles settle as individual entities, and there is no significant interaction with neighboring particles.

自由沉淀：指在低颗粒浓度的悬浮液中，颗粒以恒定加速度沉淀，沉淀过程中颗粒保持稳定，不与周围颗粒相互作用。

2 Flocculent settling: Refers to a rather dilute suspension of particles that coalesce, or flocculate during the setting operation. By coalescing, the particles increase in mass and settle at a faster rate.

絮凝沉淀：指悬浮液中颗粒浓度不高但有絮凝性，或者在絮凝过程中发生沉淀。由于凝聚，颗粒的质量增加，沉淀速度加快。

3 Hindered settling: Refers to suspensions of intermediate concentration, in which inter-particle forces are sufficient to hinder the setting of neighboring particles. The particles tend to remain in fixed positions with respect to each other, and the mass of particles settles as a unit. A solids-liquid interface develops at the top of the settling mass.

区域沉淀：指悬浮液中颗粒浓度较高，每个颗粒下降都要受到周围其他颗粒的干扰，互相牵扯，形成一个整体。在沉淀物质上面形成固 - 液界面。

4 Compression settling: Refers to settling in which the particles are of such concentration that a structure is formed, and further settling can occur only by compression of the structure. Compression takes place from the weight of the particles, which are constantly being added to the structure by sedimentation from the supernatant liquid.

压缩沉淀：指高颗粒浓度沉淀时，颗粒之间形成团块结构，互相接触，互相支撑，在上层颗粒的重力作用下，下层颗粒间的水被挤出，颗粒形成的团块结构被压缩。

第二节　沉淀池
Section 2　Sedimentation Tank

粗砂　grit
初级的　primary
沉淀　sedimentation
矩形的　rectangular
圆形的　circular
逆流　countercurrent
堰　weir
平流沉砂池　horizontal-flow grit chamber
曝气沉砂池　aerated grit chamber
初沉池　primary sedimentation tank
矩形池　rectangular tank
圆形池　circular tank
斜板(管)沉淀池　inclined plate and tube settling
错流(十字流)　cross-flow

1 Inclined plate and tube settling: Inclined plate and tube settlers are shallow settling devices consisting of stacked offset trays or bundles of small plastic tubes of various geometries that are used to enhance the settling characteristics of sedimentation basins. They are based on the theory that settling depends on the settling area rather than detention time.

斜板(管)沉淀池：斜板(管)沉淀池是由波纹板或者各种形状的小塑料管捆扎成束而成的浅层沉淀设备，用于提高沉淀池的沉淀性能。其设计理论：沉淀面积对沉淀效率的影响大于停留时间对沉淀效率的影响。

2 Although they are used predominantly in water-treatment applications, plate and tube settlers are used in wastewater-treatment for primary, secondary, and tertiary sedimentation. In primary sedimentation applications, however, fine screening should be provided ahead of the settling operation to prevent plugging of the plates or tubes.

尽管斜板(管)沉淀池可用于废水处理中的一级、二级和三级沉淀，但是用于一级沉淀时，仍然要在沉淀单元前设置细格栅防止板或管的堵塞。

第三节　气浮池
Section 3　Air-Flotation Tank

浮选　flotation
凝聚剂　coagulant
絮凝剂　flocculant
挡板　baffle

进水 influent
出水 effluent
分散空气浮上 dispersed-air flotation
溶气浮上 dissolved-air flotation

1 Flotation: Flotation is a unit operation used to separate solid or liquid particles from a liquid phase. Separation is brought about by introducing fine gas (usually air) bubbles into the liquid phase. The bubbles attach to the particulate matter, and the buoyant force of the combined particle and gas bubbles is great enough to cause the particle to rise to the surface.

浮上法:浮上法是一个用于从液相中分离固体颗粒或液体粒子的处理单元。通过向液相中通入细小的气泡(通常为空气),气泡黏附到粒子上,在浮力的作用下黏附气泡的粒子上浮到表面而得以分离。

2 In wastewater treatment, flotation is used principally to remove suspended matter and to concentrate bio-solids. The principal advantages of flotation over sedimentation are that very small or light particles that settle slowly can be removed more completely and in a shorter time.

在废水处理中,浮上法主要被用于去除悬浮颗粒或浓缩生物粒子。在去除非常小或轻的粒子时,浮上法比沉淀法效果好且时间短。

第四节 过滤
Section 4 Filtration

反洗 backwash
多层介质 multimedium
深层过滤 depth filtration
浅层过滤 surface filtration
单层介质 single medium
双层介质 dual medium

1 Mechanical squeezing: Particles larger than the pore space of the filtering medium are strained out mechanically.

机械挤压:颗粒直径大于填料孔隙的需要用机械装置进行挤压。

2 Chance contact: Particles smaller than the pore space are trapped within the filter by chance contact.

偶然接触:颗粒直径小于填料孔隙的会留在过滤器内偶然接触。

3 Impaction: Heavy particles will not follow the flow streamlines.

嵌塞:较重粒子不会随水流运动。

4 Interception: Many particles that move along in the streamline are removed when they come in contact with the surface of the filtering medium.

拦截:许多颗粒会随着水流流动,接触到填料表面的颗粒会被截留住。

5 Adhesion: Particles become attached to the surface of the filtering medium as they pass by. Because of the force of the flowing water, some materials are sheared away before they become firmly attached and are pushed deeper into the filter bed. As the bed becomes clogged, the surface shear force increases to a point at which no additional material can be removed. Some material may break through the bottom of the filter, causing the sudden appearance of turbidity in the effluent.

黏附:当颗粒通过填料,会附着到填料表面。因为水的冲击,一些材料在固定之前会被冲刷下,并被冲到滤池的底部。由于滤床变得拥堵,表面剪切力增加到一个点上,不会有额外的材料被冲走。一些材料可能冲破滤床的底部,造成浑浊污水的出现。

6 Turbidity breakthrough: Unacceptable levels of turbidity are recorded in the effluent from the filter, even though the terminal head loss has not been reached. To control the buildup of effluent turbidity levels, chemicals and polymers have been added to the filter. The point of chemical or polymer addition must be determined by testing.

浊度穿透:尽管水头损失尚未达到控制标准,滤池出水浊度超过了可接受的水平。为了防止出水浊度,需要向滤池中添加化学药品和聚合物。化学药品或聚合物投加量需经过实验确定。

7 Mudball formation: Mudballs are an agglomeration of biological floc, dirt and the filtering medium or media. If the mudballs are not removed, they will grow into large masses that often sink into the filter bed and ultimately reduce the effectiveness of the filtering and backwashing operations. The

formation of mudballs can be controlled by auxiliary washing processes such as air scour or water surface wash concurrent with, or followed by, water wash.

泥团的形成：泥团由集聚生物絮体、污垢和过滤介质组成。如果泥团不去除，它们将成长为较大的泥团，并常常沉积于过滤床而最终导致过滤和反冲洗操作效率降低。辅助冲洗过程可以控制泥团的形成，如气冲或气水反冲同时后续用水反冲。

8 Development of cracks and contraction of filter bed: If the filter bed is not cleaned properly, the grains of the filtering medium become coated. As the filter compresses, cracks develop, especially at the sidewalls of the filter. Ultimately, mudballs may develop. This problem can be controlled by adequately backwashing and scouring.

滤床裂隙的产生：如果没有妥善清理，过滤介质会被覆盖。随着过滤浓缩，会导致裂隙的产生，特别沿滤池的侧壁会产生裂隙。最终导致泥团产生。这个问题可以通过充分反冲洗和冲刷来解决。

9 Loss of filter medium or media (mechanical): In time, some of the filter materials may be lost during backwashing and through the underdrain system (where the gravel support has been upset or the underdrain system has been installed improperly). The loss of the filter material can be minimized through the proper placement of washwater troughs and underdrain system. Special baffles have also been proved effective.

过滤介质的损耗(机械)：经过一段时间，一些过滤材料可能在反冲洗过程中流失，并且通过排水系统(如支撑砾石被破坏或暗渠系统安装不当)排出。通过合理安置洗涤水槽和排水系统可以减少过滤材料的流失。专用挡板经证明也是有效的。

10 Loss of filter medium or media (operational): Depending on the characteristics of the biological floc, grains of the filter material can become attached to it, forming aggregates light enough to be floated away during the backwashing operations. The problem can be minimized by the addition of an auxiliary air and/or water scoring system.

过滤介质的损耗(运行)：取决于生物絮体是否具有附着到过滤材料颗粒上的特点，使其重量变轻，并随反冲洗操作排走。这个问题可以通过增加的一个辅助气和(或)水冲洗系统来解决。

11 Gravel mounding: Gravel mounding occurs when the various layers of the support gravel are disrupted by the application of excessive rates of flow during the backwashing operation. A gravel support with an additional 2- to 3-in (50- to-75-mm) layer of high-density material, such as ilmenite or garnet, can be used to overcome this problem.

碎石堆积：在反冲洗操作中，过大的流速会导致不同层支撑砾石混杂，发生碎石堆积现象。支撑砾石通过添加2层以上或3层以上(50mm以上或75mm以上)的高密度材料，如钛铁矿、石榴石等来解决这个问题。

第二部分　废水的化学处理
Part 2　Chemical Disposal of Wastewater

第一节　混凝
Section 1　Coagulation

凝聚　coagulation
絮凝　flocculation
胶体　colloid
电荷　charge
水解　hydrolyze
双电层　electrical double layer
凝聚沉降　coagulative sedimentation
电中和　charge neutralization
吸附架桥　polymer bridge formation

1 Colloidal particles found in wastewater typically have a net negative surface charge. The size of colloids (about 0.01 to 1μm) is such that the attractive body forces between particles are considerably less than the repelling forces of the electrical charge.

废水中的胶体颗粒一般带有负表面电荷，胶体的尺寸(0.01-1μm)使颗粒间的引力远小于颗粒间的电荷斥力。

2 Under these stable conditions, Brownian motion keeps the particles in suspension.
在稳定的情况下，布朗运动保持颗粒处于悬浮状态。

3 Brownian motion is brought about by the constant thermal bombardment of the colloidal particles by the relatively small water molecules that surround them. Coagulation is the process of destabilizing colloidal particles so that particle growth can occur as a result of particle collisions.
胶体颗粒的布朗运动是由于受到环绕其周围的较小水分子的热运动碰撞所引起的。混凝是胶体颗粒脱稳的过程，颗粒的长大则是颗粒碰撞的结果。

第二节 中和
Section 2 Neutralization

调整 adjustment | 稳定性 stabilization

1 The removal of excess acidity or alkalinity by treatment with a chemical of the opposite composition is termed neutralization. In general, all treated wastewater with excessively low or high pH will require neutralization before they can be dispersed to the environment.
用与其组分相反的化学试剂去除过量酸度与碱度的方法称为中和。一般来说，所有处理后的废水的pH均过低或过高时，排放到环境之前均需进行中和。

第三节 化学沉淀
Section 3 Chemical Precipitation

沉淀 precipitation
明矾 alum
石灰 lime
氯化物 chloride
硫酸亚铁 ferrous sulfate
氯化铁 ferric chloride
重金属 heavy metal
溶解性无机物 dissolved inorganic substance

1 The technologies available for removal of heavy metals and dissolved inorganic substances: The technologies available for the removal of heavy metals from wastewater include chemical precipitation, carbon adsorption, ion exchange, and reverse osmosis. Of these technologies, chemical precipitation is most commonly employed for most of the metals.
化学沉淀用于去除重金属和溶解性无机物：从废水中去除重金属的技术有化学沉淀法、活性炭吸附法、离子交换法以及反渗透法。对于大多数金属，化学沉淀法是最常采用的一种方法。

第四节 化学氧化还原
Section 4 Chemical Oxidation-Reduction

氧化 oxidation
还原 reduction
反应 reaction
氧化还原 oxidation-reduction
中和 neutralization

1 Chemical oxidation in wastewater treatment typically involves the use of oxidizing agents such as ozone (O_3), hydrogen peroxide (H_2O_2), permanganate (MnO_2), chloride dioxide (ClO_2), chlorine (Cl), and oxygen (O_2), to bring about change in the chemical composition of a compound or a compounds.
废水处理中的化学氧化法是使用臭氧(O_3)、双氧水(H_2O_2)、二氧化锰(MnO_2)、二氧化氯(ClO_2)、氯(Cl)以及氧(O_2)等具有代表性的氧化剂，使一种或一组化合物的组分发生变化。

Dose = concentration × time for chemical agents
剂量 = 浓度 × 化学药剂作用时间
Dose = intensity × time
剂量 = 强度 × 时间

第三部分 吸附、离子交换、膜分离及其他相转移分离法

Part 3 Adsorption, Ion Exchange, Membrane Separation and Other Separation Methods

第一节 吸附

Section 1 Adsorption

吸附 adsorption
平衡 equilibrium
等温线 isotherm
吸附容量 adsorption capacity
粒状 granular
脱臭 deodorization
再生 regeneration
活性炭 activated carbon
粉末状活性炭 powdered activated carbon(PAC)
粒状活性炭 granular activated carbon(GAC)
脱色 decoloration

1 The adsorption process takes place in four more or less definable steps: (1)bulk solution transport, (2)film diffusion transport, (3)pore transport, and(4)adsorption (or sorption).
从某种意义上讲,吸附过程至少可分为下述四个可定义步骤:(1) 总体溶液内迁移,(2) 膜扩散迁移,(3) 孔隙内迁移,(4) 吸附(或吸收)。

第二节 离子交换

Section 2 Ion Exchange

树脂 resin
交换 exchange
阳离子 action
阴离子 anion
离子交换 ion exchange
强酸性阳离子树脂 strong-acid cation resin
弱酸性阳离子树脂 weak-acid cation resin
强碱性阴离子树脂 strong-base anion resin
弱碱性阴离子树脂 weak-base anion resin
交换容量 exchange capacity

1 The most widespread use of this process is in domestic water softening, where sodium ions from a cationic-exchange resin replace the calcium and magnesium ions in the treated water, thus reducing the hardness.
离子交换工艺广泛用于生活用水的软化处理。软化是阳离子交换树脂上的钠离子与被处理水中的钙、镁离子置换,从而降低水的硬度的一种工艺。

2 Ion exchange has been used in wastewater applications for the removal of nitrogen, heavy metals, and total dissolved solids.
在废水处理中,利用离子交换法从废水中去除氮、重金属和溶解性固体。

第三节 膜分离

Section 3 Membrane Separation

膜 membrane
反应器 reactor
污垢 fouling
膜组件 module
膜生物反应器 membrane biological reactor
截留分子量 molecular weight cutoff
微滤 microfiltration (MF)
超滤 ultrafiltration (UF)
纳滤 nanofiltration (NF)
反渗透 reverse osmosis (RO)
膜过滤工艺 membrane filtration process
膜材料 membrane material
驱动力 driving force
分离尺寸 size of separation
去除机理 removal mechanism
管式膜 tubular module

中空纤维膜 hollow fiber
螺旋卷式膜 spiral wound
板框式膜组件 plate and frame
膜污染 membrane fouling
渗透 permeate

1 Molecular weight cutoff(MWCO): The molecular weight of the smallest material rejected by the membrane, usually expressed in Daltons.

截留分子量:膜截留的最小物质的分子量,通常用道尔顿表示。

2 The separation of particles in MF and UF is accomplished primarily of straining. In NF and RO, small particles are rejected by the water layer adsorbed on the surface of the membrane which is known as a dense membrane.

在微滤和超滤工艺中,颗粒的分离主要是通过筛分实现的。在纳滤和反渗透工艺中,小颗粒是由吸附在被称为致密膜的表面的水层截流的。

3 Ionic species are transported across the membrane by diffusion through the pores of macromolecule comprising the membrane. Typically NF can be used to reject constituents as small as 0.001μm whereas RO can reject particles as small as 0.0001μm. Straining is also important in NF membrane, especially at the larger pore size opening.

各种离子通过扩散迁移作用穿过膜的大分子孔。典型的纳滤工艺可用于分离 0.001μm 的组分,而反渗透可分离 0.0001μm 的颗粒。在纳滤膜中的筛分作用也很重要,特别是孔径较大的纳滤膜。

第四节 其他相转移分离法
Section 4 Other Separation Methods

吹脱 blowing
气提 stripping

第四部分 废水生化处理理论基础
Part 4 The Theoretical Foundation of Wastewater Biochemical Disposal

第一节 废水处理微生物基础
Section 1 Biological Fundamentals of Treatment of Wastewater

微生物 microorganism
生长 growth

1 Microbial growth kinetic: The kinetics of microbial growth govern the oxidation of substrate and the production of biomass, which contributes to the total suspended solids concentration in a biological reactor.

微生物生长动力学:微生物生长动力学描述的是在生物反应器中,总悬浮物浓度与底物降解量和微生物增长量有关。

2 Objective of biological treatment: The overall objectives of the biological treatment of domestic wastewater are to (1) transform dissolved and particulate biodegradable constituents into acceptable end products, (2) capture and incorporate suspended and non-settleable colloidal solids into a biological floc or bio-film, (3) transform or remove nutrients, such as nitrogen and phosphorus, and (4) in some cases, remove specific trace organic constituents and compounds.

生物处理的目的:生活污水生物处理的主要目的是:(1) 将能够生物分解的溶解性和非溶解性组分转化为可接受的最终产物,(2) 将悬浮物以及非沉淀性的胶体颗粒截留并转变成生物絮体或生物膜,(3) 去除或转化营养物质如氮和磷,(4) 在有些情况下,去除特殊的微量有机组分和化合物。

3 For industrial wastewater, the objective is to remove or reduce the concentration of organic and inorganic compounds.

对于工业废水,其目的是去除或降低有机物和无机物的浓度。

第二节 酶及酶反应

Section 2 Enzyme and Enzymatic Reaction

可生物降解 biodegradable
微生物生长动力学 microbial growth kinetic
有机颗粒物 particulate organic matter
底物利用 substrate utilization
微生物增长 biomass growth
动力学系数 kinetic coefficient

1 Types of biological processes for wastewater treatment: The principal biological processes used for wastewater treatment can be divided into two main categories: suspended growth and attached growth processes.

废水生物处理的类型:废水生物处理的方式主要分为两类:悬浮生长方式和附着生长方式。

第五部分 活性污泥法

Part 5 Active Sludge Methods

第一节 基本原理及分类

Section 1 Basic Principle and Classification

污泥 sludge
悬浮生长 suspended growth
活性污泥 activated sludge
返送污泥 return activated-sludge
剩余污泥 waste sludge
活性污泥浓度 mixed liquor suspended solids (MLSS)
挥发性悬浮固体浓度 mixed liquor volatile suspended solids (MLVSS)
污泥容积指数 sludge volume index (SVI)
氧化沟 oxidation ditch
活性污泥法 activated sludge process
推流式 plug-flow
完全混合式 complete-mix
好氧工艺 aerobic (oxic) process
厌氧工艺 anaerobic process
缺氧工艺 anoxic process
兼氧工艺 facultative process
联合工艺 hybrid (combined) process
附着生长工艺 attached-growth process
氧化塘工艺 lagoon process
悬浮生长工艺 suspended-grow process
悬浮和附着生长组合工艺 hybrid(combined) suspended-and-attached-growth process
生物滤池或活性污泥法 trickling filter/activated sludge
人工湿地 constructed wetland
悬浮生长反硝化 suspended-growth denitrification
固定膜反硝化 fixed-film denitrification
厌氧接触工艺 anaerobic contact process
厌氧消化 anaerobic digestion
厌氧固定床 anaerobic fixed bed
升流式厌氧污泥床 upflow anaerobic sludge blanket
升流式污泥床或固定床反应器 upflow sludge blanket/ fixed-bed reactor
深度处理塘 maturation (tertiary) lagoon

1 The biomass solids in a bioreactor are commonly measured as total suspended solids (TSS) and volatile suspended solids(VSS). The mixture of solids resulting from combining recycled sludge with influent wastewater in the bioreactor is termed mixed liquor suspended solids(MLSS) and mixed liquor volatile suspended solids(MLVSS).

生物反应器中的生物固体通常分为总悬浮固体(TSS)和挥发性悬浮固体(VSS)。生物反应器中的污泥与流入废水形成的混合液固体分为混合液悬浮固体(MLSS)和混合液挥发性悬浮固体(MLVSS)。

第二节 活性污泥法参数

Section 2 Parameters of Active Sludge Method

截留 retention
负荷 load
食物与微生物比 F/M the ratio of food to microorganism
容积有机负荷 volumetric organic loading
污泥停留时间 sludge retention time (SRT)

1 Certain design and operating parameters distinguish one activated-sludge process from another. The common parameter used are the solids retention time(SRT), the food to biomass (F/M) rate.
活性污泥系统设计和运行参数的确定不同于其他处理方法。经常使用的参数是污泥停留时间(SRT)和生物负荷率(F/M)。

2 The two types of reactor used commonly are the complete-mix and the plug-flow reactor. From a practical standpoint, the hydraulic detention times of many of the complete-mix and plug-flow reactor in actual use are about the same. The reason is that the designs for BOD removal are generally governed by an STR sufficient to assure good settling properties and of duration longer than needed for BOD removal.
完全混合式和推流式反应器是常用的两种类型反应器。从实际运行结果看,很多的完全混合式与推流式反应器的水力停留时间在实际运行时是相同的。其原因是,在设计时保证 SRT 足够长,以获得良好的污泥沉淀性能和长期稳定的 BOD 去除效果。

第三节　曝气
Section 3　Aeration

传递　transfer
曝气　aeration
扩散器　diffuser
风机　blower
喷射　jet
泡沫　bubble
氧传递　oxygen transfer
清水　clean water
修正系数　correction factor
鼓风曝气　air aeration
机械曝气　mechanical aeration

1 Oxygen transfer, the process by which oxygen is transferred from the gaseous to the liquid phase, is a vital part of a number of wastewater-treatment processes.
氧传递过程是氧从气态传递到液相,是废水处理系统中一个非常重要的部分。

2 There are several types of aeration systems used for wastewater treatment. The systems used depend on the function to be performed, type and geometry of the reactor, and cost to install and operate the system.
用于废水处理的曝气系统有几种类型,曝气系统的选用与反应器的形状及曝气系统的自身性能以及安装、运行成本有关。

第四节　曝气池的构造与设计
Section 4　Structure and Designation of Aeration Pod

观测　observe
剩余污泥量　sludge production
需氧量　oxygen requirement
溶解氧　dissolved oxygen
耗氧量　oxygen uptake
表观产率　observed yield (Yobs)

1 The sequencing batch reactor (SBR) system has five steps in common, which are carried out in sequence as follows: (1) fill, (2) react (aeration), (3) settle (sedimentation/clarification), (4) draw (decant), and (5) idle.
序批式活性污泥法(SBR) 通常有五个阶段,先后顺序为:(1) 充水,(2) 反应(曝气),(3) 沉淀(沉降/澄清),(4) 排出,(5) 闲置。

2 Applicable flow range: The process should be matched to the expected range of flow rates. For example, stabilization ponds are not suitable for extremely large flow rates.
适当的流量范围:工艺选择需结合要处理的流量,例如,稳定塘工艺就不适合非常大的流量。

3 Applicable flow variation: Most unit operations and processes have to be designed to operate over a wide range of flow rates. Most processes work best at a relatively constant flow rate. If the flow variation is too great, flow equalization may be necessary.

适当的流量变化：大多数装置的运行需要设计成适合很大的流量变化范围。大多数工艺在相对稳定的流量下运行良好。如果流量变化太大，需对流量进行均化处理。

4 Influent wastewater characteristics: The characteristics of the influent wastewater affect the types of processes to be used (e.g., chemical or biological) and the requirements for their proper operation.

进水的特性：进水特性影响废水处理工艺的选择(例如化学的还是生物的)和正常运作的要求。

5 Climatic constraints: Temperature affects the rate of reaction of most chemical and biological processes. Temperature may also affect the physical operation of the facilities. Warm temperatures may accelerate odor generation and also limit atmospheric dispersion.

气候条件限制：温度影响大多数化学和生物工艺的反应速率，温度也影响设备的运行。较高的气温加快臭气的产生，也限制大气扩散。

6 Reaction kinetics and reactor selection: Reactor sizing is based on the governing reaction kinetics. Data for kinetic expressions usually are derived from experience, published literature, and the results of pilot plant studies.

反应动力学和反应器的选择：反应器的大小是建立在控制性反应动力学基础上的。动力学表达式的数据通常是通过经验、出版的文献数据和中试试验获得。

7 Performance: Performance is usually measured in terms of effluent quality, which must be consistent with the effluent discharge requirements.

性能：工艺性能通常是通过出水水质来评估的，出水水质必须满足废水排放标准。

8 Treatment residuals: The types and amounts of solid, liquid, and gaseous residuals produced must be known or estimated. Often, pilot plant studies are used to identity and quantify residuals.

残余物处理：产生的固体、液体和气体残余物的类型和数量必须已知或者经过估算，通常利用中试试验获得。

9 Environmental constraints: Environmental factors, such as prevailing winds and directions and proximity to residential areas, may restrict or affect the use of certain processes, especially where odors may be produced. Noise and traffic may affect selection of a plant site. Receiving waters may have special limitations, requiring the removal of specific constituents such as nutrients.

环境制约：环境因素，例如主导风向和靠近住宅区，会限制或影响某些工艺的使用，特别是产生气味的环节。噪音和交通也会影响工厂厂址的选择。受纳水体会有特殊的限制，例如要求特定成分如营养物的去除等。

第五节　运行与管理

Section 5　Operation and Management

膨胀　bulking
泡沫　foam
活性污泥膨胀　activated sludge bulking
活性污泥丝状膨胀　activated sludge filamentation bulking

1 In many cases, MLSS with poor settling characteristics has developed into what is known as a bulking sludge condition, which defines a condition in the activated-sludge clarifier that can cause high effluent suspended solids and poor treatment performance.

在很多情况下，MLSS 沉降性能变差是导致污泥膨胀的一个原因。污泥膨胀将导致活性污泥处理系统出水中悬浮物增加和处理性能变差。

2 To maintain high levels of treatment performance with the activated-sludge process under a wide range of operating conditions, special attention must be given to process control. The principal approaches to process control are (1) maintaining dissolved oxygen levels in the aeration tanks, (2) regulating the amount of return activated sludge (SRA), and (3) controlling the waste-activated sludge (WAS).

在宽范围的运行条件下，活性污泥处理系统要维持高水平的处理性能，必须特别注意运行控制。运行控制主要有以下方法：(1) 曝气池中维持一定的溶解氧，(2) 调节回流污泥的总量(SRA)，(3) 控制剩余污泥的排放(WAS)。

第六部分 生物膜法
Part 6 Biofilm Method

第一节 生物滤池
Section 1 Biofilter

生物滤池 biofilter	塑料滤料 plastic packing
参数 parameter	砂石滤料 rock packing
塑料 plastic	慢速滤池 low-rate filter
填料 packing	中速滤池 intermediate-rate filter
级配 distribution	快速滤池 high-rate filter
生物膜 biofilm	排水系统 distribution systems
附着生长 attached growth	布水系统 under-drain systems
滴滤池 tricking filter	通风口 ventilation port
塔式生物滤池 tower tricking filter	集水沟 drainage collection trench

1 The ideal filter packing is a material that has a high surface area per unit of volume, is low in cost, has a high durability, and has an enough porosity so that clogging is minimized and good air circulation can occur.

理想的滤料具有表面积大、价格低廉、耐用性好、空隙率高、不易堵塞、透气性好等优点。

2 Tricking filter have been used to provide biological wastewater treatment of municipal and industrial wastewaters for nearly 100 years. The tricking filter is a non-submerged fixed-film biological reactor using rock or plastic packing over which wastewater is distributed continuously. Treatment occurs as the liquid flows over the attached bio-film.

生物滤池作为市政废水及工业废水的生物处理法已有近百年的历史。生物滤池是一种非淹没式的固定膜生物反应器，使用碎石或塑料作填料，在其上连续布洒废水，废水流过附着的生物膜而得到处理。

3 Hydraulic loading rate: The loading rate is an empirically-derived design and operating parameter that relates to surface shearing rate, and hydraulic detention time. Usually hydraulic loading rate is reported in units of volume of wastewater, including recycle, per unit cross-sectional area per day. Because most rock-medium trickling filters are between 1 and 2 m in depth, the volumetric loading rate used in some countries is easily translated into flow per unit area-time.

水力负荷率：负荷率主要是以经验来设计和运行的参数，涉及表面剪切率和水力停留时间。水力负荷率通常以污水体积为单位计量，包括每天单位横截面积的回流量。由于大多数滴滤床的填料层是在 1 到 2 米的深度，一些国家的容积负荷率通常容易转化为单位时间单位面积的流量。

4 Organic loading rate: Wastewater loading on trickling filters is characterized by the organic loading rate in terms of kilograms of BOD_5 per cubic meter-day ($kgBOD5/m^3 \cdot d$). There is no parameter for solids loading, and solids removal in trickling filters has not been characterized in any predictive manner.

有机负荷率：滴滤床的废水负荷一般用有机负荷率表示，即 $kgBOD_5/m^3 \cdot d$。尚没有参数表达固体负荷，滴滤床的固体去除率还没有任何表达方式。

5 Oxygen transfer: Air is usually supplied to trickling filters through natural drafts resulting from temperature differences between the ambient and the internal air. Deep plastic-medium filters often require the use of compressed air to supply a forced draft. Maximum oxygen transfer rate in natural-draft trickling filters is about $28g/m^2 \cdot d$. And this corresponds to uptake rates expected in

the biofilm for applied BOD concentrations of about 400mg/L BOD, and anoxic conditions can occur within the trickling filters, which can lead to the production of rather nasty odors. Odor problems have resulted in the covering of the filters, with venting of the odorous gas to soil filters or other odor control facilities.

氧转移率：通过周围和内部温度的差异形成的空气流动来为滴滤床供氧。滴滤床深处的塑料层填料通常需要使用压缩空气强制曝气。自然通风滴滤床的最大氧转移速率约为 $28g/m^2 \cdot d$。相当于进水 BOD 浓度 400mg/L 下生物膜工艺的氧转移速率。滴滤床内可产生缺氧环境，产生难闻的异味气体。气味问题要求滤体加盖，将臭气通过固体滤层或其他气味控制装置。

6 Secondary sedimentation facilities: All the sludge from trickling filter settling tanks is removed to sludge processing facilities. The design of these tanks is similar to the design of primary settling tanks, except that the surface loading rate is based on the plant flow plus the recycle flow minus the underflow (often neglected). Suggested overflow rates and solids loading rates for settling tanks following trickling filters range from 400 to 600 $gal/ft^2 \cdot d$ and 0.6 to 1.0 $lb/ft^2 \cdot h$, respectively.

二沉池：所有来自滴滤床沉淀池的污泥都排入污泥处理设施。这类池体设计类似于初沉池，只是表面负荷率是基于污水处理厂流量加循环流量减去渗漏量（常常被忽略）。建议滴滤床后的沉淀池溢流率和固体负荷率范围分别为 400-600$gal/ft^2 \cdot d$ 和 0.6-1.0$lb/ft^2 \cdot h$。

第二节　生物转盘
Section 2　Rotating Biological Contactor

生物转盘　rotating biological contactor

1 Disk materials: High-density polyethylene is the material used most commonly for the manufacture of RBC disks, which are in different configurations or corrugation patterns. Corrugations increase the available surface area and enhance structural stability. The type of RBC disks, classified based on the total area of disks on the shaft, are commonly termed low-density, medium-density, and high-density.

圆盘的材料：制造生物转盘最常用的材料是高密度聚乙烯，圆盘可以是不同的外形或波纹形式。波纹能增加转盘的有效表面面积，并使结构稳定性增强。转盘的类型根据轴上转盘的总面积分类，通常可分为低密度、中密度和高密度三种。

第三节　生物接触氧化法
Section 3　Biological Contact Oxidation Process

生物接触氧化法　biological contact oxidation process | 微细泡曝气　fine bubble diffuser

1 Aerobic submerged fixed-film processes consist of three phases: packing, biofilm, and liquid. The BOD and/or NH_4-N removed from the liquid flowing past the bio-film is oxidized. Oxygen is supplied by diffused aeration into the packing or by being pre-dissolved into the influent wastewater. Aerobic fixed-film processes include down-flow packed-bed reactors, up-flow packed-bed reactor, and up-flow fluidized-bed reactors.

好氧浸没式固定生物膜工艺由三部分组成：填料、生物膜和液相。废水流经生物膜，液相中的 BOD 和（或）NH_4-N 得以氧化。通过散气装置向填料供氧，或者采用流入水预充氧的方式。好氧浸没式固定生物膜工艺包括：降流式填充床反应器、升流式填充床反应器和升流式流化床反应器。

第四节　生物流化床
Section 4　Fluidized-Bed Bioreactor (FBBR)

流化　fluidize | 生物流化床　fluidized-bed bioreactor (FBBR)

第七部分　厌氧生化法
Part 7　Anaerobic Biochemistry

第一节　基本原理
Section 1　Basic Principle

厌氧　anaerobic
低生物量　lower biomass yield
高容积负荷　higher volumetric loading
长链脂肪酸　long chain fatty acid
酸性发酵　acid fermentation
营养物质需要量少　fewer nutrients required
需要深度处理　need for further treatment

第二节　影响因素
Section 2　Affecting Factors

碱度　alkalinity
废水性质　characteristics of the wastewater
流量和负荷　flow and loading
废水的碱度　wastewater alkalinity
有机物浓度和温度　Organic concentration and temperature
非溶解性有机物质的比例　fraction of non-dissolved organic material

第三节　厌氧工艺和设备
Section 3　Anaerobic Processes and Devices

接触　contact
向上流　upflow
厌氧接触法　anaerobic contact process
完全混合式　complete-mix process
升流式厌氧污泥反应器　up-flow anaerobic sludge blanket (UASB)

1 One of the most notable developments in anaerobic treatment process technology was the up-flow anaerobic sludge blanket (UASB) reactor in the late 1970s in the Netherlands by Lettinga and coworkers.
在厌氧处理工艺中，一个引人注目的成就是在 20 世纪 70 年代末由荷兰科学家 Lettinga 和他的助手设计的升流式厌氧污泥反应器。

第四节　厌氧消化过程动力学
Section 4　Anaerobic Consuming Kinetics

甲烷　methane
甲烷气理论产量　expected methane gas production

1 Anaerobic treatment processes are capable of high COD conversion efficiency to methane with minimal production costs. At SRT values greater than 20 to 50d, maximum conversion of solids may occur at temperatures above 25℃. However, high-effluent suspended solids (50 to 200mg/L) are common for anaerobic processes.
厌氧处理系统可以有效地将高浓度的 COD 转化为甲烷生产成本很小。SRT 时间较长约为 20 ～ 50d，最佳温度为 25℃。但是通常厌氧处理的出水悬浮物浓度都很高(50 ～ 200mg/L)。

第八部分　脱氮除磷
Part 8　Denitrification and Phosphorus Removal

生物除磷法　process for biological phosphorus removal
生物脱氮法　process for biological nitrogen removal
硝化　nitrification
反硝化　denitrification
分段式　staged process

1 Nitrification is the term used to describe the two-step biological process in which ammonia (NH_4-N) is oxidized to nitrite(NO_2-N) and nitrite is oxidized to nitrate (NO_3-N).
硝化作用就是指两步生化反应过程，第一步是 NH_4-N 被氧化成 NO_2-N；第二步是 NO_2-N 被氧化成 NO_3-N。

2 In the biological removal of phosphorus, the phosphorus in the influent wastewater is incorporated into cell biomass, which subsequently is removed from the process as a result of sludge washing.
生物除磷是将进入水中的磷合成细胞质，然后通过污泥洗涤的方式从系统中去除。

第九部分 污泥处理与处置
Part 9 Sludge Disposal and Treatment

来源 source
浓缩 thickening
消化 digestion
调理 conditioning
脱水 dewatering
焚化 incineration
堆制肥料 composting
脱水污泥 dewatered sludge cake
带式压滤脱水 belt-press dewatering
污泥干化床 sludge drying bed
含固率 percent of solid matter
污泥提升 sludge pumping
污泥研磨 sludge grinding
污泥混合和储存 sludge blending and storage
污泥除砂 sludge degritting
重力浓缩 gravity thickening
重力带式浓缩 gravity belt thickening
池塘 lagoon
气浮浓缩 flotation thickening
离心 centrifugation
好氧消化 aerobic digestion
污泥储存池 sludge storage basin
石灰稳定 lime stabilization
热处理 heat treatment
化学调理 chemical conditioning
长期储存 long term storage
巴斯德氏杀菌 Pasteurization
带式过滤机 belt filter press
板框压滤机 plate and frame filter press
干燥机变化 dryer variation
多效蒸发 multiple effect evaporation
多炉焚烧 multiple hearth incineration
流化床焚烧 fluidized-bed incineration
与固体废物共焚烧 coincineration with solid wastes
湿式氧化 wet air oxidation
垃圾埋填 landfill
土地利用 land application

1 Composting is a process in which organic materials undergo biological degradation to a stable end product.
堆肥是有机物生物降解过程的稳定的产物。

第十部分 废水处理的主要工艺和反应器类型
Part 10 Principal Processes and Types of Reactors Used for the Treatment of Wastewater

去除 BOD BOD removal
硝化作用 nitrification
稳定化 stabilization
生物脱氮 biological denitrification
除磷 phosphorus removal

1 Batch: Flow is neither entering nor leaving the reactor. The liquid contents are mixed completely.
间歇式：水流既不进入也不离开反应器，液体被充分混合。

2 Plug flow: Fluid particles pass through the tank and are discharged in the same sequence in which they enter.
推流式：液态颗粒穿过水池，并以相同顺序排出。

3 Complete-mix, also known as the continuous flow stirred-tank: Complete mixing occurs when the particles entering the tank are dispersed immediately throughout the tank. The particles leave the tank in proportion to their statistical population. Complete mixing can be accomplished in round or square tanks if the contents of the tank are uniformly and continuous distributed.

完全混合式，也称连续流搅拌池：颗粒进入反应器时立即扩散并发生完全混合，并按统计比例排出反应器。如果反应器中物质是均一连续的，完全混合可在圆形或者方形反应器中完成。

4 Packed bed: Packed-bed reactors are filled with some type of packing media, such as rock, slag, ceramic, or plastic. With respect to flow, they can be completely filled (anaerobic filter) or intermittently dosed (trickling filter).

堆积床：填充床式反应器内填填料，如砾石、沙子、陶瓷或者塑料。水流可以完全淹没(厌氧滤池)或者间歇布水(滴滤床)。

5 Fluidized bed: The fluidized-bed reactor is similar to the packed-bed reactor in many respects, but the packing medium is expanded by the upward movement of fluid (air or water) through the bed. Porosity of the packing can be varied by controlling the flow rate of the fluid.

流化床：流化床反应器在很多方面与填充床反应器类似，但是流化床内的填充介质在流体(气或水)的升流作用下发生膨胀，堆体的空隙度可以通过调节流体流速来控制。

6 Adsorption/desorption: Many chemical constituents tend to attach or sorb onto solids. The implication for wastewater discharges is that a substantial fraction of some toxic chemicals are associated with the suspended solids in the effluent. Adsorption combined with solids settling results in the removal from the water column of constituents.

吸附/解吸：很多化学成分往往附着或者吸附到固体上，废水排放意味着有毒化学品大部分都随出水中的悬浮固体一同流出。与固体沉积伴同的吸附可去除废水中的有毒化学物质。

7 Algal synthesis: The synthesis of algal cell tissue using the nutrients found in wastewater.

藻类合成：藻细胞组织是利用废水中的营养物质合成的。

8 Bacterial conversion: Bacterial conversion (both aerobic and anaerobic) is the most important process in the transformation of constituents released to the environment. The exertion of BOD and NOD are the most common examples of bacterial conversion encountered in water quality management. The depletion of oxygen in the aerobic conversion of organic wastes is also known as deoxygenation. Solids discharged with treated wastewater are partly organic. Upon settling to the bottom, they decompose bacterially either anaerobically or aerobically, depending on local conditions. The bacterial transformation of toxic organic compounds is also of great significance.

微生物转化：微生物转化(好氧和厌氧)是释放到环境中污染物质转化的重要过程。在水质管理中，BOD(生化需氧量)和NOD(硝化需氧量)是微生物转化最好的例证。在有机废物的好氧转化中，氧气的消耗被称为脱氧氧化。随处理废水排除的固体含部分有机质。这些固体物质沉积到底部后，视环境条件不同，可进行厌氧或好氧生物降解，有毒有机化合物的微生物转化也就非常有意义了。

9 Gas absorption/desorption: The process whereby a gas is taken up by a liquid is known as absorption. For example, when the dissolved-oxygen concentration in a body of water with a free surface is below the saturation concentration in the water, a net transfer of oxygen occurs from the atmosphere to the water. The rate of transfer (mass per unit time per unit surface area) is proportional to the amount by which the dissolved oxygen is below saturation. The addition of oxygen to water is also known as re-aeration. Desorption occurs when the concentration of the gas in the liquid exceeds the saturation value, and there is a transfer from the liquid to the atmosphere.

气体吸收/解吸：气体通过液体被吸收的过程，称为吸收。例如，当具有自由水面的水体中溶解氧浓度低于饱和浓度，就会发生从大气到水体的氧转移。转移速率(单位面积单位时间内转移的氧量)与水体中低于饱和溶解氧的溶解氧量成正比。氧从大气到水体的过程称为大气复氧。液体中气体的浓度超过饱和值时就会发生解吸，气体从液相转移到气相。

10 Natural decay: In nature, contaminants will decay for a variety of reasons, including mortality in the case of bacteria and photo-oxidation for certain organic constituents. Natural decay follows first-order kinetics.

自然降解：在自然界中，污染物会有很多种降解原因，包括某种有机化合物在细菌和光氧化作用下的降解。自然降解遵照一级反应动力学。

11 Photochemical reactions: Solar radiation is known to trigger a number of chemical reactions. Radiation in the near-ultraviolet (UV) and visible range is known to cause the breakdown of a variety of organic compounds.

光化学反应：太阳辐射会引发大量的化学反应，近紫外线区域和可见光区域的辐射会引起很多有机化合物的分解。

12 Photosynthesis/respiration: During the day, algal cells in water bodies will produce oxygen by means of photosynthesis. Dissolved oxygen concentrations as high as 30 to 40 mg/L have been measured. During the evening hours algal respiration will consume oxygen. Where heavy growths of algae are present, oxygen depletion has been observed during the evening hours.

光合作用／呼吸作用：在白天，水体中的藻类细胞通过光合作用产生氧，溶解氧的浓度高达 30～40mg/L。在夜晚，藻类呼吸会消耗氧。藻类过量生长的地方，可观察到夜间数小时内的氧气耗竭。

13 Sedimentation: The suspended solids discharged with treated wastewater ultimately settle to the bottom of the receiving water body. This settling is enhanced by flocculation and hindered by ambient turbulence. In river and coastal areas, turbulence is often sufficient to distribute the suspended solids over the entire water depth.

沉淀：随处理水排出的悬浮固体最终沉积在受纳水体的底部。沉积过程受絮凝作用强化，但也受周围湍流的阻碍。在河流和沿海地区，湍流通常使悬浮固体沿整个水深分布。

14 Sediment oxygen demand: The residual solids discharged with treated wastewater will, in time, settle to the bottom of streams and rivers. Because the particles are partly organic, they can be decomposed anaerobically as well as aerobically, depending on conditions. Algae which settle to the bottom will also be decomposed, but much more slowly. The oxygen consumed in the aerobic decomposition represents another dissolved oxygen demand in the water body.

沉积物耗氧量：随处理水排出的残余固体有时会沉积在河流的底部，由于这些颗粒含部分有机物，视环境因素不同，可发生好氧分解，也可发生厌氧分解。沉积在底部的藻类也会分解，但是大多数很缓慢。好氧分解中的氧消耗代表着水体的另一种需氧量。

15 Volatilization: Volatilization is the process whereby liquid and solids vaporize and escape to the atmosphere. Organic compounds that readily volatilize are known as VOCs (volatile organic compounds). The physics of this phenomenon is very similar to gas absorption, except that the net flux is out of the water surface.

挥发：挥发是固体和液体蒸发并溢出到大气的过程。易挥发有机化合物通常称为 VOCs。这种物理现象跟气体吸收类似，只不过是从水表面的净溢出。

16 Biological nutrient removal: The term applied to the removal of nitrogen and phosphorus in biological treatment processes.

生物营养物去除工艺：用于去除氮和磷的生物处理工艺。

17 Carbonaceous BOD removal: Biological conversion of the carbonaceous organic matter in wastewater to cell tissue and various gaseous end products. In the conversion, it is assumed that the nitrogen present in the various compounds is converted to ammonia.

碳化 BOD 去除：废水中含碳有机物通过生化反应生成细胞组织和各种气体终极产物。反应过程中，假定化合物中的氮转化为氨氮。

18 Processes applicability: The applicability of a process is evaluated on the basis of past experience data, from full-scale plants, published data, and pilot plant studies. If new or unusual conditions are

encountered, pilot plant studies are essential.

工艺的适用性：工艺的适用性是基于过去的经验数据评估的，这些数据来源于生产性处理厂、发表的数据，以及中试规模的研究。如果遇到新的或者独特的情况，需通过中试获取数据。

19 Conventional plug-flow: Settled wastewater and recycled activated sludge enter the head end of the aeration tank and are mixed by diffused air or mechanical aeration. Air application is generally uniform throughout tank length. During aeration period, adsorption, flocculation, and oxidation of organic matter occurs. Activated-sludge solids are separated in a secondary settling tank.

传统推流工艺：沉淀后的废水和回流污泥从曝气池顶端进入，利用鼓风曝气或机械曝气进行曝气混合。曝气沿整个池体长度进行布设。在曝气时间，有机物发生吸附、絮凝和氧化过程。活性污泥固体分离在二沉池完成。

20 Plug-flow step feed: Step feed is a modification of the conventional plug-flow process in which the settled wastewater is introduced at several points in the aeration tank to equalize the F/M ratio, thus lowering peak oxygen demand. Generally three or more parallel channels are used. Flexibility of operation is one of the important features of this process.

推流式阶段进水工艺：阶段进水是传统推流工艺的改良，在相同的 F/M 比下，废水沿曝气池多点进水，降低了峰值氧需求。一般设三个或更多的平行通渠道，操作灵活是该工艺重要特征之一。

21 Tapered aeration: Tapered aeration is a modification of the conventional plug-flow process. Varying aeration rates are applied over the tank length depending on the oxygen demand. Greater amounts of air are supplied to the head end of the aeration tank and the amounts diminish as the mixed liquor approaches the effluent end. Tapered aeration is usually achieved by using different spacing of the air diffusers over the tank length.

渐减曝气工艺：传统推流工艺的改良，依据沿池体长度需氧率不同设计。曝气池的前端需要大量的氧气供应，而在接近出水口处需氧量大量减少。渐减曝气工艺通常靠沿池体长度设置不同间隔的空气扩散器完成。

22 Modified aeration: Modified aeration is similar to the convention plug-flow process except that shorter aeration time and higher F/M ratios are used. BOD removal efficiency is lower than other activated-sludge processes.

改良的曝气工艺：改良曝气工艺类似于传统推流工艺，只是曝气时间较短，而 F/M 比较高，BOD 去除效率低于其他活性污泥工艺。

23 Kraus process: Kraus process is a variation of the step aeration process used to treat wastewater with low nitrogen levels. Digester supernatant is added as a food source to a portion of the return sludge in a separate aeration tank designed to nitrify. The resulting mixed liquor is then added to the main plug-flow aeration system.

Kraus 工艺：Kraus 工艺是阶段曝气工艺用于处理含氮较低污水的一种变型。消化池的上清液作为营养物质加入到回流污泥中，回流污泥来源于独立设置的硝化曝气池。由此产生的混合液体进入推流式曝气池中进行处理。

24 Complete-mix: The process is an application of the flow regime of a complete-mix reactor. Settled wastewater and recycled activated sludge are introduced typically at several point in the aeration tank. The organic load on the aeration tank and the oxygen demand are uniform throughout the tank length.

完全混合工艺：完全混合工艺是完全混合反应器的流态化应用工艺。废水和回流污泥多点进入曝气池。曝气池的有机负荷和需氧量沿池长平均分布。

25 High-rate aeration: High-rate aeration is a process modification in which high MLSS concentrations are combined with high volumetric loadings. This combination allows high F/M ratios and long mean cell residence time with relatively short hydraulic detention time.

高效曝气工艺：是针对高 MLSS 浓度和高容量负荷条件下的修正工艺。这种工艺可形成相对较短水力停留时间下的高 F/M 比和长污泥龄。

26 High-purity oxygen: High-purity oxygen is used instead of air in the activated-sludge process. Oxygen is diffused into covered aeration tanks and is recirculated.

高纯氧工艺：在活性污泥工艺中用高纯度氧气代替空气。氧气扩散到密封的曝气池中并循环使用。

27 Single-stage nitrification: In single-stage nitrification, both BOD and ammonia reduction occur in a single biological stage. Reactor configurations can be either a series of complete-mix reactors or plug-flow.

一段硝化工艺：在一段硝化处理工艺中，BOD 和氨氮减少同时发生在一个生物处理阶段。反应器的构造可以是一系列完全混合反应器，也可以是推流式反应器。

28 Separate-stage nitrification: In separate-stage nitrification, a separate reactor is used for nitrification, operating on a feed waste from a preceding biological treatment unit. The advantage of this system is that operation can be optimized to conform to the nitrification needs. An external source of carbon is normally required.

独立硝化工艺：在独立硝化工艺中，使用一台单独的反应器用于硝化，处理来自前面生物处理单元处理过的废水。该系统的优点是操作最优化，并且能达到硝化需要。通常需要添加额外碳源。

29 Single-stage nitrification/denitrification: Reactor configurations can be either a series of complete-mix reactors or plug-flow.

一段硝化/反硝化工艺(同步硝化/反硝化)：反应器的构造可以是一系列完全混合反应器，也可以是推流式反应器。

30 Nitrogen and phosphorus removal: Phosphorus removal is accomplished in conjunction with processes for nitrification and denitrification by creating conditions suitable for the luxury uptake of phosphorus.

氮和磷的去除：通过创造适合的过量吸磷条件，除磷和硝化-反硝化过程同步进行。

31 Contact stabilization: Contact stabilization uses two separate tanks or compartments for the treatment of the wastewater and stabilization of the activated sludge. The stabilized activated sludge is mixed with the influent (either raw or settled) wastewater in a contact tank. The mixed liquor settled in a secondary setting tank and return sludge are aerated separately in a re-aeration basin to stabilize the organic matter. Aeration volume requirement are typically 50 percent less than convention plug flow.

接触稳定工艺：接触稳定工艺使用两个独立的池体来处理污水和稳定活性污泥。稳定化的活性污泥在接触池内与进水(包括原水和沉淀水)废水混合。混合液体在二沉池中沉降下来，回流污泥在一个曝气池内单独曝气，以稳定有机质。曝气量的要求通常比传统推流工艺小百分之五十。

32 Extended aeration: Extended aeration process is similar to the conventional plug-flow process except that it operates in the endogenous respiration phase of the growth curve, which requires a low organic loading and long aeration time. This process is used extensively for prefabricated package plants for small communities.

延时曝气工艺：延时曝气工艺类似于传统的推流过程，只是它运行在内源呼吸阶段，需要低有机负荷和长曝气时间。该工艺广泛用于小社区的成套处理设施。

33 Oxidation ditch process: The oxidation ditch consists of a ring-or oval-shaped channel and is equipped with mechanical aeration devices. Screened wastewater enters the ditch, is aerated, and circulates at about 0.8 to 1.2 ft/s (0.25 to 0.35m/s). Oxidation ditches typically operate in an extended aeration mode with long detention and solids retention times. Secondary sedimentation tanks are used for most applications.

氧化沟工艺：氧化沟由一个环形或椭圆形渠道组成，设有机械曝气装置。栅后污水进入沟内，曝气，水流速度在 0.8～1.2 ft/s(0.25～0.35m/s)。氧化沟通常以延时曝气模式运行，需要较长的污泥停留时间。大多情况下需要二沉淀池。

34 Intermittent decanted extended aeration: A single reactor in which all of the steps of the activated-sludge process occur. Flow into the reactor is continuous as compared to the sequencing

batch reactor. Because the mixed liquor remains in the reactor during all of the treatment steps, separate secondary sedimentation facilities are not required. Sludge wasting occurs during the aeration portion of the cycle.

间歇排水延时曝气工艺：活性污泥工艺所有的步骤都运行在一个单一的反应池中。相比序批式反应器，该工艺是连续进水。由于所有的处理步骤混合液都保留在反应器内，不需要单独设置二沉池，在循环周期的曝气阶段排除污泥。

35 Sequencing batch reactor: The sequencing batch reactor is a fill and draw type reactor system involving one or two complete-mixed reactors in which all steps of the activated-sludge process occur. Because the mixed liquor remains in the reactor during all of the treatment steps, separate secondary sedimentation facilities are not required. Typically, two reactors are used for sequencing.

序批式反应器：序批式反应器是一种进水和排水式反应器系统，涉及一个或两个完全混合反应器，所有活性污泥反应过程都发生在反应器中。由于所有的处理步骤的混合液仍然留在反应器内，可以不设二沉池。一般两个反应器按顺序使用。

第十一部分　例题

Part 11　Examples

Examples: Design of a Primary Sedimentation Basin

The average flow rate at a small municipal wastewater treatment plant is 0.5 Mgal/d. The highest observed peak daily flow rate is 1.25 Mgal/d. Design a rectangular (aspect ratio 4:1) primary clarifier for this wastewater treatment facility. Calculate the scour velocity, to determine if settled material will become resuspended. Estimate the BOD and TSS removal at average and peak flow. Use a surface overflow rate (SOR) of 900 gal/ft^2 · d at average flow and a side water depth of 12 ft.

例：初沉池的设计

一个小型城市废水处理厂的平均流量为 0.5 Mgal/d，最高流量为 1.25 Mgal/d。为这个废水处理厂设计一个矩形的初沉池(长宽比为 4:1)。计算水流速度，判断沉淀物质是否发生再悬浮现象。估计平均流量和最高流量下的 BOD 和 TSS 去除率。平均流量下的表面过流率(SOR)为 900 gal/ft^2 · d，水深 12ft。

Solution

解：

1. Calculate the required surface area. For average flow conditions, the required area is

计算所需表面积。平均水流条件下所需面积为

$$A = \frac{Q}{SOR} = \frac{0.5 \times 10^6 \text{ gal/d}}{900 \text{ gal/ft}^2 \cdot \text{d}} = 556 \text{ ft}^2$$

2. Determine the tank surface dimensions. For an aspect ratio of 4 to 1,

计算池体表面尺寸，根据长宽比 4 : 1，则有

$$4L^2 - 556 \text{ ft}^2$$

$L = 11.8$ ft, so the idealized surface dimensions are 11.8 ft by 47.2 ft. However, for the sake of convenience, the surface dimensions will be rounded to 12 ft by 48 ft.

$L = 11.8$ ft，所以实际宽为 11.8 ft，长为 47.2 ft。为了简便起见，将宽修正为 12 ft，长修正为 48 ft。

3. Determine the detention time and overflow rate at average flow. For the assumed side water depth of 12 ft,

平均流量下确定停留时间和过流率。假定水深 12 英尺，则有

$$\text{Tank volume (池体容积)} = 12 \text{ ft} \times 48 \text{ ft} \times 12 \text{ ft} = 6912 \text{ ft}^3$$

$$\text{Overflow rate(过流率)} = \frac{Q}{A} = \frac{0.5 \times 10^6 \text{ gal/d}}{12 \text{ ft} \times 48 \text{ ft}} = 868 \text{ gal/ft}^2 \cdot \text{d}$$

$$\text{Detention time(停留时间)} = \frac{V}{Q} = \left(\frac{6912 \text{ ft}^3}{0.5 \times 10^6 \text{ gal/d}}\right)\left(\frac{7.48 \text{ gal}}{\text{ft}^3}\right)\left(\frac{24 \text{ h}}{\text{d}}\right) = 2.48 \text{ h}$$

4. Determine the detention time and overflow rate at peak flow：
确定峰值流量下的停留时间和过流率

$$\text{Overflow rate(过流率)} = \frac{Q}{A} = \frac{1.25\times10^6\ \text{gal/d}}{12\ \text{ft}\times48\ \text{ft}} = 2170\ \text{gal/ft}^2\cdot\text{d}$$

$$\text{Detention time(停留时间)} = \frac{V}{Q} = \left(\frac{6912\ \text{ft}^3}{1.25\times10^6\ \text{gal/d}}\right)\left(\frac{7.48\ \text{gal}}{\text{ft}^3}\right)\left(\frac{24\ \text{h}}{\text{d}}\right) = 0.99\ \text{h}$$

5. Calculate the scour velocity，using the following values：
利用以下参数计算冲蚀流速

Cohesion constant	内聚力常数	$k = 0.05$
Specific gravity	比重	$s = 1.25$
Acceleration due to gravity	重力加速度	$g = 32.2\ \text{ft/s}^2$
Diameter of particles	颗粒直径	$d = 100\ \mu\text{m} = 3.28\times10^{-4}\ \text{ft}$
Darcy-Weisbach friction factor	达西-韦斯巴赫阻力常数	$f = 0.025$

$$\nu_H = \left[\frac{8k(s-1)\,gd}{f}\right]^{1/2} = \left[\frac{(8)(0.05)(0.25)(32.2)(3.28\times10^{-4})}{0.025}\right]^{1/2} = 0.21\ \text{ft/s}$$

6. Compare the scour velocity calculated in the previous step to the peak flow horizontal velocity (the peak flow divided by the cross-sectional area through which the flow passes). The peak flow horizontal velocity through the setting tank is

将上一步计算的冲蚀速度与峰值流量水平流速进行比较(峰值流量除以它通过的横断面积)，通过沉淀池的峰值流量水平速度为：

$$\nu = \frac{Q}{A_x} = \left(\frac{1.25\times10^6\ \text{gal/d}}{12\ \text{ft}\times12\ \text{ft}}\right)\left(\frac{\text{ft}^3}{7.48\ \text{gal}}\right)\left[\frac{1}{(24\ \text{h/d})(3600\ \text{s/h})}\right] = 0.013\ \text{ft/s}$$

The horizontal velocity value，even at peak flow，is substantially less than the scour velocity. Therefore，settled matter should not be resuspended.

即使在峰值流量下，水平流速也远低于冲蚀速度，因此，沉淀物质不会产生再悬浮。

7. Use accompanying coefficients to estimate the removal rates for BOD and TSS at average and peak flow.

根据相应的系数来估计在平均值和峰值流量下的 BOD 和 TSS 的去除率。

a. At average flow：
在平均流量下：

$$\text{BOD removal(BOD 去除率)} = \frac{t}{a+bt} = \frac{2.48}{0.018+(0.020)(2.48)} = 37\%$$

$$\text{TSS removal(TSS 去除率)} = \frac{t}{a+bt} = \frac{2.48}{0.0075+(0.014)(2.48)} = 59\%$$

b. At peak flow：
在峰值流量下：

$$\text{BOD removal(BOD 去除率)} = \frac{t}{a+bt} = \frac{0.99}{0.018+(0.020)(0.99)} = 26\%$$

$$\text{TSS removal(TSS 去除率)} = \frac{t}{a+bt} = \frac{0.99}{0.0075+(0.014)(0.99)} = 46\%$$

第十二部分 摘要及全文
Part 12 Abstract and Full Paper

摘要翻译示例 1：

Enhancement of polycyclic aromatic hydrocarbons removal during anaerobic treatment of urban sludge

Abstract：Anaerobically stabilized sewage sludge has potential to partially substitute synthetic fertilizers. The main risk with the recycling of urban sludge on agricultural soil is the accumulation of unwanted products, such as trace metals and organic micropollutants. In this context, the polycyclic aromatic hydrocarbons (PAHs) are particularly monitored because of their toxic properties at low concentrations and their high resistance to biological degradation. The aim of the present study was to optimize PAHs removal during anaerobic digestion of contaminated sewage sludge. Thirteen PAHs were monitored in laboratory scale anaerobic bioreactors under mesophilic (35℃) and thermophilic (55℃) conditions. Abiotic losses were statistically significant for the lightest PAHs, such as fluorcne, phenanthrene and anthracene. It was shown that PAHs removal was due to a specific biological activity. Biological PAHs removal was significantly enhanced by an increase of the temperature from 35℃ to 55℃, especially for the heaviest PAHs. Bioaugmentation experiment was also performed by addition of a PAHs-adapted bacterial finally shown that PAHs removal efficiencies and methanogenic performances were closely linked. The rate of biogas production may be used as an indicator of bacterial activity on PAHS removal.

Keywords: Bioaugmentation; biogas yield; mesophilic conditions; methanogenesis; sewage sludge; thermophilic conditions

城市污泥厌氧消化中 PAHs 的强化去除

摘要:厌氧稳定污泥可以作为肥料使用。将污泥回用于农业土壤的主要风险是一些有害物质的积累,例如,痕量金属和微量有毒有机物。由于多环芳烃(PAHs) 在低浓度下的毒性特征和不易生物降解的特点,本研究对其进行主要监控。本研究的目的就是在污泥厌氧消化中优化 PAHs 的去除。小试规模的厌氧反应器运行条件分别为中温(35℃) 和高温(55℃),监控 13 种 PAHs。对于低分子量的 PAHs,如芴、菲、蒽,非生物减少具有统计意义。结果表明,PAHs 去除归因于比对生物活性。温度从 35℃ 升高到 55℃,PAHs 的生物去除明显加强,尤其是对多环的 PAHs。通过向反应器中投加 PAHs 适应菌群考察生物强化的效果,表明 PAHs 去除明显增加。PAHs 去除率和产甲烷的性能密切相关。甲烷产量可以作为细菌去除 PAHs 活性的指标。

关键词:生物强化;甲烷产量;中温条件;甲烷生成;污水污泥;高温条件

摘要翻译示例 2:

Environmental technology education in Portugal: analysis and perspectives

Abstract: The scope of the present communication is the current status of environmental engineering in Portugal. The different approaches concerning the environmental engineering courses available at undergraduate levels as well as some post-graduation studies are described. Furthermore, an analysis based on strengths, weakness, opportunities and trend issues regarding the environmental engineering profession in Portugal is presented.

Keywords: Environmental engineering; education; curriculum; Portugal

葡萄牙的环境工程教育:分析与展望

摘要:本文主要涉及葡萄牙的环境工程现状,介绍了环境工程本科课程设置以及部分研究生课程;同时,就葡萄牙环境工程专业的优缺点、就业机会以及发展趋势进行分析。

关键词:环境工程;教育;课程;葡萄牙

全文翻译示例

Onsite Wastewater Treatment Systems Technology: Continuous-Flow, Suspended Growth Aerobic Systems (CFSGAS)

原位废水处理技术:连续流悬浮生长好氧系统(CFSGAS)

Description　系统描述

The activated sludge process in an aerobic suspended-growth process maintains a relatively high population of micro-organisms (biomass) by recycling settled biomass back to the treatment process. The biomass converts soluble and colloidal biodegradable organic matter and some inorganic compounds into cell mass and metabolic end product. The biomass is separated from the wastewater through settling in a clarifier for recycling or wasting to sludge handing processes.

好氧悬浮生长系统中的活性污泥工艺靠回流污泥来维持曝气池中相对较高的微生物浓度。微生物将溶解态和胶体态的可生物降解有机物以及一些无机物转移到细胞体内，同时形成新陈代谢终极产物。微生物在沉淀池中进行泥水分离，部分污泥回流，剩余部分进入污泥处理工艺。

Preliminary treatment to remove settleable solids and floatable materials is usually provided by a septic tank or other primary treatment device. Most onsite designs are capable of providing significant ammonia oxidation and effective removal of organic matter.

通常，可利用化粪池或其他初沉设施对可沉淀固体物质或漂浮物质进行预处理，进而得以去除。大多数原位污水处理系统设计具备氨氧化和有机物去除功能。

The basic system consists of a number of interrelated components (as shown in Fig. 6-1):

原位污水处理系统一般包括如下互相联系的部分(见图 6-1)：

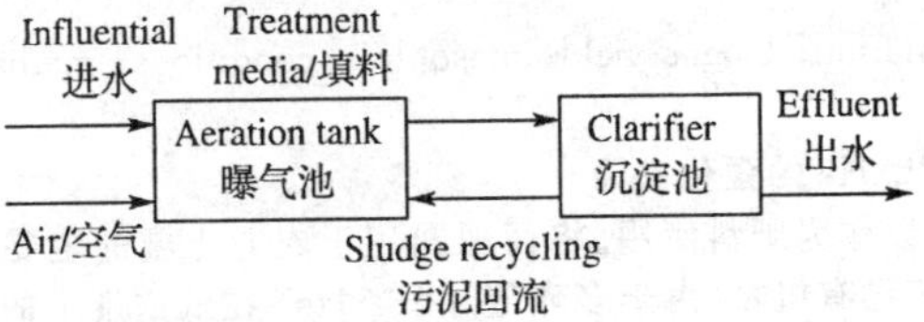

Fig. 6-1 A basic CFSGAS configuration

图 6-1 连续好氧悬浮生长系统基本工艺流程

- An aeration tank.
- An oxygen source and equipment to disperse atmospheric or pressurized air or oxygen into the aeration tank at a rate sufficient to always maintain positive dissolved oxygen.
- A means to appropriately mix the aeration basin and ensure suspension of the biomass (usually accomplished by the aeration system).
- A clarifier to separate the biomass from the treated effluent and collect settled biomass for recycling to the aeration basin.
- 曝气池
- 氧源及供氧设备，将空气或压缩空气或纯氧以一定流量供给曝气池，维持充足的溶解氧浓度。
- 对曝气池进行混合的设备，以确保污泥絮体呈悬浮状态(通常由曝气系统完成)。
- 在沉淀池中进行泥水分离，收集回流所需的沉淀污泥。

Several modifications of this basic process are commercially available. These include different aeration devices; different means of sludge collection and recycling to the aerator; the use of coarse membrane filters in lieu of, or in addition to, the clarifier; and process enhancement through the addition of an inert media area on which biofilms can grow. The addition of surfaces where biota can become attached and grow increase the capacity of the system (increased organic loading possible). This last modification is the most significant enhancement and is described below.

上述基本工艺在应用中存在几种变换形式，包括不同的曝气装置，不同的污泥收集和回流方法，除了沉淀池外采用的膜滤器，添加惰性填料强化生物膜附着，等等。高比表面积填料的添加能够更多地附着微生物，提高系统的处理能力(可能提高有机负荷)。后者起到最有效的强化作用，下面具体描述。

The combined fixed-film/suspended growth process is sometimes referred to as class of treatment processes called coupled contact aeration, enhanced, or high biomass systems. To enhance performance and increase the capacity of the aeration tank, an inert support medium is added to the aeration tank. This allows a fixed film of biomass to attach and grow on the medium to augment the suspended microbial population, providing more biomass to feed on wastewater constituents (Fig. 6-2). Synthetic trickling filter media, loops of fiber bundles, and a variety of different plastic surface configuration can be suspended in the aeration tank. Advantages include increased active microbial mass per unit volume, enhanced potential for nitrification, reduced suspended solids loading to the clarifier, improved solids separation characteristics, reduced sludge production, and resilience under variable influent conditions.

固定膜-悬浮生长联合工艺有时称为复合接触氧化-强化或高生物量处理工艺。为了强化和提高曝气池的处理能力，可将惰性材料投加到曝气池中。微生物在填料表面形成附着的生物膜，提高了悬浮微生物的数量，因此，有更多的微生物以废水中的营养物为食(图 6-2)。合成滴滤床填料，纤维填料以及不同的塑料表面结构能够悬浮在曝气池中。由此带来的优点是，提高了单位体积的活性微生物量，强化了硝化作用的潜力，减少了沉淀池中的悬浮固体，改善了固体分离特征，减少了污泥产量，适应不同的进水条件。

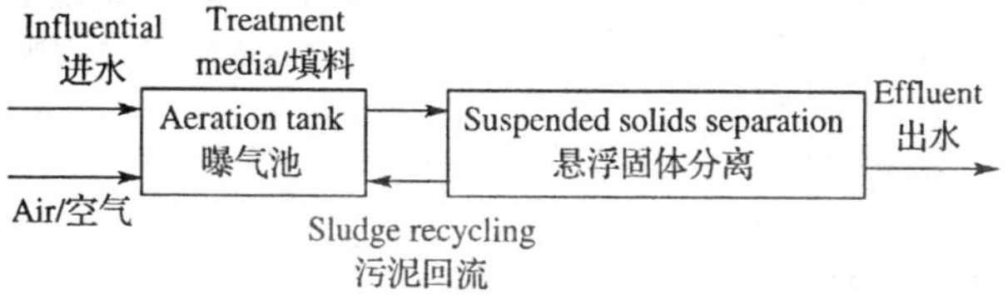

Fig. 6-2　An enhanced CFSGAS or high biomass system

图 6-2　强化的好氧悬浮生长系统或高生物量系统

Typical Application　典型应用

These systems are usually preceded by a septic tank and followed by a subsurface wastewater infiltration system (SWIS). Despite some claims of reduced SWIS sizing when compared to the conventional septic tank pretreatment, the designer is cautioned to consider ground water protection. These systems should be applied only where onsite system management services are available. For surface water discharge, the system must be followed by disinfection at a minimum to consistently meet discharge standards. However, some subsurface (non-human-contact) reuse may be implemented without further treatment. High biomass systems can be a low-cost means of upgrading existing overloaded CFSGAS units that currently do not meet BOD or nitrification goals. They can also complete directly with conventional designs because they have greater stability in handling highly variable loadings.

上述这些系统通常设在化粪池之后，后续潜流废水过滤系统。相对于传统的化粪池预处理，虽然可减少潜流废水过滤系统的规模，但考虑到地下水保护，设计者在应用时趋于保守设计。这些系统仅用于可提供原位系统管理设施的区域。如果处理水排向地表水体，排放前必须进行消毒，以满足排放标准的需要。但回用于非人类直接接触的地下回灌则无需进一步处理。对于目前不能满足 BOD 或硝化要求的连续流悬浮生长好氧系统，用高生物量系统对其升级改造是最经济的方法。高生物量系统具有较强的抗冲击稳定性，因此，也可以直接按照传统的设计实施。

Design Assumptions　设计假定

The extended aeration type of CFSGAS is the most commonly used design. At present there is no generic information on design parameters for fixed film activated sludge systems. Package plants are delivered based on design flow rates. A conservative design approach for extended aeration systems is presented in table 4-1. The inert medium should support additional biomass and add to the total system microbial mass. Because the increase in microbial population is difficult to measure, any "credits" for this addition would have to be based on empirical observation. Claims for significantly decreased sludge production, increased oxygen transfer efficiency, and improved settleability of the sludge have not been universally proved. However, a number of successful installation for onsite and small municipal systems have been in operation throughout the world for more than 10 years.

延时曝气连续流悬浮生长好氧系统是最普遍的设计形式。目前，尚没有固定膜活性污泥系统的通用设计信息。可根据设计流量配送成套污水处理设施。延时曝气系统的保守设计参数列于表 4-1。惰性填料可提供额外的微生物量，增加系统的总生物量水平。由于增加的微生物数量难以计量，一般都是基于经验观察获得。该系统是否明显降低污泥产量、增加氧传质效率、改善污泥的沉降性等尚未得到普遍认可。但是，一些原位处理和小型市政污水处理系统已经在世界各地成功地运行了10年。

Tab. 4-1 Design parameters for CFSGAS extended aeration package plants
表4-1 连续流悬浮生长好氧系统延时曝气成套设备设计参数

Parameter 参数	Extended aeration 延时曝气
Pretreatment (if needed) 预处理(如果需要)	Septic tank or equivalent 化粪池或其他类似设施
Mixed Liquor Suspended Solids (mg/L) 混合液悬浮固体浓度(mg/L)	2000-6000
F/M Load (1b BOD/d/MLVSS) F/M (1b BOD/d/MLVSS)	0.05-0.15
Hydraulic Retention Time (h) 水力停留时间(h)	24-120
Solids Retention Time (days) 固体停留时间(天)	20-40
Mixing Power Input 混合动力输入	0.2-3.0 hp/1000ft
Clarifier Overflow Rate (gpd/ft) 沉淀池溢流率	200-400 avg., 800 peak 200～400(平均), 800(峰值)
Clarifier Solids Loading (lb/d/ft) 沉淀池固体负荷	30 avg., 50 peak 30(平均), 50(峰值)
Dissolved Oxygen (mg/L) 溶解氧	>2.0
Residuals Generated 剩余污泥产量	0.6-0.9 lb TSS/lb BOD removed 0.6～0.9 lb TSS/lb BOD(去除)
Sludge Removal 污泥清除	3-6 months as needed 视需要3～6个月

Onsite package treatment units should be constructed of noncorrosive materials, such as coated concrete, plastic, fiberglass, or coated steel. Units may be stand-alone or manufactured to drop into a compartmented septic tank. Some units are installed aboveground on a concrete slab with proper housing to protect against severe climatic conditions. Units may also be buried underground as long as easy access is provided to all mechanical parts, electrical control systems, and water surfaces. All electrical components should follow NEC code and be waterproof and/or housed from the elements. If airlift pumps are used, large-diameter units should be provided to avoid clogging. Blowers, pumps, and other mechanical devices should be designed for continuous use because they will be abused by climatic conditions and the corrosive atmosphere within the treatment environment. Easy access to all moving parts should be provided for routine maintenance. An effective alarm system should be employed. Typical land area requirements for package plants are modest.

原位处理设备应该使用不锈钢材料，例如，涂层水泥、塑料、玻璃纤维或涂层钢。这些单元可以是独立的，或建在化粪池的隔间内。一些设备可安装在地上的混凝土板上，为防止极端天气的影响，需建造房屋进行保护。处理设备也可埋于地下，但须保证人员便于接触机械部件、电子控制系统和水面。所有的电子部件需遵循NEC(National Economic Council，美国全国经济委员会)标准，并进行防水处理设置隔离间。如果使用气泵，为避免堵塞，应使用大管径设备。由于气候条件和处理环境的腐蚀性气体的影响，鼓风机、泵以及其他机械设备应设计成连续使用，以防止事故的发生。日常维护需要所有移动部件，便于人员进行接触和维护操作。同时，需要设置有效的预警系统。此外，需要适量的土地供污水处理厂建设使用。

For engineered package plants, final clarifier designs should be conservative for high MLSS and poor settleability of biomass. Because of the potential for bulking sludge, secondary clarifiers should be equipped with surface skimming devices to remove greases and floating solids, as well as efficient screens.

对于工程化的成套污水处理系统，考虑到高 MLSS 浓度污泥和沉降性差污泥的沉淀效果，设在工艺最后的沉淀池设计需保守一些。由于存在污泥膨胀的风险，二沉池应该安装浮渣清理设备，以去除油脂和漂浮固体，并进行有效的过滤。

Performance　性能

Well-operated CFSGAS extended aeration units that are well operated can achieve BOD concentrations ranging from 10 to 50 mg/L and TSS concentrations ranging from 15 to 60 mg/L. Some studies have indicated poorer performance owing to surge flows, variable loading, and inadequate maintenance. Nitrification can also be significant in these aeration units during warmer periods. Some nitrogen removal can be achieved by denitrification, which can remove 30 to 40 percent of the total nitrogen (TN) under optimum conditions. Average total nitrogen effluent concentrations in residential extended aeration units range from 17 to 40 mg/L.

运行良好的连续流悬浮生长好氧系统中，如果延时曝气单元也运行良好的话，出水 BOD 浓度可达到 10 ～ 50 mg/L，TSS 浓度在 15 ～ 60 mg/L。但是，有些研究表明其效果并不理想，其中原因主要是大的流量波动、负荷变化以及缺乏维护等。在气候温暖的季节，曝气池中会发生明显的硝化作用，部分氮通过反硝化作用得以去除，运行条件适宜的情况下，可以去除大约 30% ～ 40% 的总氮。延时曝气池的出水总氮浓度大约在 17 ～ 40 mg/L。

High biomass systems have produced BOD and TSS effluents 5 to 40 mg/L. Although they are less dependent on temperature than the extended aeration CFSGAS, temperature does have an impact on their seasonal capability to nitrify the influent ammonium-nitrogen to nitrate-nitrogen. All CFSGAS systems do an excellent job of removing toxic organics and heavy metals. Most CFSGAS systems do not remove more than a small percentage of phosphorus (10 to 20 percent) and nitrogen(15 to 25 percent).

高生物量系统的出水 BOD 和 TSS 浓度在 5 ～ 40 mg/L。虽然该系统受温度的影响较延时曝气连续流悬浮生长好氧系统弱，但温度的确对氨氮转化为硝酸盐氮的硝化作用产生季节性影响。所有的连续流悬浮生长好氧系统在去除有毒有机物和重金属方面效果极佳，但大多数系统的除磷和脱氮能力不足，除磷率只有 10% ～ 20%，脱氮率 15% ～ 25%。

Management Requirements　管理要求

CFSGAS systems must be managed and maintained by trained personnel rather than homeowners to perform acceptably. Power requirements vary from 2.5 to 10 kWh/day. They should be inspected at least every 2 to 3 months. During these inspections, excess solids pumping should be based on the mixed liquor measurements. It is estimated that an effective program will require between 12 and 28 person-hours annually, in addition to analytical testing of the effluent, where require. Management contracts should be in place for the life of the system. Common operational problems with extended aeration systems are provided in table 6-1. Residuals generated will vary from 0.6 to 0.9 lb TSS per lb BOD removed, over and above the normal septic tank sludge produced.

连续流悬浮生长好氧系统要想运行良好，需由经过培训的专业人员进行维护和管理，而不是普通业主。其运行所需的电耗为 2.5 ～ 10 kWh/d。系统至少 2 ～ 3 个月检查一次，检查期间，剩余污泥量需根据测量的混合液浓度确定。据估算，除了必需的出水水质检测分析外，一般每年检查需要 12 ～ 28 个工时。管理合同的期限应设定为系统的整个使用期。延时曝气系统普遍存在的操作问题列于表 6-1。剩余污泥的产生量为每去除 1 磅 BOD 产生 0.6 ～ 0.9 磅 TSS，超过一般化粪池的污泥产量。

Tab. 6-1 Common operational problems of extended aeration package plants

表 6-1 延时曝气成套设备的运行问题

Observation 现象	Cause 原因	Remedy 解决办法
Excessive local turbulence in aeration tank 曝气池中出现大量局部涡流	Diffuser plugging Pipe breakage Excessive aeration 扩散器堵塞 管道破裂 过量曝气	Remove and clean Replace as required Throttle blower 去除、清理 需要时更换 节流阀风机
White, thick, billowy foam on aeration tank 曝气池表面出现大量很厚的白色泡沫	Insufficient MLSS MLSS 过低	Avoid wasting solids 避免剩余污泥排放
Thick, scummy, dark tan foam on aeration tank 曝气池表面出现很厚的深棕褐色浮渣	High MLSS MLSS 过高	Waste solids 排放剩余污泥
Dark brown/black foam and mixed liquor in aeration tank 曝气池中出现深棕色/黑色泡沫混合液	Anaerobic conditions Aerator failure 厌氧条件 曝气器问题	Check aeration system, aeration tank 检查曝气系统和曝气池
Billowing sludge washout in clarifier 大量污泥流出沉淀池	Hydraulic or solids overload Bulking sludge 水力负荷或固体负荷过高 污泥膨胀	Waste sludge; check flow to unit 排放剩余污泥,检查单元流量
Clumps of rising sludge in clarifier 沉淀池中污泥上浮	Denitrification Septic conditions in clarifier 反硝化 沉淀池中存在腐化条件	Increase sludge return rate to decrease sludge retention time in clarifier Increase return rate 提高污泥回流率,减少沉淀池中的污泥停留时间 提高回流率
Fine dispersed floc, turbid effluent 污泥絮体破碎,出水浑浊	Turbulence in aeration tank Sludge age too high 曝气池中紊动过高 污泥龄过长	Reduce power input Waste sludge 减少能力输入 排放剩余污泥
Poor TSS and/or BOD removal TSS/BOD 去除率下降	Excess flow and strength variations 流量过大以及负荷过高	Install flow smoothing system 安装流量调节系统
Poor nitrification 硝化作用不理想	Low temperatures Excessive biocide use 温度过低 过量使用杀虫剂	Insulate, upgrade to high biomass, etc. Reduce biocide loading 保温,升级为高生物量系统 减少杀虫剂负荷

Risk Management Issues 风险管理问题

CFSGAS systems require effluent disinfection at a minimum to meet surface discharge or any surface reuse water quality requirement. They are quite sensitive to temperature, interruption of electric supply, influent variability, or shock loadings of toxic chemicals. The septic tank helps protect these units from the latter problems. Aesthetically, noise from the blowers is the major irritant, while odors can be significant during power outages or organic overloading periods. High biomass units are more resistant to the above impacts. The systems are not well suited to seasonal use because of long start-up times.

为符合地表排放或回用水质要求，连续流悬浮生长好氧系统的出水需进行消毒处理。该系统对温度非常敏感，也会受断电、进水波动以及有毒化学品负荷的影响。对于有毒化学品的影响问题，化粪池对后续单元可起到缓解作用。美学方面，鼓风机的噪声是主要问题，而由于断电或有机负荷过高也会引起明显的气味问题。高生物量系统受上述问题的影响小些。由于启动时间较长，该系统不适合季节性运行。

Costs 费用

The installing costs of package plants are highly variable but are usually less than ＄10,000. Operation and maintenance (O/M) costs are primarily dependent on local power and labor costs, varying from ＄400 to ＄600 per year in most cases.

成套设备的安装费变化幅度较大，但一般不超过 10 000 美元。运行和维护费用主要取决于当地的电费和劳动力成本，一般每年在 400 ～ 600 美元。

第七章 大气污染控制工程
Chapter 7 Air Pollution Control Engineering

第一节 绪论
Section 1 Introduction

大气 atmosphere
环境空气 ambient air
大气组成 atmospheric composition
大气水 atmospheric water
溶胶 sol
气溶胶 aerosol
凝结型气溶胶 condensation aerosol
分散型气溶胶 dispersion aerosol
胶体 colloid
颗粒 particle
胶粒 colloidal particle
悬浮颗粒物 suspended particulate
一次颗粒物 primary particulate
二次颗粒物 secondary particulate
液滴 droplet
尘粒 grit
粉尘 dust
飘尘 airborne dust
降尘 dust fall
飞灰 fly ash
烟 fume
烟雾 fog
黑烟 smoke
烟灰 soot
气温 atmospheric temperature
气压 atmospheric pressure
热平衡 heat balance
标准大气 standard atmosphere
大气污染 air pollution
大气污染源 atmospheric pollution source
林火 forest fire
火山 volcano
地热流 geothermal stream
油田 oil field
天然气田 natural gas field
工艺装置 process unit
焚烧 incineration
排放 emission
排放控制 emission control
排放标准 emission standard
室外空气 outdoor air
环境空气质量标准 ambient air quality standard
大气环境质量标准 atmospheric environmental quality standard
大气环境影响评价 assessment of atmospheric environmental influence
大气环境容量 atmospheric environmental capacity
大气质量评价 atmospheric quality assessment
大气污染监测 air pollution monitoring
总悬浮颗粒物(TSP) total suspended particulate
可吸入颗粒物(IP) inhalable particulate
臭氧 ozone
铅 lead
苯并[a]芘 benzo[a]pyrene
氟化物 fluoride
空气污染指数 air pollution index(API)
大气污染物 atmospheric pollutant
无机污染物 inorganic pollutant
有机污染物 organic pollutant
光离解 photodissociation
光分解 photolysis
光氧化 photooxidation
化学氧化 chemical oxidation
温室效应 greenhouse effect
温室气体 greenhouse gas
全球变暖,地球暖化 global warming
臭氧层破坏 ozone layer destruction
光化学反应 photochemical reaction
查普曼方程 Chapman Equation
保护臭氧层 ozone layer conservation
氟利昂 Freon
制冷剂 refrigerant
喷雾剂 spraying agent
洗涤剂 detergent
隔热材料 heat insulating material

酸雨　acid rain
酸性降水　acid precipitation
硫酸　sulfuric acid
硫酸盐　sulfate
硝酸　nitric acid
硝酸盐　nitrate
抑制剂　inhibitor
湖泊酸化　lake acidification
土壤酸化　soil acidification
森林死亡　forest death
土地荒漠化　land desertification
二氧化硫　sulfur dioxide
硫氧化物　sulfur oxides
氮氧化物　nitrogen oxides
一氧化氮　nitrogen monoxide
烟雾　smog
光化学烟雾　photochemical smog
硫酸烟雾　sulfurous smog
机理　mechanism
光敏剂　photosensitizer
反应性烃　reactive hydrocarbon
光化学氧化剂　photochemical oxidant
过氧乙酰硝酸酯　peroxyacetyl nitrate
居室空气污染　indoor air pollution
废气治理　waste gas treatment
净气　gas purification
大气污染综合防治　Integrated prevention and control of air pollution

1 Atmosphere: The entire mass of air which surrounds the Earth.
大气是指环绕地球的全部空气。

2 Ambient air: Outdoor air to which people, plants, animals and structures are exposed.
环境空气是指人类、植物、动物和建筑物暴露于其中的室外空气。

3 Air pollution may be defined as any atmospheric condition in which substances are present at concentrations high enough above their normal ambient levels to produce a measurable effect on man, animals, vegetation, or materials.
大气中某些物质的浓度，超过了正常的环境水平，以致对人类、动物、植物和其他物质产生可测量的影响，这种情况就是大气污染。

4 Human are putting gaseous materials into the atmosphere that may cause the earth's average temperature to rise. This is called global warming or the greenhouse effect.
人类不断排放气态物质进入大气中，引起地球平均温度升高，这种效应称为全球变暖或温室效应。

5 The group consisting of CO_2, CH4, N_2O, and CFCs are collectively called greenhouse gases.
温室气体包括二氧化碳、甲烷、氧化亚氮和氟利昂等。

6 Global air pollution problems include global warming, destruction of the ozone layer, and acid rain problems.
全球性大气污染问题包括全球变暖、臭氧层破坏和酸雨等三大问题。

7 The normal technical measure of acidity is pH, generally, any rain with a pH less than the 5.6 is considered acidic.
酸度的常规技术测量方法是测定 pH 值，一般情况下，pH 值小于 5.6 的雨称为酸雨。

第二节　燃烧
Section 2　Combustion

燃料　fuel
燃料发热量　heating value of fuel
燃料消耗率　fuel rate
锅炉热损失　heat loss of boiler
天然气　natural gas
石油　petroleum
型煤　briquette coal
褐煤　lignite
煤工业分析　proximate analysis of coal
煤元素分析　ultimate analysis of coal
非常规燃料　unconventional fuel
无污染燃料　pollution-free fuel
燃烧　combustion
不完全燃烧　incomplete combustion
燃烧热　heat of combustion
热力氮氧化物　thermal nitrogen oxide
瞬时氮氧化物　prompt nitrogen oxide
燃料氮氧化物　fuel nitrogen oxide

空气燃料比 air-to-fuel ratio
理论空气量 theoretical air
空气过剩系数 excess air number
烟气状态参数 parameter of fuel gas state
燃烧装置 firing device
排气 exhaust gas
烟气 fuel gas
烟气分析 analysis of fuel gas
奥氏烟气分析仪 Orsat analyzer
流速 fluid velocity

1 There are two common definitions of heat of combustion. The higher heating value definition assumes that the water produced by combustion is condensed, thus giving up its latent heat of condensation. The lower heating value assumes that the water leaves the combustor as a gas, and hence the lower value is less than the higher heating value.

关于燃烧热的定义有两种：高位发热量是指包括燃烧生成物中水蒸气的汽化潜热；低位发热量是指燃烧产物中的水蒸气仍以气态存在时完全燃烧过程所释放的热量，故低位发热量小于高位发热量。

2 Example 1: C: 77.2%, H: 5.2%, N: 1.2%, S: 2.6%, O: 5.9% and ash: 7.9% by weight. Determine the normalized molar composition.

例 1：某燃料的化学组成以质量计为：C:77.20%；H:5.20%；N:1.20%；S:2.60%；O:5.90%；灰分 7.90%。确定其摩尔组成。

Element	Wt %	mol/100g	mol/mol(C)
元素	质量计 %	mol/100g	mol/mol(碳)
C	77.2	÷12 = 6.43	÷6.43 = 1.00
H	5.20	÷1 = 5.20	÷6.43 = 0.809
N	1.20	÷14 = 0.0857	÷6.43 = 0.013
S	2.60	÷32 = 0.0813	÷6.43 = 0.013
O	5.90	÷16 = 0.369	÷6.43 = 0.057
Ash(灰分)		7.9	÷6.43 = 1.23 g/mol(C)

The normalized molar composition: CH0.808N0.013S0.013O0.057

其摩尔组成可表示为：CH0.808N0.013S0.013O0.057

3 Example 2: A "typical Pittsburgh seam coal" has the following ultimate analysis by weight: hydrogen, 5.0%; carbon, 75.8%; nitrogen, 1.5%; sulfur, 1.6%; oxygen, 7.4%; ash, 8.7%. It is burned with 20% excess air, with humidity 0.0116 mol H_2O/mol dry air, and combustion is complete. Determine the amount and composition of the gas produced.

例 2：一"典型的匹兹堡层煤"的元素分析按重量百分比表示，结果如下：氢 5.0%；碳 75.8%；氮 1.5%；硫 1.6%；氧 7.4%；灰 8.7%，燃烧条件为空气过量 20%，空气湿度为 0.0116 mol H_2O/mol 干空气，并假定完全燃烧，试计算烟气的量和组成。

第三节 大气扩散

Section 3 Atmospheric Diffusion

大气圈 atmospheric sphere
大气层 atmospheric layer
洁净空气 clean air
层结作用 stratification
对流层 troposphere
平流层 stratosphere
臭氧层 ozone layer
中间层 mesosphere
热层 thermosphere
电离层 ionosphere
外逸层 exosphere
气团 air mass
暖气团 warm air mass
冷气团 cold air mass
气团变性 air mass modification
锋 front
暖锋 warm front
冷锋 cold front
大气边界 atmospheric boundary layer
近地层 surface layer
混合层 mixing layer
混合层厚度 mixing depth

自由大气　free atmosphere
大气环流　general circulation of atmosphere
局地环流　local circulation
山谷风　mountain-valley breeze
峡谷风　gorge wind
海陆风　sea-land breeze
向岸风　coastward wind
离岸风　offshore wind
风场　wind field
风向　wind direction
风速　wind speed
风力等级　wind scale
气象要素　meteorological element
云状　cloud form
云量　cloud amount
烟囱　chimney
排气管　exhaust pipe
高烟囱排放　high stack emission
有效排放高度　effective height of mission
热力抬升　buoyancy rise
动力抬升　plume rise due to momentum
烟羽　plume
烟羽型　plume type
翻卷型烟羽　looping plumes type
锥型烟羽　conning plumes type
扇型烟羽　fanning plumes type
熏烟型烟羽　fumigating plumes type
上升型烟羽　lofting plumes type
烟羽抬升　plume rise
湍流　turbulence
湍流强度　turbulence intensity
风洞模拟　tunnel simulation
危险风速　critical wind of menace
大气稳定度　atmospheric stability
温度廓线　temperature profile
直减率，递减率　lapse rate
气温直减率　temperature lapse rate
干绝热直减率　adiabatic lapse rate
绝热过程　adiabatic process
逆温　inversion
辐射逆温　radiation inversion
下沉逆温　subsidence inversion
平流逆温　advection inversion
干绝热过程　dry adiabatic process
大气选择吸收　atmospheric selective absorption
对流过程　convective process
半衰期　half-life
拉格朗日法　method of Lagrange
欧拉法　method of Euler
大气扩散　atmospheric diffusion
大气扩散相似理论　similarity theory of atmospheric diffusion
大气扩散统计理论　statistical theory of atmospheric diffusion
大气扩散估算方法　method of atmospheric diffusion estimating
大气扩散模　atmospheric diffusion model
高斯烟羽模式　The Gaussian Plume Model
大气扩散参数　atmospheric diffusion parameter
扩散系数公式　diffusion coefficient equation
帕斯奎尔-吉福德扩散曲线　Pasquill-Gifford diffusion curve
双向风标　bidirectional vane
高斯分布(或正态分布)　Gaussianor normal distribution
污染范围　pollution scale
全球的　global
区域的　regional
城市的　urban
局部的　local
本底　background

1 Meteorological services regularly prepare wind roses like that in Fig. 3-2 on page 68. These summarize the frequency of winds varying velocities and direction from which it comes.
气象服务用的风玫瑰图，如 68 页图 3-2 所示，它们概括了不同风向的频率、风速的变化情况。

2 Most diffusion models use the Gaussian plume idea, which is also a material balance model. In it one considers a point source such as a factory smokestack (which is not really a point but a small area that can be satisfactorily approximated as a point) and attempts to compute the downwind concentration resulting from this point source. The schematic representation and nomenclature are shown in Fig. 4-2, where the origin of the coordinate system is placed at the base of the smokestack, with the x axis aligned in the downwind direction. The contaminated gas stream (normally called a plume) is shown rising from the smokestack and then leveling off to travel in the x direction and spreading in the y and

z directions as it travels.

绝大部分的扩散模式都应用高斯烟羽概念，高斯烟羽模式也是一种物料衡算模式。它把工厂烟囱视为点源(实际上并不是点源，但由于其区域较小可近似为一点源)，然后计算点源下风向污染物浓度。示意图如图 4-2 所示，坐标系原点为烟囱基点，x 轴为下风向，污染气流(一般称为烟羽)从烟囱处升起后沿 x、y、z 方向扩散。

第四节 颗粒物控制技术

Section 4 Particulate Control Technology

细微颗粒物 fine particulate
颗粒物特征 particulate characteristics
颗粒物粒度 particulate size
颗粒物比表面积 specific surface of particle
颗粒物比电阻 dust electrical resistivity
颗粒物安息角 repose angle of particle
颗粒物含水率 water content of particle
流体雷诺数 Reynolds number of fluid
颗粒雷诺数 Reynolds number of particle
努森数 Knudsen number
曳力系数 coefficient of drag
坎宁汉因数 Cunningham factor
最终沉降速度 terminal setting velocity
颗粒扩散系数 diffusion coefficient of particle
热泳现象 thermophoresis
光泳现象 photophoresis
惯性碰撞 inertial impaction
斯托克斯准数 Stokes number
拦截 direct interception
絮凝 flocculation
聚集 aggregation
凝聚 coagulation
扩散沉淀 diffusion deposition
除尘效率 particle collection efficiency
透过滤 penetration
分割粒径 cut diameter for particles
压力损失 pressure drop
损失系数 pressure drop coefficient
除尘 dust removal
除尘装置 dust removal device
沉降室 settling chamber
挡板集尘室 baffle chamber
旋风分离器 cyclone separator
袋滤器 bag filter
洗涤式除尘器 scrubber
静电除尘器 electrostatic precipitator
机械除尘器 mechanical collector
重力沉降室 gravity settling chamber
惯性除尘器 inertial dust separator
旋风除尘器 cyclone collector
回流式旋风除尘器 reverse-flow cyclone collector
直流式旋风除尘器 straight-through cyclone collector
多管旋风除尘器 multiple cyclone collector
过滤除尘器 filter
袋式除尘器 bag house
滤料 filtration media
合成纤维滤料 synthetic fiber
玻璃纤维滤料 glass fiber
覆膜滤料 teflon lined filter
透气性 permeability
抗拉强度 tensile strength
气布比 air-to-cloth ratio
过滤速度 superficial face velocity
机械振动清灰袋式除尘器 bag house with shake cleaning
逆气流清灰袋式除尘器 bag house with reverse flow cleaning
脉冲喷吹清灰袋式除尘器 bag house with pulse jet cleaning
扁布袋除尘器 rectangular bag house
颗粒层除尘器 granular bed filter
静电除尘器 electrostatic precipitator
电晕放电 corona discharge
驱进速度 drift velocity
集尘极 collecting electrode
比集尘面积 specific collection area
板间距 distance between collecting electrodes
气体流速 gas velocity
气流分布 distribution of gas velocity
烟气调质 flue gas condition
电极清灰 removal of collected particle from electrodes
宽间距静电除尘器 wide space electrostatic precipitator

高压脉冲静电除尘器 pulse charging electrostatic precipitator
湿式静电除尘器 wet electrostatic precipitator
双区静电除尘器 two-stage electrostatic precipitator
湿式除尘器 wet collector of particulates
重力喷雾洗涤器 gravitational spray scrubber
旋风洗涤器 centrifugal scrubber
旋筒式水膜洗涤器 horizonal cyclone water-film scrubber
中心喷雾旋风洗涤器 cyclone spray scrubber
自激喷雾洗涤器 impingement-entrainment scrubber
泡沫塔洗涤器 foam tower scrubber
填料床洗涤器 packed bed scrubber
湍球塔洗涤器 fluidized packed bed scrubber
文丘里洗涤器 venturi scrubber
分割-功率关系 cut and power relationship
分割直径法 cut diameter method
除雾器 demister

1 Gravity settling chambers, cyclones, and ESPs work by driving the particles to a solid wall where they form agglomerates that can be collected. Although these devices look different from one another, they all use the same idea and are described by the same general design equations.
重力沉降室、旋风除尘器和静电除尘器的工作原理都是通过颗粒运动到固体壁面凝聚成块被捕集下来。虽然这三种除尘器看起来各不相同,但他们有着相同的原理,并且都可用相同的设计方式来描述。

2 Example: A 1μ diameter particle of a material with a dielectric constant of 6 has reached its equilibrium charge in an ESP at a place where the field strength is 300 kV/m. How many electronic charges does it have? Calculate the drift velocity for the particle
例:直径为1微米的粒子置于电晕电场,粒子介电常数为6,场强为300 kV/m,求粒子荷电量,计算粒子的驱进速度。

第五节 硫氧化物控制
Section 5 Control of Sulfur Oxides

煤的综合利用 comprehensive utilization of coal
燃料脱硫 desulfurization of fuel
重油脱硫 desulfurization of heavy oil
煤脱硫 desulfurization of coal
物理法脱硫 physical desulfurization
化学法脱硫 chemical desulfurization
重力洗选脱硫 gravity floating desulfurization
磁力脱硫 magnetic desulfurization
细菌脱硫 bacterial desulfurization
微波脱硫 microwave desulfurization
煤的气化 coal gasification
煤的液化 coal liquefaction
流化床燃烧脱硫 desulfurization by fluidized bed combustion
酸碱反应 acid-base reaction
烟气脱硫 flue gas desulfurization (FGD)
焙烧 roasting
烧结 agglutination
石膏 gypsum
浆液 slurry
亚硫酸钠 sodium sulfite
氧化镁 magnesium oxide
湿法脱硫 wet process of FGD
石灰 lime
石灰-石灰石法脱硫 desulfurization by lime and limestone
石灰石湿式洗涤器 limestone wet scrubber
氧化镁法脱硫 desulfurization by magnesium oxide
双碱流程 dual-alkali process
亚硫酸钠法脱硫 sodium sulphite of FGD
氨吸收法脱硫 ammonia process of FGD
干法脱硫 dry process of FGD
喷雾干燥法脱硫 spray-dryer process of FGD
活性炭吸附法脱硫 activated carbon process of FGD
电子束辐照脱硫 desulfurization by electron beam irradiation
海水脱硫 desulfurization by sea water
干式催化氧化法脱硫 dry catalytic oxidation process of FGD
液相催化氧化法脱硫 liquid phase catalytic oxidation process of FGD

1 Fig. 8-1 Page 304 shows in part how sulfur moves in the environment as a result of human activities. It does not include the large amounts of sulfur emitted by volcanic eruptions nor the movement of sulfur into growing plants and then back out of decaying plants. Sulfur is the 16-most abundant element in the earth's crust, with an abundance of about 260 ppm. The vast majority of this sulfur exists in the form of sulfates, mostly as gypsum, $CaSO_4.2H_2O$, or anhydride $CaSO_4$. Gypsum is a chemically inert, nontoxic, slightly water-soluble mineral, found widely throughout the world.
在 304 页图 8-1 中显示了由于人为活动造成的硫在自然环境中的循环。其中，由于火山爆发引起的大量的硫排放以及由于植物生长吸收及植物腐烂释放的硫没有包括在内。硫是地壳中第十六丰富的元素，其丰度约为 260×10^{-6}（质量分数）。在地壳中硫主要是以硫酸盐的形式存在，其中大部分是石膏 $CaSO_4.2H_2O$ 或硬石膏 $CaSO_4$。石膏是一种化学惰性、无毒、微溶于水的矿物质，在全球范围内广泛存在。

第六节　氮氧化物控制
Section 6　Control of Nitrogen Oxides

烟气脱氮　flue gas denitrogenation
燃料型氮氧化物　fuel nitrogen oxide
热力型氮氧化物　thermal nitrogen oxide
瞬时型氮氧化物　prompt nitrogen oxide
催化还原法脱氮　control of NO_x with catalytic reduction process
吸收法控制氮氧化物　control of NO_x by absorption
水吸收法脱氮　control of NO_x by absorption-process with water
酸吸收法脱氮　control of NO_x by absorption process with acid
碱吸收法脱氮　control of NO_x by absorption process with alkali
吸附法控制氮氧化物　control of No_x by absorption
分子筛吸附法脱氮　control of NO_x by absorption process with molecular sieve
硅胶吸附法脱氮　control of NO_x by absorption process with silica gel
活性炭吸附法脱氮　NO_x control with activated carbon adsorption
低氮氧化物燃烧技术　low nitrogen oxide combustion technology
烟气循环　flue gas recirculation
二段燃烧法　two-stage combustion

1 Scientists classify the nitrogen oxides found in combustion gases as thermal, prompt, and fuel nitrogen oxides. Thermal nitrogen oxides, which are generally the most significant, are formed by the simple heating of oxygen and nitrogen. Prompt refers to the nitrogen oxides that form very quickly as a result of the interaction of nitrogen and oxygen with some of the active hydrocarbon species derived from fuel in the fuel-rich parts of flames. Fuel nitrogen oxide is formed by conversion of some of the nitrogen originally present in the fuel to NO_x.
科学家们把燃烧过程中形成的氮氧化物分为热力型、瞬时型和燃料型氮氧化物。热力型氮氧化物占绝大部分，一般是由空气中氮和氧反应生成。瞬时型氮氧化物是指由氮气、氧气和燃料中碳氢化合物在富燃条件下快速反应生成的氮氧化物。燃料型氮氧化物是指由燃料中含氮有机物直接氧化而成的氮氧化物。

第七节　挥发性有机物控制
Section 7　Control of Volatile Organic Compounds (VOCS)

气态污染物净化　control of gaseous pollutants
吸收　absorption
物理吸收　physical absorption
双膜理论　two-film theory
气膜控制　gas film control
液膜控制　liquid film control
液气比　liquid-to-gas flow ratio
吸收操作线　operating line of absorption process
吸收装置　absorber
吸附　adsorption
吸附法净化气态污染物　control of gaseous pollutants by adsorption

物理吸附 physical adsorption
化学吸附 chemical adsorption
交换吸附 exchange adsorption
吸附等温线 adsorption isotherm
弗里德里胥吸附等温线 Freundlich adsorption isotherm
弗里德里胥方程 Freundlich equation
朗格缪尔方程 Langmuir equation
BET 方程 BET equation
活性炭 activated carbon
吸附系数 adsorption coefficient
吸附容量 adsorption capacity
吸附剂比表面积 specific surface area of adsorbent
吸附选择性 adsorptive specificity
吸附平衡 adsorption equilibrium
吸附等温线 adsorption isotherms
吸附热 heat of adsorption
固定床吸附器 fixed bed adsorber
蜂窝回转轮吸附器 honeycomb rotor concentration adsorber
希洛夫方程 Schieloeff Equation
吸附负荷曲线 absorptive load curve
穿透点 break-through point
穿透曲线 break-through curve
保护作用时间 protection time
移动床吸附器 moving bed adsorber
升温脱附 desorption by heating
降压脱附 desorption by evacuating
汽提脱附 desorption by stripping
置换脱附 desorption by displacing
冷凝 condensation
冷凝法净化气态污染物 control of gaseous pollutants by condensation
直接接触冷凝器 direct contact condenser
表面冷凝器 surface condenser
燃烧 combustion
催化剂 catalyst
催化剂活性 catalytic activity
催化剂选择性 catalytic selectivity
催化剂中毒 poisoning of catalyst
催化剂选择 selection of catalyst
催化氧化 catalytic oxidation
催化还原 catalytic reduction
含氟废气控制 control of fluoric gas
吸收法控制含氟废气 control of fluoric gas with absorption
水吸收含氟废气 absorption of fluoric gas by water
氟铝酸法回收氟化氢 hydrogen fluoride recovery by fluoaluminic acid process
氨法回收氟硅酸 fluosilicic acid recovery by ammonia process
碱吸收含氟废气 absorption of fluoric gas by alkali
硫化氢控制 control of hydrogen sulfide
干式氧化法控制硫化氢 control of hydrogen sulfide by dry oxidation process
吸附法控制硫化氢 control of hydrogen sulfide by adsorption
吸收法控制硫化氢 control of hydrogen sulfide by absorption
弱碱溶液吸收控制硫化氢 absorption of hydrogen sulfide by weak alkali solution
碱性盐溶液吸收控制硫化氢 absorption of hydrogen sulfide by basic salt solution
有机溶剂吸收控制硫化氢 control of hydrogen sulfide by absorption with organic solvent
吸收氧化法控制硫化氢 control of hydrogen sulfide by absorption oxidation process
垃圾焚烧尾气净化 control of flue gas from MSW incineration
有机废气治理 control of organic gas
有机废气催化燃烧 catalytic combustion of organic gas
催化燃烧催化剂 catalyst for catalytic combustion
有机废气直接燃烧 direct combustion of organic gas
有机废气吸附 adsorption of organic gas
气体生物净化 biotreatment of gaseous pollutant
生物过滤器 biofilter
生物滴滤器 biotrickling filter
气体膜分离 membrane separation of gaseous pollutant
光催化氧化 photo-catalytic oxidation
等离子净化 plasma destroy
生物涤气器 bioscrubber

1 Volatile organic compounds (VOCs) are a large family of compounds. VOCs are liquids or solids that contain organic carbon, which vaporized at significant rates. Some are toxic and carcinogenic, and are

regulated individually as hazardous pollutants to humans.

挥发性有机物是一类有机化合物统称，它指含有有机碳的液态或固态物质，它们蒸发速率很大。有些挥发性有机物是有毒的、致癌的，是对人类具有危险性的污染物，必须加以控制。

第八节 汽车尾气污染控制

Section 8 Control of Automobile Engine Emission

汽车污染控制 automotive pollution control

汽油 gasoline

柴油 diesel oil

无铅汽油 unleaded gasoline

尾气 tail gas

烷烃 alkane

汽油机尾气污染控制 control of gasoline engine emission

柴油机尾气污染控制 control of diesel engine emission

内燃机 internal combustion(IC)engine

汽轮发动机 gas turbine engine

四冲程内燃汽油发动机 four-stroke IC gasoline engine

延迟点火 spark timing

电子喷射 electronic fuel injection

液化石油气汽车 LPG vehicle liquefied petroleum gas

压缩天然气汽车 CNG vehicle compressed natural gas

汽车尾气净化催化剂 catalyst of mobile exhaust control

三效催化反应器 three-way catalytic reactor

1 For control pollution, every country raises the control requirement about the emission pollutant, constitute the compelling emission standards to control the emission from automobile.

为了治理环境污染，各国相继针对车辆排放制定强制性排放标准，以控制汽车污染物的排放量。

2 Presently, catalytic converter is the most effective equipment to emission control, which can reduce harmful ingredient of automobile emission.

催化转化器能减少汽车排放物中的有害成分，是目前最有效的汽车排放控制装备。

第九节 臭气控制

Section 9 Odor Control

臭气浓度 odor concentration

臭气强度 odor intensity

臭气吸附 adsorption of odor

臭气吸收 absorption of odor

臭气火焰氧化 flame oxidation of odor

臭氧化学氧化 chemical oxidation of odor

臭氧催化氧化 catalytic oxidation of odor

臭氧掩蔽 masking of odor

臭氧抵消 counteraction of odor

脱臭 deodorization

恶臭 offensive odor

生物脱臭 biotreatment of odor

室内空气污染 indoor air pollution

厨房油烟净化 abatement of the oil mist from cooking

1 Fig. 13-1 shows, in very simplified form, a cross-sectional view of a typical auto engine. It shows only one piston and cylinder; most auto engines have four such pistons and cylinders, though some have six or eight. In operation, the crankshaft rotates, causing the piston to move up and down, driven by the crank, connecting rod, and wrist pin. To begin a cycle, with the piston at the top during the first stroke the piston moves downward while the intake value is open, so that an air-fuel mixture is sucked into the combustion chamber.

图 13-1 为典型汽车发动机的结构示意图，它为一个缸体，大部分汽车发动机装有 4 个缸、6 缸或 8 缸。在工作过程中，发动机推动活塞做上下往复运动，通过连杆、曲轴柄带动曲轴旋转。在循环开始时，活塞位于上止点，进气门打开，活塞向下移动，空气和燃料的混合物通过进气门进入缸体。

第八章　环境噪声控制工程
Chapter 8　Environmental Noise Control Engineering

第一节　绪论
Section 1　Introduction

声学　acoustics; theory of sound
环境声学　environmental acoustics
环境光学　environmental optics
物理声学　physical acoustics
波动声学　wave acoustics
统计声学　statistical acoustics
几何声学　geometrical acoustics
建筑声学　architectural acoustics
噪声　noise
噪声源　noise source
环境噪声　ambient noise, environmental noise
无规噪声　random noise
背景噪声　background noise
可听声　audible sound
次声　infrasound; infrasonic sound
噪声污染　noise pollution
噪声污染监测　noise pollution monitoring
噪声水平　level of noise
噪声危害　noise hazard
噪声控制　noise control
自然界噪声　noise in nature
机动车辆噪声　vehicle noise
轮胎噪声　tyre noise
轨道交通噪声　railroad noise
地铁噪声　underground noise
轮轨噪声　wheel/rail noise
机车噪声　locomotive noise
飞机和机场噪声　aircraft and airport noise
船舶噪声　ship noise
工业噪声　industrial noise
建筑施工噪声　construction noise
社会生活噪声　noise of social activities
交通噪声　traffic noise

1 We will begin with an introduction to the basic concepts of acoustic and acoustic measurement. It is important for the designer to understand the nomenclature and physical principles involved in sound transmission in order to suggest a rational procedure for noise reduction.
我们从介绍声学和声学测量的基本原理开始，对于设计者来说，了解声音传播过程的有关术语和原理对提出合理的降噪程序是至关重要的。

2 Noise is usually defined as any perceived sound that is objectionable or damaging for a human.
噪声一般指对人产生讨厌的或破坏性的可察觉的声音。

3 Generally, acoustics system consists of the source of sound, the transmission path of sound, or the receiver of sound. Noise may be controlled by modification of the source of sound, the transmission path of sound, or the receiver of sound.
声学系统一般由声源、传播途径和接收器三环节组成。我们可以通过噪声源、传播途径、接收器的调整来进行噪声控制。

4 Traffic noise of city roads is a major source of environmental noise pollution in urban areas. Rational planning of city roads is of great significance to the future noise control in the urban areas.
道路交通噪声是城市环境噪声的主要污染源。合理的城市道路规划，对未来的城市噪声控制具有重要的意义。

第二节　声波基本性质
Section 2　Character of Sound Wave

声波　sound wave
反射　reflection
散射　scattering
衍射(绕射)　diffraction

光辐射 visible radiation
光反射 reflection
相互反射 inter-reflection
光透射 transmission
反射比 reflectance
透射比 transmittance
光吸收 absorption
光折射 refraction
折射率 refractive index
漫反射 diffuse reflection
漫透射 diffusion transmission
漫射因数 diffusion factor
镜像法 image sound source
纵波 longitudinal wave
波阵面 wave front
球面波 spherical wave
平面波 plane wave
柱面波 cylindrical wave
弯曲波 bending wave
冲击波 shock wave
静压 static pressure
声压 sound pressure
质点位移 particle displacement
质点速度 particle velocity
质点加速度 particle acceleration
体积速度 volume velocity
声速 sound velocity
声速梯度 sound velocity gradient
阻抗 impedance
声阻抗 acoustic impedance
声阻 acoustic resistance
声抗 acoustic reactance
声质量 acoustic mass
声阻抗率 specific acoustic impedance
声特性阻抗 acoustic characteristic impedance
辐射阻抗 radiation impedance
声强 sound intensity
声源强度 strength of a sound source
声能密度 sound energy density
声功率 sound power
声源功率 sound power of a sound source
声线描绘法 sound ray tracing technique
频谱 frequency spectrum
线谱 line spectrum
连续谱 continuous spectrum
声源 sound source
点声源 point source
线声源 line source
面声源 plane source
单极子声源 monopole source
偶极子声源 dipole source
干涉 interference
相干性 coherence
级 level
分贝(dB) decibel
声级 sound level
声压级 sound pressure level
声强级 sound intensity level
声功率级 sound power level
频带声功率级 band sound power level
声线 sound ray
传声媒质 sound bearing medium
声像 sound image
声影区 sound shadow region
声桥 sound bridge
大气声衰减 sound attenuation in air
逾量(附加)声衰减 excess attenuation
发散衰减 divergence attenuation
绿化声衰减 sound attenuation in landscape
指向性 directivity
指向性因数 directivity factor
指向性指数 directivity index

1 In many situations, the size of the source of sound is relatively small, and the sound is radiated from the source uniformly in all directions. In this case, the sound wave would not be planar; instead, the sound waves are called spherical waves. For spherical waves, we see that the acoustic pressure varies inversely with the distance from the sound source.
在许多情况下,声源的尺寸相对很小,声音向各个方向均匀辐射,在这种情形下,声波不是平面的,称为球面波。球面波的声压随传播距离 r 的增加而减少。

2 The sound pressure (p) is defined as the instantaneous difference between the local pressure (P) and the ambient pressure (Po) for sound wave in the material. The acoustic pressure for the plane simple harmonic sound wave moving in the positive x-direction may be represented by the following, $p(x,t)$ = pmax cos(wt-kx).

声压定义为媒质中压力与净压的差值。沿 x 正方向传播的平面声波的声压的表达式如下：$p(x,t)$ = Pmax cos(wt-kx)。

3 The acoustic intensity(I) is defined as the average energy transmitted through a unit area per unit time, or the acoustic power (W) transmitted per unit area.

声强(I)定义为在单位面积单位时间内通过的声能或单位面积的声功率(W)。

第三节 噪声评价
Section 3 Noise Assessment

环境影响评价 environmental impact assessment
噪声控制标准 criteria for noise control
噪声污染级 noise pollution level (NPL)
交通噪声指数 traffic noise index
噪声主观评价 subjective assessment of noise
噪声预测 noise forecast
响度 loudness
噪度 noisiness
宋 sone
响度级 loudness level
等响曲线 equal-loudness contour
听阈 threshold of audibility
感觉阈 threshold of feeling
听觉区域 auditory sensation area
感觉噪声级 perceived noise level
有效感觉噪声级 effective perceived noise level
等噪线 equal-noisiness contour
计权有效连续感觉噪声级 weighted effective continuous perceived noise level
语言干扰级 speech interference level
噪声冲击 noise impact
噪声冲击指数 (NII)noise impact index
语言听力损失 hearing loss speech
计权 weighting
计权网络 weighting network
A 计权声级 A-weighted sound pressure level
C 计权声级 C-weighted sound pressure level
D 计权声级 D-weighted sound pressure level
E 计权声级 E-weighted sound pressure level
掩蔽 masking
纯音 pure tone
音质设计 acoustics design
暴露声级 sound exposure
等效声级 equivalent sound pressure level
昼夜等效[连续 A]声级 day-night equivalent [continuous A-weighted]sound pressure level
累计百分数声级 percentile level

1 The assessment of environmental noise involves two main stages. The first stage in carrying out an environmental noise assessment is the quantification of noise. This involves the measurement of existing sources or the prediction of noise from future sources. The second stage involves evaluating the actual or predicted value against some criteria.

环境噪声评价包括两个主要阶段：第一个阶段是噪声的量化，这个过程包括对现有噪声现状的测量和噪声预测。第二个阶段是评价现有噪声值和预测值超标情况。

第四节 噪声测量
Section 4 Noise Measurement

噪声测量设备 noise measurement equipment
噪声计 noise meter
噪声暴露 noise exposure
噪声暴露计 noise exposure meter
噪声暴露预测 noise exposure forecast (NEF)
噪声本底 noise background
噪声比 noise ratio
噪声源声功率测量 noise source power measurement
传声器 microphone
动态范围 dynamic range
测量传声器 measuring microphone
扬声器系统 loudspeaker system
标准声源 reference sound source
声级计 sound level meter
积分声级计 integrating sound level meter
活塞发生器 piston phone
声校正器 sound calibrator
滤波器 filter

频谱分析仪 spectrum analyzer
级记录仪 level recorder
声强仪 sound intensity analyzer
防风罩 windscreen，windshield
消声室 anechoic room，anechoic chamber
半消声室 semi-anechoic room
混响室 reverberation room
隔声实验室 sound insulation
驻波管 standing wave tube
声屏障插入损失测量 insertion loss measurement of barriers
声强测量 sound intensity measurement
数字信号处理 digital signal processing (DSP)
数字测量技术 digital measurement technique
声功率级测量 sound power level measurement
环境噪声测量 environmental noise measurement
噪声源鉴别 noise source identification
护耳器 ear protectors

1 Noise control programs often require measurements of various acoustic quantities to determine the effectiveness of the noise control procedure. Noise measurement must be made to determine compliance with noise regulations. Noise measurement may be needed for diagnostic purposes of locate the source (or sources) of noise in a piece of machinery. The transmission path for noise in a system may be identified through acoustic measurements.

噪声控制过程中经常需要测量声音的量级来确定噪声控制的效果。噪声测量必须和噪声传播规律相一致，通过测量仪器可以确定当地噪声源，通过噪声测量可以鉴别系统中噪声的传播路径。

2 The basic parts of most sound level meters include a microphone amplifier, weighting networks, and a display indicated decibels. Typical sound level meters are shown in Table 4-1 on page 65.

大部分声级计包括传声器、放大器、计权网络、显示器等基本部件，典型的声级计列于65页表4-1中。

第五节 吸声

Section 5 Sound Absorption

吸声 sound absorption，acoustic absorption
多空吸声材料 porous absorbing material
直达声场 direct sound field
混响声场 reverberant sound field
室内声学 room acoustics
声学测量 acoustical measurement
频程 frequency interval
倍频程 octave
1/3倍频程 1/3 octave
混响 reverberation
声疲劳 acoustic fatigue
疲劳寿命 fatigue life
声失效 acoustics malfunction
传播系数 propagation coefficient
衰减系数 attenuation coefficient
菲涅耳数 Fresnel number
声场 sound field
自由声场 free sound field
半自由声场 half free field
无规入射 random incidence
扩散声场 diffuse sound field
近场 near field
远场 far field
空气声 air-borne sound，air-borne noise
固体声(结构声) structure-borne noise
脉冲声 impulsive sound
直达声 direct sound
类比 analogy
混响时间 reverberation time
混响半径 reverberation radius
平均自由程 mean free path
房间常数 room constant
房间吸声量 room absorption
吸声系数 sound absorption coefficient
正入射吸声系数 Vertical incidence sound absorption coefficient
统计吸声系数 statistical absorption coefficient
平均吸声系数 average sound absorption coefficient
降噪系数 noise reduction coefficient
吸声量(等效吸声面积) equivalent absorption area
吸声能力 acoustic absorptivity
边缘效应 edge effect
孔隙率 porosity
流阻 flow resistance

声反射系数 sound reflection coefficient
声透射系数 sound transmission coefficient
吸声板 acoustic board
吸声材料 sound absorption material
吸声结构 sound absorption structure
吸声带宽 bandwidth of sound absorption
吸声降噪 noise reduction by absorption
吸声测量 acoustic absorptivity measurement
吸声砖 acoustical tile
亥姆霍兹共振器 Helmholtz resonator
薄板吸声结构 panel absorption structure
薄膜吸声结构 membrane absorption structure
穿孔板吸声结构 absorption structure of perforated panel
微穿孔共振吸声结构 resonance absorption structure of micro-perforated panel
吸声体；吸声器 acoustic absorber
空间吸声体 suspended absorber
吸声尖劈 wedge absorber
喷涂吸声结构 spray absorption structure

1 Sound-absorbing materials are used to reduce the sound levels in a room or to reduce the reverberation. The direct sound field which consists of the acoustic energy associated with sound waves that have not struck surface in the room. The reverberant sound field refers to the sound field which consists of the remainder of the energy.
吸声材料用于降低室内噪声级或降低室内混响。直达声场指从声源直接到达受声点的声波能量形成的声场，混响声场指剩余的能量形成的声场。

第六节 隔声
Section 6 Sound Insulation

降噪量 noise abatement
传声损失 sound transmission loss
声屏障 sound barrier
隔声间 acoustic booth
隔声板 acoustic septum
隔声室 sound insulation room
隔声门 sound insulation door
隔声窗 sound insulation window
隔声罩 acoustic shield
隔声纤维板 sound-proof fiber board
隔声标准 sound insulation standard
隔声评价 sound insulation assessment
隔声频率特性曲线 sound insulation frequency characteristic curve
隔声材料 deafening material
隔声效果 deafening effect
隔声指数 sound insulation index
计权隔声量（隔声指数） weighted sound isolation factor
平均隔声量 average sound isolation factor
侧向传声 flanking transmission
插入损失 insertion loss
质量定律 mass law
劲度控制 stiffness control
质量控制 mass control
吻合效应 coincidence effect

1 Sound insulation is the prevention of transmission of sound or alternatively, a reduction of sound energy transmitted into an adjoining air space. Two types of sound insulation are to be dealt with in building construction. The ideal material for good sound insulation has a very high mass and low stiffness but some of the building materials have low mass and relatively high stiffness. Details of some of these are shown in Table 2.
隔声是指在传播途径中使声音受阻或声能由于传至相邻的空间而降低的措施。在建筑物中可以采取两种隔声方法。理想的隔声材料要有较大的质量和较低的劲度，但是一些建筑材料的质量却很小，相对劲度较大，关于它们的详细描述见表2。

第七节 消声
Section 7 Sound Reduction

气流噪声（空气动力噪声） flow noise
稳态噪声 steady noise
非稳态噪声 nonsteady noise
低噪声路面 road of low noise

消声器 muffler, noise silencer
消声室 anechoic chamber
消声系数 extinction coefficient
消声材料 sound damping material
消声装置 muffler device/sound damping device
阻性消声器 resistive
抗性消声器 reactive muffler
阻抗复合型消声 impedance composite muffler
共振式消声器 resonance muffler
微穿孔板消声器 micro-perforated panel muffler
消声弯头 lined bend
小孔消声器 micropore muffler
排气喷流消声器 exhaustion jet muffler
蒸汽消声器 steam muffler
低噪声喷口 low noise nozzle
插管共振式消声器 pipe resonator muffle

1 A silencer is an important noise control element for reduction of machinery exhaust noise, fan noise, and other noise sources involving flow of a gas. In general, a silencer may be defined as an element in the flow duct that acts to reduce the sound transmitted along the duct while allowing free flow or the gas through the flow passage.

消声器是一种很重要的噪声控制元件,主要用于减少机械噪声、鼓风机噪声和其他噪声源噪声,包括气流噪声等。一般情况下,消声器是一种既能降低管道内噪声传播又能允许气流顺利通过的管道中的元件。

第八节 隔振

Section 8 Vibration Isolation

振动 vibration
振动污染 vibration pollution
振动位移 vibration displacement
振动速度 vibration velocity
振动加速度 vibration acceleration
功率谱密度 power spectrum density
声致振动 acoustically excited vibration
振动位移级 vibration displacement level
振动速度级 vibration velocity level
振动加速度级 vibration acceleration level
基频 fundamental frequency
主频 basic frequency
简正频率 normal frequency
固有频率 natural frequency
自由度 degrees freedom
简正振动方式 normal mode of vibration
无阻尼固有频率 undamped natural frequency
阻尼固有频率 damped natural frequency
稳态振动 steady-state vibration
简谐振动 simple harmonic vibration
瞬态振动 transient vibration
受迫振动 force vibration
自由振动 free vibration
冲击振动 ballistic vibration
无规振动 random vibration
共振 resonance
反共振 anti-resonance
传递率 transmissibility
临界速度 critical speed
阻尼 damping
损失因数 loss factor
黏滞阻尼 viscous damping
干(摩擦)阻尼 dry-friction damping
弛豫阻尼 hysteresis damping
非线性阻尼 nonlinear damping
临界阻尼 critical damping
阻尼比 damping ratio
阻尼系数 damping coefficient
结构阻尼 structural damping
机电类比 electro-mechanical analogy
隔振 vibration isolation
减振 vibration reduction
有源减振 active vibration reduction
减振器 dashpot
冲击 shock
隔振器;隔振装置 vibration isolator
隔振材料 vibration isolation material
隔振垫 vibra-damping apacer
隔振基座 vibra-damping mount
隔振效率 vibration isolation efficiency
弹簧隔振器 spring isolator
橡胶隔振器 rubber isolator
动力吸振器 dynamic vibration absorber
阻尼材料 damping material

黏弹性材料　viscoelastic material
阻尼结构　damping construct
阻尼控制　damping control
振动阻尼器　vibration damper
摩擦阻尼器　friction damper
空气阻尼器　air damper
磁阻尼器　friction damper
隔振效率　vibration isolation efficiency
统计能量分析　statistical energy analysis
有限元法　finite-element method
模态分析　modal analysis
加速度计　accelerometer
振动计　vibrometer
振动评价标准　vibration criteria
城市区域环境振动标准　vibration standard of city-zone environment

1 One of the major sources of noise in mechanical equipment is the noise produced by energy radiated from vibrating solid surface in the machine. In addition, noise may be produced when vibratory motion or forces are transmitted from the machine to its support structure, through connecting piping, etc. Noise reduction may be achieved by isolating the vibrations of the machine from the connected elements.

机械设备中噪声的主要来源之一是由于振动固体的表面产生的能量辐射。另外，在振动或振动力中产生的噪声可以通过相连的管道从机械传递给支撑结构，我们可以通过隔离机械振动的传递来降低噪声。

2 Many aspects of vibration isolation theory may be understood by examination of a single-degree-of-freedom (SDOF) system consisting of a mass and a linear spring, as shown in Fig. 10-6 on page 208.

隔振原理的许多方面从研究单自由度振动系统着手，单自由度振动系统包括质量块和弹簧两部分，如 208 页图 10-6 所示。

第九节　噪声控制实例
Section 9　Example of Noise Control

撞击声　impact sound
撞击声压级　impact sound pressure level
撞击声指数　impact sound index
撞击声改善　improvement of impact sound
涡流声　vortex tone
喷注噪声　jet noise
湍流噪声　turbulent noise
阀门噪声　valve noise
管道噪声　tube noise
再生噪声　regenerated noise
齿轮噪声　gear noise
螺旋桨噪声　propeller noise
湍流附面层噪声　turbulent boundary layer noise
有源噪声控制　active noise control
自适应系统　adaptive system
流体管道系统噪声控制　noise control of the fluid pipes system
木工机床噪声　noise of the timber working machine
纺织机噪声　noise of the mechanical loom
冲床噪声　noise of the punching machine
球磨机噪声　noise of the globe mill
压缩机噪声　noise of the compressor
风机噪声　noise of the fun
民用建筑噪声控制设计　noise control of civil architecture design
通风机噪声控制途径的探讨　researches on noise reduction of ventilator

1 We will present some case studies illustrating the application of noise control procedures that have been discussed in the previous chapters of this book. Additional case studies may be found in the literature.

本章将介绍一些实例研究，举例说明本书前面几章讨论的噪声控制过程，其他的实例研究可从文献中获得。

2 Control of the noise of civil architecture is one of the major measures to make the environment quiet. Based on the GB standards, the measures are offered in this paper through analyzing the source of noises. It is a kind of reference articles for the designers of environment protection and architecture acoustics.

民用建筑噪声控制是实现安静化的主要措施之一。文章以标准规范为依据，通过噪声源分析，提出控制对策，以供环境保护和建筑声学设计人员参考。

3 Noise pollution has been a public nuisance, its control technique is a new branch of borderline subject. The paper describes the characteristic and the cause of noise, researches the noise reduction of ventilator and makes it clear that it is necessary for production and environmental protection to consider the measure of reducing noise.

噪声污染已成为举世瞩目的一大公害，其控制技术是一门新兴的边缘科学。本文对通风机的噪声特性及产生原因进行了描述，对通风机噪声控制的途径进行了探讨，表明了在工程设计中需要综合考虑通风机的降噪措施，以适应生产和环保的需要。

第九章　环境监测
Chapter 9　Environmental Monitoring

第一节　绪论
Section 1　Introduction

环境科学　environmental science
环境污染　environmental pollution
环境质量　environmental quality
环境化学　environmental chemistry
环境标准　environmental standard
环境规划　environmental planning engineering
环境紧急事故　environmental emergency
环境保护　environmental protection
污染事故　pollution accident
污染防治　pollution control
环境系统工程　environmental system

1 Article 20　The competent administrative department for environmental protection under the State Council shall establish a system for monitoring environmental noise, establish monitoring standards, and set up a monitoring network with the relevant departments.

第二十条　国务院环境保护行政主管部门应当建立环境噪声监测制度,制定监测规范,并同有关部门一起组织监测网络。

2 Article 22　The administrative department of environmental protection under the State Council shall set up a monitoring system for atmospheric pollution, organize a monitoring network and work out unified monitoring measures.

第二十二条　国务院环境保护行政主管部门建立大气污染监测制度,组织监测网络,制定统一的监测方法。

第二节　水和废水监测
Section 2　Water and Waste Water Monitoring

水污染　water pollution
水资源　water resource
废水处理　wastewater treatment
污水　sewage
浊度　turbidity
过滤　filtration
絮凝　flocculation
生活废水　domestic sewage
色度计　chromaticity meter
废水流量　wastewater flow
地表水　surface water
采样器　sampler
采样方法　sampling method
监测　monitoring
监测方法　monitoring method
监测站　monitoring station
常规监测　conventional monitoring
原子吸收光谱法　atomic absorption spectrometry
分光光度法　spectrophotometry
等离子发射光谱法　plasma emission spectroscopy
电化学　electrochemistry
离子色谱法　ion chromatography
监测断面　monitoring section
工业废水　industrial waste water
城市污水　urban sewage
瞬时水样　instantaneous water sample
混合水样　mixed water sample
综合水样　comprehensive water sample
消解　digestion
流量计　flow meter
流量测量　flow measurement
溶解氧　dissolved oxygen
蒸馏　distillation
稀释　dilution
浓缩　concentrate
萃取　extract
富集　enrichment
分离　separation

吸附 adsorption
离子交换 ion exchange
残渣 residue
悬浮物 suspended solid
电导率 electric conductance
电导仪 conductivity meter
酸度 acidity
酸度计 acidity meter/pH meter
生物需氧量(BOD) biological oxygen demand
化学需氧量(COD) chemical oxygen demand
总有机碳(TOC) total organic carbon
总需氧量(TOD) total oxygen demand
高锰酸盐指数 potassium permanganate index
挥发酚 volatile phenol
挥发性有机污染物 volatile organic pollutants

1 The major pollution indicators are dissolved oxygen, potassium permanganate, BOD, volatile phenol and petroleum.
主要污染指标为溶解氧、高锰酸盐指数、生化需氧量、挥发酚和石油类。

2 Establishing the value function graphs for each individual parameter and seeing whether the environmental quality is improved or not (0-10 can be used to show the degree of the environmental quality, where 0 that indicates the environment quality is the worst, and 10 the best).
建立诸环境因子的环境变化程度(情况)与环境质量的函数图,以研究环境质量是否提高。环境质量可采用 0 ～ 10,0 表示环境质量最差,10 表示环境质量最好。

第三节 空气和废气监测
Section 3 Air and Waste Gases Monitoring

空气 air
气压计 barometer
空气污染 air pollution
空气污染物 air pollutant
大气层 aerosphere
空气污染源 air pollution source
大气扩散 atmospheric diffusion
气溶胶 aerocolloid
粉尘 dust
空气组成 air composition
空气监测 air monitoring
空气监测网 air monitoring network
空气监测站 air monitoring station
空气污染指数 air pollution index
空气质量标准 air quality standard
烟 smoke
雾 fog
灰尘 dust
气态污染物 gaseous pollutant
氮氧化物 nitrogen oxide
颗粒污染物 particulate pollutant
光化学气溶胶 photochemical aerosol
烟囱高度 chimney height
光化学现象 photochemical phenomenon
光化学烟雾 photochemical smog
光化学反应 photochemical reaction
光化学氧化剂 photochemical oxidant
总烃 total hydrocarbon
非甲烷烃 non-methane hydrocarbon
总悬浮颗粒物 total suspended particulate (TSP)
一次大气污染物 primary air pollutant
二次大气污染物 secondary air pollutant
城市空气污染源 urban air pollution source
可吸入颗粒物 inhalable particulate
气体吸收器 gas absorber
袋式除尘器 bag precipitator
静电除尘器 electrostatic precipitator
旋风分离器 cyclone separator
玻璃纤维 fiber glass
皮托管 Pitot tube
烟道 chimney

1 The particulate remains the key pollutant affecting the air quality in China. The annual concentration of the urban particulate in 64.1% of the cities exceeded Grade II.
颗粒物仍是影响我国城市空气质量的主要污染物,64.1% 的城市颗粒物年均浓度超过国家空气质量二级标准。

2 Above measures to cut the gaseous pollutants are effective means for controlling the production of secondary particles and ozone pollution.

采取以上削减气态污染物的措施，均有利于控制二次颗粒物的产生和臭氧污染。

3 The pollutants often occur with high levels of other toxins such as lead, zinc, and ground-level ozone, a component of smog formed by chemical reactions between car exhausts and sunlight.
这些污染物常常与其他有害物质，如铅、锌和地面臭氧一起出现。地面臭氧是汽车排气与太阳光一起发生光化学反应形成的一种烟雾成分。

第四节 噪声与放射性监测
Section 4 Noise and Radioactivity Monitoring

声波 sound wave
声压 sound pressure
声压级 sound pressure level
声级计 sound level meter
声强 sound intensity
声强级 sound intensity level
声功率 sound power
背景噪声 background noise
噪声频带 noise band
噪声过滤器 noise filter
噪声截止频率 noise cut-off frequency
等效连续声级 equivalent continuous sound level
消音器 muffler
分贝 decibel
响度 loudness
响度级 loudness level
放射性 radioactivity
放射性污染 radioactivity pollution
放射性物质 radioactive matter
剂量 dose
放射性同位素 radioisotope/radioactive isotope
放射性衰变 radioisotope decay
剂量当量 dose equivalent
半衰期 half life
核素 nuclide
正比计数器 proportional counter
盖革计数器 geiger counter

1 Law of the People's Republic of China on Prevention and Control of Pollution from Environmental Noise
中华人民共和国环境噪声污染防治法

2 Among 322 cities that were under road noise monitoring program, 5 percent had serious noise pollution; as for the 315 cities that were under area noise monitoring program, 52.1 percent of the cities had less noise pollution than the set standard for urban living areas.
对道路噪声监测的 322 个城市中噪声污染严重的占 5%，对区域环境噪声监测的 315 个城市中 52.1% 的区域环境噪声优于城市居住区声环境质量标准。

3 There are many toxic (poisonous) pollutants, but the best studied are radioactive elements and certain chemical compounds used to kill insects.
污染物质有许多是有毒物质，其中对放射性元素和一些特定的用来杀灭昆虫的化学化合物的污染研究得最多。

第五节 环境监测质量保证
Section 5 Environmental Monitoring Quality Assurance

蒸馏水 distilled water
化学试剂 chemical reagent
化验报告 analysis report
化学纯 chemical pure
分析纯 analytical pure
生物试剂 biological reagent
误差 error
误差分布曲线 error distribution curve
偏差 deviation
方差 variance
计量认证 metrology accreditation
质量体系 quality system
去离子水 deionized water
标准偏差 standard deviation
平均数 average
样本 sample
正态分布 normal distribution
置信区间 confidence interval

空白试验　blank test
标准曲线　standard curve
检测限　detection limit
稳定性　stability
精密度　precision
量筒　graduated flask
反滴定　back titration
缓冲剂　buffer
可见紫外吸收光谱　visible and ultraviolet absorption spectroscopy
红外光谱　infrared spectroscopy(IR)
原子吸收光谱　atomic absorption spectroscopy (AAS)
荧光分析法　fluorescence analysis(FA)
流动注射分析法　flow injection analysis(FIA)
气相色谱　gas chromatography(GC)
高效液相色谱　high performance liquid chromatograph (HPLC)
核磁共振波谱法　nuclear magnetic resonance (NMR)

1 The accuracy and precision meet the requirements of analysis of the pharmaceutical preparations.
其准确度和精密度能满足制剂分析的要求。

2 Accreditation for Laboratories and Environmental Monitoring Quality Assurance.
实验室认可与环境监测质量保证。

第十章　环境规划与管理
Chapter 10　Environmental Planning and Management

第一节　环境规划概述
Section 1　Introduction of Environmental Planning

环境规划　environmental planning
社会调查　social survey
发展状况　status of development
可持续发展　sustainable development
政策规划　policy planning
财政资助　financial assistance
土地利用规划　land use planning
环境管理指标　environmental management indicator
环境质量指标　environmental quality indicator
典型应用　typical application
空间规划　spatial planning
城市规划　urban planning
河流水域的规划　rivers and watersheds planning
沿海地区规划　coastal zones planning

1 Searching Google for "Environmental planning" gives 89 900 000 hits.
在 Google 中以"环境规划"为关键词,可搜索到 8 990 万条相关记录

2 Purpose 1: Environmental planning to coordinate private sector activity — make sure that private sector activity is consistent with environmental targets.
目标 1:协调私有部门的活动 —— 确保私有部门的活动与环境目标相一致。

3 Purpose 2: Environmental planning to integrate environmental concerns into other public planning processes — make sure that the environmental implications of other plans do not violate environmental standards and targets.
目标 2:将环境观念融入其他公共规划的决策过程中 —— 确保其他规划对环境的影响不会违反环境标准和目标。

第二节　环境规划的理论基础
Section 2　The Theoretical Basis of Environmental Planning

公平性原则　fairness principle
持续性原则　sustainability principle
多元目标下的新价值观　the new values under diversified targets
共同性原则　common principle
生态位理论　niche theory
生态学理论　ecology theory
环境容载力　environmental bearing capacity
减量化再循环再利用(3R)　reduce, recycle, reuse
复合生态系统　complex ecological system
世界环境与发展委员会(WCED)　World Commission on Environment and Development
我们共同的未来　our common future
联合国环境与发展大会(UNCED)　UN Conference on Environment and Development
国际自然和自然资源保护联盟(IUCN)　International Union for Conservation of Nature and Natural Resources
生态的永续性　ecological sustainability
生态服务　eco-service

1 He has developed a great interest in the scientific innovation strategies of colleges and universities in light of the niche theory.
生态位理论指导下的高校科学创新策略引起了他很大兴趣。

2 The bearing capacities of four major air contaminants (SO_2, TSP, NO_x, descendent dust) and the carrying capacities for economy and society were figured out based on the concept of air environmental bearing capacity.
在环境容载力概念的基础上,计算了四种主要大气污染物(二氧化硫、总悬浮颗粒物、氮氧化物和降尘)的环境容载力及经济、社会承载力。

3 Niche theory is one of the most important ecological theories. It is widely applied to analyzing such phenomena as competition among, and evolution of, urban ecosystem functional modules.
生态位理论是最重要的生态理论之一。它被广泛应用于一些现象的分析，例如城市生态系统功能模块的竞争和演变。

第三节　环境系统识别与评价
Section 3　Identifying and Assessment of Environmental System

层次分析法　analytic hierarchy process (AHP)
生态暴露评价　ecological exposure assessment
生态影响评价　ecological effect assessment
生态风险评价　ecological hazard/risk assessment
生态系统医学　eco-system medicine
生态足迹　ecological footprint
生物生产面积　biologically productive area
生态生产性面积　ecologically productive area
化石能源地　fossil energy land
可耕地　arable land
牧草地　pasture
建成地　built-up areas
产量因子　yield factor
生态承载力　ecological capacity
生态赤字/生态盈余　ecological deficit/ecological remainder
产量因子　yield factor
均衡因子　equivalent factor
战略环境影响评价　strategic environmental impact assessment

1 Environmental impact assessment shall be prepared for the preparation of the Olympic Games.
为开展奥运活动要进行环境影响评价。

2 The environmental impact assessment can be classified into two categories according to the temporal and spatial dimensions.
环境影响评价的种类可依据工程的时、空关系分为两大类。

3 An environmental impact assessment now under way will determine the best combination of treatment level and outfall location.
现在正在进行环境影响评价将决定处理水平及排污渠位置的最佳组合。

第四节　环境规划目标的确定与决策技术
Section 4　Determination and Decision-making Technology of Environmental Planning Objectives

规划层次　plan level
环境预测　environmental prediction
费用效益分析　cost benefit analysis
动态规划　dynamic programming
贝尔曼最优化原理　Bellman's principle of optimality
国家交通规划　national transport plan
长期规划　long term plan
大量运用长期经济学模型　heavy use of long term economic model
经济模型　economic model
部门预测模型　sector forecast model
差距分析　gap analysis
专家采访　interview
对利益相关方进行调研对话　participatory methods/dialogue

1 Regulatory Impact Assessment (or Analysis) (sometimes called cost benefit analysis in Europe)
监管影响评估(或分析)(在欧洲有时称之为费用效益分析)

2 A document created before a new government regulation is introduced.
在执行一项新的政府规定之前所作的报告。

3 Produced in many countries, although their scope, content, role and influence on decision-making vary.
在很多国家都实施，尽管范围、内容、侧重点以及对决策的影响力各不相同。

4 Provide a detailed appraisal of the potential impacts of a new regulation in order to assess whether the regulation is likely to achieve the desired objectives.
此报告详细评估了新规定的环境影响，以评价此规定是否可能达到预期目标。

5 Includes detailed cost benefit assessments.
包括详细的费用效益分析评估。

第五节 环境管理概论
Section 5 Introduction to Environmental Management

环境计划管理 environmental planning management
环境质量管理 environmental quality management
环境技术管理 environmental technology management
环境监督管理 environmental supervision management
基本职能 basic function
清洁空气法案 clean air act
发电厂汞排放规定 mercury rule for power plants
环境标准 environmental standard
源削减 source reduction
废物减量化 waste minimization

1 Planning function designs future activities in advance. It includes research activities, decision-making in different conditions and making plans.
计划职能会对未来活动进行预先谋划,其中包括研究活动、条件决策、编制计划.

2 Organizing function regulates reasonable coordination relationships at work for every member to reach the goal of the organization. It includes framework design, personnel deployment and organization supervision.
组织职能为实现组织目标,为每个组织成员规定在工作中形成的合理分工协作关系,其中包括组织结构设计、人员配备、组织监督。

3 Environmental Quality Management, Inc. (EQM) is an environmental engineering and remediation company headquartered in Cincinnati, Ohio that has been active in providing environmental remediation support in response to terrorist attacks, the space shuttle disaster, superfund site cleanup, hazardous chemical spills and natural disasters.
环境质量管理公司(EQM)是总部设在美国俄亥俄州辛辛那提的关于环境工程和修复的公司,该公司一直积极为应对恐怖袭击、航天飞机灾难、超级基金场地清理、危险化学品溢漏和自然灾害的环境修复提供支持。

第六节 环境政策和管理制度
Section 6 Environmental Policies and Management Systems

改进环境绩效 improved environmental performance
增强法规遵从性 enhanced compliance
污染防治 pollution prevention
保护(节约)资源 resource conservation
提高效率/降低成本 increased efficiency/reduced cost
提高员工士气 enhanced employee morale
环境问题和责任 environmental issues and responsibilities

1 A theoretical study of the total quantity control of pollutants (TQCP) has been made in the paper. TQCP has aimed at environmental quality and linked up all the environmental management systems and measures formed in the past years. TQCP has also represented the new stage of environmental management and has been the main means to strengthen environmental management.
本文就污染物排放总量控制作了理论思考。总量控制瞄准了环境质量,衔接、关联着我国多年来已实施的诸项环境管理制度、措施,代表着我国环境管理的新水平,是强化环境管理的工作主线。

2 The environment and development synthetic decision-making system was one of the new environmental management systems, and was the inevitable choice to perfect the government's decision-making methods under socialist market economy situation.
环境与发展综合决策制度是新的环境保护管理制度之一,是社会主义市场经济条件下完善政府决策方式的必然选择。

第十一章　环境土壤学
Chapter 11　Environmental Soil Science

第一节　地质学基础
Section 1　Geological Foundation

地球的一般特征　general features of the Earth
地质作用　geologic process
地貌　topography
晶体矿物　crystallization mineral
非晶体矿物　no-crystallization mineral
原生矿物　primary mineral
次生矿物　secondary mineral
土壤矿物　soil mineral
铝硅酸盐　aluminosilicate
层状硅酸盐　phyllosilicate，layer silicate
黏土矿物　clay mineral
黏粒矿物　clay-sized mineral
间层黏土矿物　interstratified clay mineral
四面体片　tetrahedral sheet
八面体片　octahedral sheet
二八面体片　dio-ctahedral sheet
三八面体片　trio-ctahedral sheet
单元晶层　unit layer
1∶1型矿物　1∶1 type mineral
2∶1型矿物　2∶1 type mineral
底面间距　basal spacing
同晶置换　isomorphous substitution
高岭石　kaolinite
埃洛石　halloysite
蒙皂石　smectite
蒙脱石　montomorillonite
贝得石　beidellite
绿脱石　nontronite
蛭石　vermiculite
绿泥石　chlorite
过渡矿物　intergradient mineral
水云母　hydromica
伊利石　illite
伊毛缟石　imogolite
水铝英石　allophane
三水铝石　gibbsite
水铁矿　ferrihydrite
针铁矿　goethite
赤铁矿　hematite
游离氧化铁　free iron oxide
矿物风化序列　weathering sequence of mineral
断键　broken bond
脱钾作用　depotassication
楔形带　wedge zone
形位效应　steric effect

1 Soil：The naturally occurring unconsolidated material on the surface of the earth.
土壤：土壤就是陆地表面能够生产植物收获物的疏松表层。

2 Parent material：The unconsolidated and more or less chemically weathered mineral or organic matter from which the solum of a soil has developed by pedogenic processes.
母质：风化作用使岩石破碎，理化性质改变，形成结构疏松的风化壳，其上部可称为土壤母质。

第二节　岩石风化和土壤形成
Section 2　Rock Weathering and Soil Formation

风化过程　weathering process
风化产物的类型　type of weathering product
土壤形成　soil formation
土壤剖面及形态特征　soil profile and shape characteristic
物理风化　physical weathering
化学风化　chemical weathering
生物风化　biological weathering
风化产物　weathering product
风化残余物　weathering residue
风化强度　weathering intensity
风化指数　weathering index
硅铝率　silica-alumina ratio
硅铝铁率　silica-sesquioxide ratio

风化淋溶系数 ba value
半风化体 saprolite
风化壳 weathering crust
碎屑风化壳 clastic weathering crust
含盐风化壳 salic weathering crust
碳酸盐风化壳 carbonated weathering crust
硅铝风化壳 siallitic weathering crust
铁铝风化壳 ferrallitic weathering crust
母质 parent material
土壤发育 soil development
土壤发育序列 soil development sequence
原始土壤 primitive soil/initial soil
幼年土壤 young soil
成熟土壤 mature soil
土壤年龄 soil age
土壤绝对年龄 absolute age of soil
土壤相对年龄 relative age of soil
土壤地球化学 soil geochemistry
土壤生物地球化学 soil biogeochemistry
土壤发生 soil genesis
土壤形成 soil formation
土壤形成因素 soil-forming factor
土壤形成过程 soil-forming process
淋溶作用 eluviation
淋洗作用 leaching
螯合淋溶作用 cheluviation
淋移作用 mechanical eluviation/lessivage
淀积作用 illuviation
淋淀作用 eluviation-illuviation
生物积累作用 biological accumulation
腐殖质积累作用 humus accumulation
泥炭形成 peat formation
盐化作用 salinization
碱化作用 solonization
次生盐化 secondary salinization
脱盐作用 desalinization
脱碱作用 solodization/solotization
石膏聚积作用 gypsum accumulation
钙积作用 calcification
复钙作用 recalcification
脱钙作用 decalcification
潜育作用 gleyization
假潜育作用 pesudogleyization
次生潜育化 secondary gleyization
硅化作用 silicification
脱硅[作用] desilicification
灰化[作用] podzoliation
棕化[作用] brownification/braunification
红化[作用] rubification
硅铝化[作用] siallitization
铁铝化[作用] ferrallitization
富铝化[作用] allitization
铁质化[作用] ferruginization
黏化[作用] clayification
残积黏化[作用] residual clayification
次生黏化[作用] secondary clayification
淀积黏化[作用] illuvial clayification/argillification
土壤扰动作用 pedoturbation
单个土体 pedon
聚合土体 polypedon
土壤剖面 soil profile
侏儒剖面 nanoprofile
巨型剖面 giant profile
扰动剖面 disturbed profile
剖面构型 profile pattern
土体层 solum
土壤发生层 soil genetic horizon/soil horizon
发生层分化 horizon differentiation/horizonation
发生层界线 horizon boundary
发生层命名 horizon designation
基本发生层 master horizon/principle horizon
有机层 organic horizon
腐殖质层 humus horizon
淋溶层 eluvial horizon
淀积层 illuvial horizon
母质层 parent material horizon
母岩 parent rock
残积物 residual deposit
崩积物 colluvial deposit
坡积物 slope deposit/slope wash
洪积物 diluvial deposit
冲积物 alluvial deposit
湖积物 lacustrine deposit
浅海沉积物 shallow marine deposit/neritic deposit
冰川沉积物 glacial deposit
冰水沉积物 glaciofluvial deposit
风积物 aeolian deposit
黄土 loess
黄土状物质 loess-like material
第四纪红色黏土 quaternary red clay

珊瑚砂 coral sand
弱分解有机层 slightly decomposed organic horizon
半分解有机层 intermediate decomposed organic horizon
高分解有机层 highly decomposed organic horizon
草毡层 sod layer
耕作层 cultivated horizon
犁底层 plow pan
灰化层 podzolized horizon
漂洗层 bleached horizon
次生黏化层 secondary clayified horizon
网纹层 reticulated mottling horizon
泥炭层 peat horizon
潜育层 gley horizon
潴育层 waterloggogenic horizon
渗育层 percogenic horizon
荒漠砾幂 desert pavement
荒漠漆皮 desert varnish
土壤层次 soil layer
表土层 surface soil layer
心土层 subsoil layer
底土层 substratum
埋藏层 buried horizon
表土 surface soil/top soil
心土 subsoil
底土 bottom soil
诊断层 diagnostic horizon
诊断表层 diagnostic surface horizon
有机表层 histic epipedon
垫熟表层 plaggen epipedon
暗沃表层 mollic epipedon
暗瘠表层 umbric epipedon
淡薄表层 ochric epipedon
黑色表层 melanic epipedon
人为表层 anthropic epipedon
灌淤表层 siltigic epipedon
堆垫表层 cumuli epipedon
肥熟表层 fimic epipedon
水耕表层 anthrostagnic epipedon
诊断表下层 diagnostic subsurface horizon
漂白层 albic horizon
舌状层 glossic horizon
雏形层 cambic horizon
耕作淀积层 agric horizon
淀积黏化层 argillic horizon
耕作淀铅层 argic horizon
黏磐 clay pan
高岭层 kandic horizon
氧化层 oxic horizon
薄铁磐层 placic horizon
铁磐 iron pan
硅质硬磐 duripan
脆磐 fragipan
腐殖质淀积层 sombric horizon
灰化淀积层 spodic horizon
钠质层 natric horizon
盐积层 salic horizon
盐磐 salt pan
石膏层 gypsic horizon
石化石膏层 petrogypsic horizon
钙积层 calcic horizon
石化钙积层 petrocalcic horizon
含硫层 sulfuric horizon
诊断特性 diagnostic characteristics
质地突变 abrupt textural change
漂白物质 albic material
漂白物质指间状延伸 interfingering of albic material
火山灰土壤特性 andic soil property
潮湿特征 aquic condition
线胀度 linear extensibility
硅质硬结核 durinode
可辨认次生碳酸盐 identifiable secondary carbonate
石质接触面 lithic contact
准石质接触面 paralithic contact
准石质物质 paralithic material
石化铁质接触面 petroferic contact
致密接触面 densic contact
致密物质 densic material
聚铁网纹体 plinthite
聚铁网纹层 plinthic horizon
淋淀层段 sequum
双淋淀层段 bisequum

[1] Weathering: The breakdown of rocks and minerals at and below the Earth's surface by the action of physical and chemical processes.
风化:地表的岩石在外界因素的作用下,发生形态、组成和性质变化的过程。

2 Biogeochemical cycle: The movement of chemical elements from organism to physical environment to organism in more or less circular pathways.
生物地球化学循环:环境中各种元素沿着特定的路线运动,由周围环境进入生物体,最后回到环境中的循环运动过程,称为生物地球化学循环。

3 Eluviation: The transportation of soil material in suspension or in solution within the soil by the downward or lateral movement of water.
淋溶作用:淋溶作用是指土壤中的土壤物质以悬浮液或水溶液形式向下或通过水侧向运动进行转移的过程。

第三节 土壤生物学
Section 3 Soil Biology

土壤动物 soil animal
土壤微生物 soil microorganism
植物根系及微生物的联合 plant root system and combination from the microorganism
土壤酶 soil enzyme
土壤微生物生物量 soil microbial biomass
土壤生物量 soil biomass
土壤微生物区系 soil microflora
土壤动物学 soil zoology
土壤动物区系 soil fauna
土壤微动物区系 soil microfauna
寄生现象 parasitism
共生现象 symbiotism
捕食现象 predation
土著性细菌 autochthonous bacteria
发酵性细菌 zymogenic bacteria
土传植物病菌 soil-borne plant
根圈 rhizosphere
根圈微生物 rhizosphere microorganism
有害根圈微生物 deterious rhizosphere microorganism (DRMO)
内生菌 endophyte
附生菌 epiphyte
根圈效应 rhizosphere effect
根土比 R/S ratio
根区 root zone
根面 rhizoplane
根瘤菌根瘤 rhizobial nodule
根瘤菌剂 rhizobium inoculant
放线菌根瘤 actinorhizal nodule
茎瘤 stem nodule
菌根 mycorrhiza
外生菌根 ectomycorrhiza
内生菌根 endomycorrhiza
丛枝菌根 arbuscular mycorrhiza
类菌体 bacteroid
侵入线 infection thread
结瘤 nodulation
竞争结瘤 competitive nodulation
豆血红蛋白 leghaemoglobin
固氮酶 nitrogenase
固氮基因 nitrogen fixation gene
固氮作用 nitrogen fixation
生物固氮作用 biological nitrogen fixation
固氮酶活性 nitrogenase activity
联合固氮作用 associative nitrogen fixation
共生固氮作用 symbiotic nitrogen fixation
非共生固氮作用 non-symbiotic nitrogen fixation
硝化作用 nitrification
亚硝化作用 nitrosification
反硝化作用 denitrification
硝化抑制剂 nitrification inhibitor
植物凝集素 lectin
硫化作用 sulfuricaion
反硫化作用 desulfurication
氨化作用 ammonification
纤维素分解作用 cellulose decomposition
矿化作用 mineralization
生物固持作用 biological immobilization
毒性有机物的去毒作用 detoxication of toxic organic compound
有机物的毒化 toxication of organic compound
共代谢 co-metabolism
土壤环流术 soil perfusion technique
根瘤菌 rhizobia
自生固氮菌 free-living nitrogen fixing bacteria
甲烷细菌 methane bacteria
铁细菌 iron bacteria
硫细菌 sulfur bacteria
硝化细菌 nitrifying bacteria

反硝化细菌　denitrifying bacteria
纤维分解菌　cellulose-decomposing bacteria
黏细菌　myxobacteria
生物降解　biodegradation
土壤生物活性　soil biological activity
好气分解　aerobic decomposition
嫌气分解　anaerobic decomposition
熏蒸法，熏蒸剂　fumigant
生物可降解性　biodegradability
土地生物处理　land biological treatment
土壤消毒　soil disinfection
土壤源温室气体　soil-borne greenhouse gas
放线菌根瘤共生　actinorhizal symbiosis
细菌　bacterium
噬菌体　bacteriophage
微生物根部定植　colonization of root by microorganism
蓝细菌　cyanobacteria
真菌　fungus
杀真菌剂　fungicide
遗传工程微生物　genetically engineered microorganism(GEM)
接种物　inoculum
接种密度　inoculum density
生物发光技术　bioluminescence technique
微生物群落　microbial community
菌根圈　mycorrhizosphere
根瘤素　nodulin
结瘤基因　nodulation gene
贫营养性　oligotrophy
贫营养微生物　oligotrophic microorganisms
质粒　plasmid
共生质粒　symplasmid
原生动物　protozoa
植物圈　phytosphere
植物促长根圈细菌　plant growth promoting rhizobacteria(PGPR)
趋化性　chemotaxis
根瘤菌毒素　rhizobiotoxin
土壤-根界面　soil-root interface
铁载体　siderophore
孢子　spore
微生物营养群　trophic groups of microorganism
非生物物质　xenobiotic substance
氮素有效性比率　nitrogen availability ratio
乙炔还原检测　acetylene reduction assay
微生物积累　microbial accumulation
微生物除草剂　microbial herbicide
微生物杀虫剂　microbial pesticide
微生物接种剂　microbial inoculant
生物转化　bioconversion
生物治疗　bioremediation
生物变质　biodeterioration
纤维素分解活性　cellulolytic activity
群落　community
岩内微生物群落　endolithic microbial community
内共生体　endosymbiont
低营养流生境　less nutrient flux habitat
生境　habitat
微生境　microhabitat
生态位　ecological niche
光合微生物　photosynthetic microorganism
先锋群落　pioneer community
外源污染物　xenobiotic pollutants
氨氧化细菌　ammonia-oxidizing bacteria
腐解作用　decay
细菌生理群　physiological group of bacteria
土壤局部灭菌　partial sterilization of soil
食虫真菌　predacious fungi
土壤富集法　soil enrichment method
嗜热微生物　thermophilic microorganism
干土效应　effect of soil drying
冻土效应　effect of soil freezing
碳循环　carbon cycle
氮循环　nitrogen cycle
磷循环　phosphorus cycle
硫循环　sulfur cycle
有机氮库　organic nitrogen pool
土壤酶学　soil enzymology
土壤酶　soil enzyme
土壤酶活性　soil enzyme activity
土壤酶促反应　soil enzymatic reaction
土壤酶抑制剂　soil enzyme inhibitor
土壤非活体酶　soil abiotic enzyme
土壤脲酶　soil urease
土壤酶保护容量　enzyme protection capacity of soil
胞外酶　extracellular enzyme
胞内酶　endocellular enzyme/endoenzyme
土壤酶激活　activation of soil enzyme
土壤酶稳定性　stability of soil enzyme
生化强度　biochemical capacity

水解酶类　hydrolase
氧化还原酶类　oxidoreductase
裂解酶类　lyase
转移酶类　transferase
土壤蔗糖酶　soil saccharase

1 Decomposition: The breakdown of complex organic molecules of dead protoplasm and cell walls into simpler organic and inorganic molecules, which may be used again by other organisms.
分解作用:有机物质经过代谢降解变成简单的有机和无机物质的过程。

2 Plant residues: Organic compounds in the form of leaf, stem, and root tissues remaining after plants are harvested.
植物残体:未分解的植物茎、叶、根等植物组织及其部分分解的有机物。

第四节　土壤有机质
Section 4　Soil Organic Matter

土壤有机质的来源、组成　source and composition of soil organic matter
植物凋落物　litter
根系分泌物　root exudate
动物粪便　animal fecal
植物残体　plant residue
动物残体　dead animal
根茬　root
秸秆还田　straw return
土壤原有机质　native soil organic matter
土壤有机质分解率　decomposition rate of soil organic matter
激发效应　priming effect
土壤腐殖质　soil humus
轻组腐殖物质　humic substance in light fraction
重组腐殖物质　humic substance in heavy fraction
腐殖化作用　humification
腐殖酸;胡敏酸　humic acid
A 型胡敏酸　A type humic acid
B 型胡敏酸　B type humic acid
绿色胡敏酸　green humic acid
灰色胡敏酸　grey humic acid
富里酸　fulvic acid
胡敏素　humin
活性腐殖质　active humus
腐殖物质组合　fraction of humic substance
腐殖物质发色基团　chromophoric group of humic substance
胡敏酸 A4/A6 比值　A4/A6 ratio of humic acid
胡敏酸-富啡酸比值　HA/FA ratio
腐殖物质生理效应　physiological effect of humic substance
腐殖化系数　humification coefficient
腐殖化程度　degree of humification
金属-腐殖物质络合物　metal humic substances complex
动植物残体　plant and animal residue
土壤多糖　soil polysaccharide
土壤有机碳-有机氮比值　organic carbo-nitrogen ratio in soil

1 Soil organisms (biota): Microbes, plants and animals occupying an area together.
土壤生物:土壤中活的有机体,主要指生活在土壤中的微生物、动物和植物等总称为土壤生物。

2 Metabolism: The total of all the chemical reactions that occur within a living organism.
新陈代谢:生物体从环境摄取营养物转变为自身物质,同时将自身原有组成转变为废物排出到环境中的不断更新的过程。

2 N-compounds: Organic compounds containing N. e. g, proteins.
含氮的化合物:土壤中含有 N 的有机物,例如蛋白质。

3 Ash: The substance remaining after the ignition of plant material.
灰分:植物燃烧后的剩余物质。

第五节　土壤物理
Section 5　Soil Physics

土壤物理性质　soil physical property
土壤三相　soil three phases
土壤总容积　soil total volume
固相容积　solid phase volume

液相容积 liquid phase volume
气相容积 air phase volume
土壤颗粒 soil particle
土壤颗粒大小分布 particle-size distribution
土粒有效直径 effective diameter of soil
斯托克斯定律 Stokes' law
土壤颗粒大小分析 soil particle size analysis
石块 stone
砾 cobble
石块质 stony
砾质 cobbly
砾石 cobblestone
石砾 gravel
粗屑体 coarse fragment
砂粒 sand
粉[砂]粒 silt
黏粒 clay
物理性砂粒 physical sand
物理性黏粒 physical clay
土壤分散[作用] soil dispersion
土壤分散剂 soil dispersing agent
土粒沉降累积曲线 particle sedi-mental accumulated curve
土壤质地 soil texture
壤砂土 loamy sand
砂壤土 sandy loam
壤土 loam
粉[砂]壤土 silt loam
砂质黏壤土 sandy clay loam
黏壤土 clay loam
粉[砂]质黏壤土 silty clay loam
砂黏土 sand clay
粉[砂]黏土 silty clay
粗砂土 coarse sand
细砂土 fine sand
轻壤土 light loam
中壤土 medium loam
重壤土 heavy loam
轻黏土 light clay
中黏土 medium clay
重黏土 heavy clay
土壤质地剖面 soil texture profile
土壤结构 soil structure
土粒胶结[作用] soil particle cementation
土粒凝聚[作用] soil particle coagulation
干湿交替[作用] alternation of drying and wetting
冻融交替[作用] alternation of freezing and thawing
团聚[作用] aggregation
碎裂[作用] fragmentation
土壤结构分类 soil structure classification
土壤结构类型 soil structure type
结构形态 structure morphology
块状结构 blocky structure
角块状结构 angular blocky structure
亚解块状结构 subangular blocky structure
粒状结构 granular structure
柱状结构 columnar structure
棱状结构 prismatic structure
片状结构 platy structure
土壤结构度 soil structure grade
大结构 macrostructure
微结构 microstructure
团聚体 aggregate
大团聚体 macroaggregate
微团聚体 microaggregate
原生颗粒 primary particle
生颗粒 secondary particle
结构单位 structure unit
结构剖面 structure profile
原状土 undisturbed soil
结壳 crust
稳定性团聚体 stable aggregate
水稳性团聚体 water stable aggregate
力稳性团聚体 mechanical stable aggregate
非稳定性团聚体 instable aggregate
团聚体消散 aggregate slaking
团聚体崩解 aggregate disintegration
结构退化 structure degradation
结构指标 structure index
分散系数 dispersive coefficient
结构系数 structure coefficient
破裂系数 rupture modulus
土壤改良剂 soil amendment
土壤结构改良剂 soil conditioner
土壤比表面 soil specific surface area
土壤密度 soil density
土粒密度 soil particle density
孔径分布 pore-size distribution
土壤孔隙 soil pore space
土壤孔隙度 soil porosity

当量孔隙　equivalent pore
充气孔隙　air-filled pore
持水孔隙　vater-holding pore
团聚体内孔隙　intra-aggregate pore
团聚体间孔隙　inter-aggregate pore
通气孔隙度　air porosity/aeration porosity
土壤比容　soil specific volume
浸水土壤密度　water immersed soil density
孔隙比　pore space ratio
土壤水[分]　soil water
土壤水形态　soil water form
土壤固态水　soil water in solid phase
土壤液态水　soil water in liquid phase
土壤气态水　soil water in vapor phase
束缚水　confined water
薄膜水　film water
触点水　pendular water
毛管水　capillary water
毛管凝结水　capillary condensation water
毛管悬着水　capillary suspending water
毛管支持水　capillary supporting water
重力水　gravitational water
吸湿水　hygroscopic water
最大吸湿量　maximum hygroscopicity
土壤水[分]常数　soil water constant
饱和含水量　saturated water content
田间持水量　field capacity
萎蔫含水量　wilting point
土壤含水量　soil water content
绝对含水量　absolute water content
相对含水量　relative water content
质量含水量　mass water content
体积含水量　volumetric water content
土壤水[分]状况　soil water regime
风干土　air-dry soil
烘干土　oven dry soil
土壤水能态　energy state of soil water
土水势　soil water potential
基质势　matric potential
重力势　gravitational potential
渗透势　osmotic potential
压力势　pressure potential
总势　total potential
水力势　hydraulic potential
土壤[水]吸力　soil water suction
基质吸力　matric suction
渗透吸力　osmotic suction
土壤水分特征曲线　soil water characteristic curve
脱水曲线　water desorption curve
吸水曲线　water sorption curve
土壤水分滞后现象　soil water hysteresis
进气吸力值　value of air-entry suction
比水容量　specific water capacity
张力计　tensiometer
土壤水流　soil water flow
土壤饱和水流　saturated soil water flow
土壤非饱和水流　unsaturated soil water flow
稳态水流　steady state water flow
非稳态水流　nonsteady state water flow
优先流　preferential flow
毛管水活动带　zone of capillary flow
水汽运动　vapor movement
土壤水分扩散率　soil water diffusivity
土壤导水率　soil hydraulic conductivity
饱和导水率　saturated hydraulic conductivity
非饱和导水率 unsaturated hydraulic conductivity
水势梯度　water potential gradient
湿度梯度　moisture gradient
水力梯度　hydraulic gradient
吸力梯度　suction gradient
达西定律　Darcy's law
水通量　water flux
水通量密度　water flux density
溶质运移　solute transfer
溶质对流　solute convection
溶质扩散　solute diffusion
溶质弥散　solute dispersion
扩散-弥散　diffusion-dispersion
扩散系数　diffusion coefficient
弥散系数　dispersion coefficient
扩散-弥散系数　diffusion-dispersion coefficient
混合置换　miscible displacement
水文循环　hydrologic cycle
土壤水平衡　soil water balance
降水　precipitation
有效降水　effective precipitation
土壤水入渗　soil water infiltration
侧渗　lateral seepage
渗滤　percolation
入渗通量　infiltration flux
初始入渗率　initial infiltration rate

稳定入渗率 stable infiltration rate
内通透性 intrinsic permeability
湿润锋 wetting front
土壤水分再分布 soil water redistribution
空间变异 spatial variability
土壤排水 soil drainage
地面排水 surface drainage
地下排水 underground drainage
暗管排水 tile drainage
明沟排水 ditch drainage
透水层 permeable layer
不透水层 impermeable layer
临界水位 critical groundwater table
蓄水层 water storage layer
渍水 waterlogging
淹水 water flooding
渗滤水采集器 lysimeter
土壤水分蒸发 soil water evaporation
土壤水汽扩散 soil vapor diffusion
土壤潜在蒸发 soil potential evaporation
蒸腾 transpiration
蒸腾系数 transpiration coefficient
蒸散作用 evapotranspiration
潜在蒸散 potential evapotranspiration
实际蒸散 actual evapotranspiration
潜在蒸腾 potential transpiration
相对蒸腾 relative transpiration
干燥度 aridity
旱农 dryland farming
雨养农业 rainfed farming
水分胁迫 water stress
土壤-植物-大气连续体 soil-plant-atmosphere continuum (SPAC)
主动吸收水分 active uptake water
被动吸收水分 passive uptake water
有效水 available water
灌溉 irrigation
灌溉效益 irrigation efficiency
节水灌溉 water saving irrigation
节水农业 water saving agriculture
土壤空气状况 soil air regime
土壤空气 soil air
土壤空气容量 soil air capacity
土壤空气组成 soil air composition
土壤通气性 soil aeration
土壤空气交换 soil air exchange
土壤空气扩散 soil air diffusion
土壤空气扩散系数 diffusion coefficient of soil air
土壤空气相对扩散系数 relative diffusion coefficient of soil air
气通量 air flux
氧扩散率 oxygen diffusion rate (ODR)
土壤呼吸 soil respiration
土壤呼吸强度 intensity of soil respiration
呼吸商 respiratory quotient (RQ)
土壤温度 soil temperature
土壤温度状况 soil temperature regime
土壤温度梯度 soil temperature gradient
土壤热 soil heat
土壤热状况 soil heat regime, soil thermal regime
土壤热交换 soil heat exchange, soil thermal exchange
土壤热容量 soil heat capacity, soil thermal capacity
质量热容量 mass heat capacity
容积热容量 volume heat capacity
土壤热传导 soil heat conduction, soil thermal conduction
湿润热 wetting heat
导热率 heat conductivity, thermal conductivity
土壤热扩散率 soil heat diffusivity, soil thermal diffusivity
土壤热流 soil heat flow
土壤热平衡 soil heat balance, soil thermal balance
热通量密度 heat flux density
土壤力学 soil mechanics
土壤动力学性质 dynamic property of soil
耕作动力学 tillage dynamics
土壤结持度 soil consistency
土壤结持性 soil consistence
阿特贝限 Atterberg limit
流限 liquid limit
塑性上限 upper plastic limit
塑性下限 lower plastic limit
塑性距 plasticity range
可塑性 plasticity
塑性指数 plastic index
塑性流定律 law of plastic flow
土壤黏度 soil viscosity

土壤表观黏度　soil apparent viscosity
黏着点　sticky point
壤流变学　soil rheology
土壤黏着力　soil adhesion
土壤黏结力　soil cohesion
黏粒活度　activity of clay
土壤变形　soil deformation
土壤剪切　soil shear
剪切强度　shear strength
土壤摩擦力　soil friction
土壤膨胀　soil swelling/soil expansion
线胀系数　coefficient of linear extensibility
土壤膨胀压　soil swelling pressure
土壤膨胀指数　soil swelling index
土壤收缩　soil shrinkage
结构收缩　structure shrinkage
土壤固结　soil consolidation
土壤坚实度　soil hardness
土壤压实　soil compaction
土壤黏闭　soil puddling
可耕性　tillability
土块　clod
土块分布　clod distribution
土壤电磁性　soil electromagnetism

1 Soil water potential: The difference between the energy of water in soil and of pure, free water at the same temperature. The water potential of pure water is zero, so that of a solution will be negative.
土壤水势:土壤势是所相同温度下,土壤水、孔隙水和自由水之间的能量之差。纯水的水势规定为 0,土壤水势一般为负值。

2 Capillary water: Water that is left in soil, alone with hygroscopic moisture and water vapor, after the gravitational water has drained off. Capillary water is held by surface tension as a film of moisture on the surface of soil particles and peds, and as minute bodies of water filling part to the pore space between particles.
毛管水:受毛管压力作用而保持在土壤孔隙中的水分。运动方向和速度依毛管压力的大小而定。毛管水上升高度与毛管半径成反比,毛管水运动速度与毛管半径平方成正比。

3 Hygroscopic water: Water absorbed from the atmosphere and held very tightly by the soil particles, so that it is unavailable to plants in amounts sufficient for them to survive. It is the water lost from an air-dry soil when it is heated to 105°C.
吸湿水:由于固体土粒表面分子引力和静电引力对空气中的水汽分子产生的吸附力而紧密保持的水分称吸湿水。当温度为 105℃ 时,土壤干燥过程失去的水分。

第六节　土壤化学
Section 6　Soil Chemistry

土壤矿物化学　soil mineral chemistry
土壤腐殖质化学　soil humus chemistry
土壤物理化学　soil physical chemistry
土壤胶体化学　soil colloid chemistry
土壤表面化学　soil surface chemistry
土壤电化学　soil electrochemistry
土壤溶液化学　soil solution chemistry
土壤养分化学　soil nutrient chemistry
土壤污染化学　soil pollution chemistry
土壤矿物　soil mineral
水铁矿　ferrihydrite
针铁矿　goethite
赤铁矿　hematite
游离氧化铁　free iron oxide
矿物风化序列　weathering sequence of mineral
断键　broken bond
脱钾作用　depotassication
楔形带　wedge zone
形位效应　steric effect
土壤胶体　soil colloid
土壤有机胶体　soil organic colloid
土壤矿质胶体　soil mineral colloid
有机矿质复合体　organo-mineral complex
土壤酸胶体　soil acidoid
土壤碱胶体　soil basoid
叠胶　tactoid
两性胶体　ampholytoid
钙质黏粒　calcium clay
钠质黏粒　sodium clay
氢铝质黏粒　H-Al-clay
层间表面　interlayer surface
硅氧烷表面　siloxane surface

硅烷醇基 silanol group
铝醇基 aluminol group
铁醇基 ferrol group
羟基表面 hydroxyl surface
羟基铝间层 hydroxyl-aluminum interlayer
羟基化作用 hydroxylation
质子化作用 protonation
恒电荷表面 constant charge surface
土粒电荷 soil particle charge
表面电荷 surface charge
结构电荷 structure charge
层电荷 layer charge
单位化学式电荷 charge per formula unit
固有电荷 intrinsic charge
永久电荷 permanent charge
可变电荷 variable charge
质子电荷 proton charge
表面电荷密度 surface charge density
扩散双电层 diffused double layer
决定电位离子 potential determining ion
阳离子交换 cation exchange
阳离子交换量 cation exchange capacity
阳离子交换位 cation exchange site
层间交换位 interlayer exchange site
吸附态离子 adsorbed ion
交换性阳离子 exchangeable cation
交换性阴离子 exchangeable anion
交换性盐基 exchange base
实际阳离子交换量 effective cation exchange capacity (ECEC)
盐基饱和度 base saturation percentage
盐基饱和度效应 effect of degree of base saturation
阴离子交换 anion exchange
阴离子交换量 anion exchange capacity
阴离子吸持 anion retention
阴离子吸附 anion adsorption
阴离子渗入 anion penetration
离子选择性 ion selectivity
专性吸附 specific adsorption
非专性吸附 nonspecific adsorption
选择吸附 selective adsorption
优先吸附 preference adsorption
负吸附 negative adsorption
选择系数 selectivity coefficient
反荷离子 counter ion
互补离子 complementary ion
互补离子效应 effect of complementary ion
嵌插作用 intercalation
铁解作用 ferrolysis
吸附作用 adsorption
解吸作用 desorption
固定作用 fixation
表面络合物 surface complex
内圈络合物 innersphere complex
外圈络合物 outersphere complex
谐溶 congruent dissolution
不谐溶 incongruent dissolution
配位体交换 ligand exchange
土壤溶液电导率 electrical conductivity of soil solution
土壤酸、碱平衡 soil acid-base equilibrium
土壤反应 soil reaction
土壤 pH soil pH
土壤酸度 soil acidity
土壤碱度 soil alkalinity
土壤活性酸度 soil active acidity
土壤潜性酸度 soil potential acidity
土壤残留酸度 soil residual acidity
土壤残留碱度 soil residual alkalinity
土壤交换性酸度 soil exchangeable acidity
土壤盐置换性酸 salt-replaceable acid of soil
土壤水解性酸度 soil hydrolytic acidity
土壤水浸液酸 water-extractable acid of soil
土壤盐浸液酸 salt-extractable acid of soil
土壤表面酸度 soil surface acidity
总酸度 total acidity
土壤酸化作用 acidification
土壤碱化作用 alkalinization
土壤钠质化作用 sodication
酸性土壤 acid soil
中性土壤 neutral soil
碱性土壤 alkaline soil
石灰性土 calcareous soil
石灰需要量 lime requirement
土壤缓冲性 soil buffering
土壤缓冲容量 soil buffer capacity, buffer power
土壤缓冲物质 soil buffer compound
铝聚合作用 aluminum polymerization
铝毒 aluminum toxicity
土壤氧化还原体系 soil redox system
土壤氧化还原电位 soil redox potential

土壤氧化还原状况 soil redox status
氧化还原耦 redox couple
动电电位 electrokinetic potential
临界电位 critical potential
液接电位 liquid junction potential
悬液效应 suspension effect
电荷零点 zero point charge (ZPC)
等电点 isoelectric point
离子选择电极 ion selective electrode
土壤化学固定 soil chemical fixation
化学反硝化作用 chemical denitrification
钾固定 potassium fixation
铵固定 ammonium fixation
磷酸固定 phosphate fixation
磷酸吸附 phosphate adsorption
磷酸吸持 phosphate retention
化学位 chemical potential
离子交换动力学 dynamics of ion exchange
化学风化动力学 dynamics of chemical weathering
氧化还原动力学 dynamics of oxidation-reduction
土壤含盐量 soil salt content
土壤盐渍度 soil salinity
土壤可溶盐 soil soluble salt
土壤测定 soil test
土壤溶液 soil solution
土壤浸提剂 soil extractant
土壤浸出液 soil extract
土壤饱和浸出液 soil saturation extract

1 Soil adsorption: The process by which atoms, molecules, or ions taken up from the soil solution or soil atmosphere and retained on surface of soils by chemical or physical binding.
土壤吸附作用:土壤溶液和土壤空气中的原子、分子通过物理、化学作用于土壤矿物表面的过程。

2 Soil adsorption competence: A collection of various organic and inorganic substances in soil that is capable of adsorbing ions and molecules.
土壤吸附能力:土壤具有吸附无机物和有机物的能力。

3 Total soil acidity: The total soil acidity is the sum of active and reserve acidity.
土壤总酸度:土壤总酸度是指活性酸度和潜性酸度的和。

4 Exchangeable cation: A positively charged ion held on or near the surface of a soild other positively charged, hydroxy ions in the soil solution.
交换性阳离子:土壤中带负电荷的胶体所吸附的阳离子,在静电引力、离子本身的热运动或浓度梯度的作用下,可以和土壤溶液或其他胶体表面的阳离子进行交换。能互相交换的阳离子就称为交换性阳离子。

第七节 土壤养分
Section 7 Soil Nutrients

植物营养学 plant nutrition
植物矿质营养 mineral nutrition of plant
植物有机营养 organic nutrition of plant
有机营养学物 theory of organic nutrition
最低因子律 law of the minimum
矿质营养学说 theory of mineral nutrition
米采利希定律 Mitscherlich's law
报酬递减律 law of diminishing returns
归还学说 theory of returns
根际营养 rhizosphere nutrition
根氧化力 oxidizing force of root
根系分泌物 root exudate
根系阳离子交换量 cation exchange capacity of root
根外营养 exoroot nutrition
阶段营养期 different nutritional stage
营养临界期 critical period of nutrition
肥料最大效率期 maximum efficiency stage of fertilization
选择吸收 selective absorption
交换吸收 exchange absorption
主要吸收养分 active uptake nutrient
被动吸收养分 passive uptake nutrient
奢侈吸收 luxury absorption
接触交换 contact exchange
吸附学说 theory of adsorption
载体学说 carrier theory
阴离子吸收学说 theory anion absorption
自由空间 free space
道南平衡 Donnan equilibrium

能斯特方程 Nernst equation
离子泵 ion pump
腺苷三磷酸酶 ATPase
离子载体 ionophore
胞饮作用 pinocytosis
稀释效应 dilution effect
养分交互作用 nutrient interaction
离子拮抗作用 ion antagonism
离子协和作用 ion synergism
竞争性抑制 competitive inhibition
非竞争性抑制 noncompetitive inhibition
维茨效应 Viet's effect
根系截获 root interception
质流养分 mass flow nutrition
扩散养分 diffusion nutrition
植物体内养分运输 nutrients transport in plant
质外体运输 apoplast transport
共质体运输 symplast transport
短距离运输 short distance transport
长距离运输 long distance transport
木质部运输 xylem transport
韧皮部运输 phloem transport
源-汇关系 source-sink relationship
植物激素 phytohormone
生长抑制剂 growth inhibitor
大量元素养分 macronutrient
中量元素养分 middle element nutrient
微量元素养分 micronutrient
必需元素 essential element
有益元素 beneficial element
大量元素 macroelement
微量元素 microelement, trace element
灰分元素 ash element
植物养分比例 plant nutrients ration
养分有效性 nutrient availability
养分生物有效性 nutrient bioavailability
养分再利用 nutrient reutilization
养分富集 nutrient enrichment
养分胁迫 nutrient stress
养分效率 nutrient efficiency
营养条件 nutritional condition
营养水平 nutritional level
营养物质 nutritive material
营养液 nutrient solution
氮同化作用 nitrogen assimilation
氮平衡 nitrogen balance
氮代谢 nitrogen metabolism
氨毒 ammonia toxicity
钙调素 calmodulin (CAM)
表面迁移 surface migration
土壤养分 soil nutrient
有效性养分 available nutrient
无效养分 non-available nutrient
土壤养分形态 form of soil nutrient
全氮 total nitrogen
全磷 total phosphorus
全钾 total potassium
可溶性养分 soluble nutrient
水溶性养分 water soluble nutrient
酸溶性养分 acid soluble nutrient
难溶性养分 difficultly soluble nutrient
缓效性养分 slowly available nutrient
交换性养分 exchangeable nutrient
非交换性养分 non-exchangeable nutrient
活性养分 active nutrient
养分循环 nutrient cycle
养分平衡 nutrient balance
养分固定 nutrient fixation
养分吸附 nutrient adsorption
养分释放 nutrient releasing
养分解吸 nutrient loss
养分损失 nutrient loss
养分淋失 leaching loss of nutrient
养分转化 nutrient transformation
碳氮比 C/N ratio
有机氮 organic nitrogen
无机氮 inorganic nitrogen
氮矿化 nitrogen mineralization
氮转化 nitrogen transformation
水解氮 hydrolyzable nitrogen
铵态氮 ammonium nitrogen
硝态氮 nitrate nitrogen
亚硝态氮 nitrite nitrogen
碱解氮 alkali-hydrolyzable nitrogen
酸解氮 acid-hydrolyzable nitrogen
酰胺态氮 amide nitrogen
矿化速率 mineralization rate
活性磷 labile phosphorus
枸溶性磷 citric acid soluble phosphorus
有机磷 organic phosphorus
无机磷 inorganic phosphorus
钾平衡 potassium balance

缓效钾　slowly available potassium
矿物态钾　mineral potassium
层间钾　interlayer potassium
单因子试验　unifactor experiment
多因子试验　multifactor experiment
随机区组试验　randomized block experiment
完全区组设计　complete block design
不完全区组设计　incomplete block design
拉丁方设计　Latin square design
裂区设计　split plot design
回归设计　regression design
正交设计　orthogonal design
回归正交设计　regression-orthogonal design
旋转设计　rotation design
最优设计　optimum design
试验误差　experimental error
统计检验　statistical test
示踪核素　tracer nuclides
稳定性同位素示踪技术　stable isotope tracer technique
放射性同位素示踪技术　radio isotope tracer technique
放射自显影术　autoradiography
示踪测定　isotopic tracer

1 Soil nutrient: Elements or compounds essential as raw materials for organism growth and development.
土壤养分:土壤中植物生长发育所需要的营养元素。

2 Plant nutrient: An element which is absorbed by plants and is necessary for the completion of the normal life cycle.
植物养分:植物完成生命周期所需的基本化学组分。

3 Macronutrient: A chemical element necessary in large amounts (usually greater than 1 part per million in the plant) for the growth of plants and usually applied artificially in fertilizer or liming materials.
大量元素:指含量占生物总重量万分之一以上的化学元素,对植物体内的生物氧化还原过程起着重要作用,主要通过肥料和矿物质中获得。

4 Micronutrient: A chemical element necessary only in small amounts (usually less than 1part per million in the plant)for the growth of plants and the health of animals.
微量元素:指含量占生物总重量小于万分之一以上的化学元素,对植物的生长和动物的健康起着重要作用。

第八节　土壤质量
Section 8　Soil Quality

土壤环境　soil environment
土壤环境因子　soil environment factor
土壤环境容量　soil environment capacity
土壤环境质量　soil environment quality
土壤环境质量评价　soil environment quality assessment
土壤环境监测　soil environment monitoring
土壤环境指示物　soil environment indicator
土壤环境保护　soil environment protection
土壤环境工程　soil environment engineering
土壤元素背景值　background value of soil element
土壤污染　soil pollution
农药污染　pesticide pollution
肥料污染　fertilizer pollution
土壤放射性污染　soil radioactive contamination
土壤污染物　soil pollutant
生化需氧量　biochemical oxygen demand (BOD)
化学需氧量　chemical oxygen demand (COD)
酸雨　acid rain
温室气体　greenhouse gasses
温室效应　greenhouse effect
土壤污染负荷　soil pollution load
土壤污染指数　soil pollution index
土壤单项污染指数　individual pollution index of soil
污染物累积指数　accumulation index of pollutant
污水灌溉　sewage irrigation
土壤自净　soil self-purification
土壤自净能力　soil self-purification capacity

1 Soil pollution: When many harmful substances in soil exceed the self-purification capacity of the soil, can cause changes in the composition, structure, microbial activities and function of the soil. Harmful material or its decomposition products in soil in the gradually accumulation through "soil → plant → human", or through "soil → water → human" indirect is human absorption, reached against human health of degree, is soil pollution.

土壤污染:当土壤中含有害物质过多,超过土壤的自净能力,就会引起土壤的组成、结构和功能发生变化,微生物活动受到抑制,有害物质或其分解产物在土壤中逐渐积累通过"土壤 → 植物 → 人体",或通过"土壤 → 水 → 人体"间接被人体吸收,达到危害人体健康的程度,就是土壤污染。

2 Soil aggregate: A group of soil particles cohering in such way that they behave mechanically as a unit.

土壤团聚体:土壤团聚体指土粒通过各种自然过程的作用而形成的良好的土壤结构体。

3 Lithosphere: The upper (oceanic and continental) layer of the solid Earth, comprising all crustal rocks and the brittle part of the uppermost mantle.

岩石圈:石圈是地球上部相对于软流圈而言的坚硬的岩石圈层,包括地壳的全部和上地幔的上部,由花岗质岩、玄武质岩和超基性岩组成。

第十二章　生态学
Chapter 12　Ecology

第一节　生态学概述
Section 1　Introduction to Ecology

生态学　ecology
环境生态学　environmental ecology
组织层次　levels of organization
个体　individual
种群　population
群落　community
生态系统　ecosystem
分子生态学　molecular ecology
个体生态学　autecology
种群生态学　population ecology
群落生态学　synecology
生态系统生态学　ecosystem ecology
景观生态学　landscape ecology
全球生态学　global ecology
动物生态学　animal ecology
昆虫生态学　insect ecology
微生物生态学　microbial ecology
人类生态学　human ecology
淡水生态学　fresh-water ecology
海洋生态学　marine ecology
河口生态学　estuary ecology
湿地生态学　wetland ecology
热带生态学　tropical ecology
陆地生态学　terrestrial ecology
城市生态学　urban ecology
污染生态学　pollution ecology
渔业生态学　fishery ecology
放射生态学　radio ecology
资源生态学　resource ecology
数学生态学　mathematical ecology
化学生态学　chemical ecology
物理生态学　physical ecology
地理生态学　geographic ecology
生理生态学　physiological ecology
进化生态学　evolutionary ecology
行为生态学　behavioral ecology
遗传生态学　genetic ecology
经济生态学　economic ecology
环境问题　environmental issues
环境污染　environmental pollution
生态破坏　ecological damage
酸雨　acid rain
温室效应　green house effect
温室气体　green house gases
臭氧空洞　ozone hollow/hole
负反馈效应　negative feedback effect
环境质量　environmental quality
生态资源　ecological resource
缓冲能力　buffer capacity
自净能力　self-purification capacity
可持续发展　sustainable development
环境运动　environmental movement
生物带　biotic belt
受损生态系统　damaged ecosystem
位移　displacement
恢复　restoration
生态重建　ecological restoration
生态规划　ecological planning

1 Ecology is the study of organisms in relation to the surroundings in which they live.
生态学是研究生物体与其环境相互作用的科学。

2 There are four identifiable subdivisions of scale which ecologists investigate: (i)individual, considering the response of individuals to their environments; (ii)population, examining the response of populations of a single species to the environment, and considering processes such as abundance and fluctuations; (iii)community, the composition and structure of communities; (iv) ecosystem, the processes occurring within ecosystems, such as energy flow, food webs and the cycling of nutrients.
生态学者进行了四个可辨层次的研究:(i) 个体,探究个体对其环境的反应;(ii) 种群,研究单个物种的种群对于环境的反应和探讨诸如多度及波动等过程;(iii) 群落,群落的组成和结构;(iv) 生态系统,生态系统内的各种过程,例如能量流动、食物网和物质循环等。

第二节　生物与环境
Section 2　Organisms and Environment

环境　environment
生境　habitat
宇宙环境　space environment
地球环境　global environment
区域环境　regional environment
微环境　micro-environment
小气候　microclimate
生物气候　bioclimate
密度制约因子　density-dependent factor
非密度制约因子　density-independent factor
最小因子定律　law of the minimum
耐受定律　law of tolerance
广适性生物　eurytropic organism
狭适性生物　stenotropic organism
限制因子　limiting factor
稳态　homeostasis
阳生植物　sun plant
阴生植物　shade plant
长日照植物　long-day plant
短日照植物　short-day plant
光周期　photoperiod
水生植物　aquatic plant
陆生植物　terrestrial plant
水生动物　aquatic animal
漂浮植物　floating plant
浮叶植物　emerging plant
沉水植物　submerged plant
土壤质地　texture of soil
土壤结构　structure of soil
腐殖质　humus
矿质营养　mineral nutrient
生物群系　biomes
盐生植物　halophyte
氧化土　oxisol
新成土　entisol
灰土　spodosol
淋溶土 alfisol
终成土　ultisol
始成土　inceptisol
砖红壤　laterite soil
盐碱土　saline-alkali soil
钙土植物　calciphyte
土层　soil horizon
限制因子　limiting factor
Liebig 最小因子定律　Liebig's law of minimum
Shelford 耐性定律　Shelford's law of tolerance
生态幅　ecological amplitude
生态价　ecological valence
生物因子　biotic factor
非生物因子　abiotic factor
气候因子　climate factor
耐冷菌　psychrotrophs
嗜温微生物　mesophile
纬度地带性　latitude zonality
垂直地带性　vertical zonality
因子补偿作用　factor compensation
驯化　acclimation
适应　adaptation
适应组合　adaptive suites
光补偿点　light compensation point
光饱和点　light saturate point
光周期现象　photoperiodism
稳定性选择　steady selection
单向性选择　directional selection
分裂性选择　disruptive selection
大气圈　atmosphere
水圈　hydrosphere
岩石圈　lithosphere
土壤圈　pedosphere
人为环境　man-made environment
水土流失　soil and water loss
再生性资源　regenerative resource
无脊椎动物　invertebrate
脊椎动物　vertebrate
新陈代谢　metabolism
趋同适应　convergent adaptation
趋异适应　divergence adaptation
辐射适应　adaptive radiation
保护色　protective coloration
警戒色　warning coloration
拟态　mimesis
迁移　migration
生物钟　biological clock
休眠　dormancy
春化作用　vernalization

1 Biosphere: the thin layer of air (atmosphere), water (hydrosphere), and soil and rock (lithosphere and pedosphere) that contains the conditions to support life.
生物圈是指具有支持生命体条件的大气圈、水圈、土壤及岩石圈。

2 Adaptation: any heritable trait possessed by an organism which aids survival or reproduction is an adaptation. Such traits may be physiological, morphological or behavioral. Adaption is the result of natural selection.
适应:有机体所具有的有助于生存和生殖的任何可遗传特征都是适应。适应性特征可以是生理的或行为的特征。适应是自然选择的结果。

2 Genotype and phenotype: the genotype is the genetic composition of an individual. The phenotype is the individual organism, a product of the interaction between its genotype and its environment. The ability of the phenotype to vary due to environmental influences on its genotype is known as phenotypic plasticity.
基因型和表型:基因型是个体的遗传组成。表型是各个有机体,它是基因型与环境的相互作用的产物。由于环境对基因型的影响,表型发生变化的能力叫做表型可塑性。

3 Resources: anything which the organism uses up or depletes is a resource for that organism. Thus, nectar is a resource for honey bees, and light is a resource for all green plants.
资源:有机体消耗的任何东西,对该有机体而言,就是资源。例如,蜜是蜜蜂的资源,光是一切绿色植物的资源。

4 Environment variation: most organisms have to cope with a continually changing external environment over a range of timescales. Some environmental factors may change over seconds or minutes (e.g. sunlight intensity when there is patchy cloud) whilst others may change daily or seasonally or over a much longer period (e.g. glaciations cycle).
环境变化:大多数有机体都必须应付在一定时间尺度范围内不断变化着的外界环境。某些环境因子的变化以秒或分计(如当有云块时的阳光强度),另一些因子的变化以日或季计,甚至为更长的时期(如冰河周期)。

5 Internal regulation: biological cells cannot function with a wildly fluctuating environment and organisms therefore take steps to limit their internal variation.
内调节:生物细胞不可能在剧烈的变动环境中生存,因此,有机体要采取行动以限制其内环境的变化。

6 Homeostasis: the maintenance of a relatively constant internal environment by an organism in a variable external environment is called homeostasis. All organisms adopt a degree of homeostatic control.
稳态:有机体在可变动的外部环境中维持一个相对恒定的内部环境,称为稳态。一切有机体都采取一定程度的稳态控制。

7 Negative feedback: most biological homeostatic mechanisms act in a broadly similar way: if the current internal level of a factor (e.g. temperature or osmolarity) is too high, the mechanism will reduce it; if the level is too low the mechanism will increase it. This process, by which the response is opposite to the signal, is known as negative feedback.
负反馈:大多数生物稳态机制以大致一样的方式起作用:如果一个因子的内部水平(如温度或渗透性)太高,该机制将减少它;如果水平太低,就增加它。这个过程叫做负反馈。负反馈反应的方向与信号相反。

8 Tolerance: organisms can cope with variation in their external environment (though different species may differ markedly). The upper and lower extremes of environmental conditions which members of a species can survive are the species limits of tolerance.
耐受性:有机体能够应付其外部环境的变化(虽然不同物种表现很不相同)。物种的成员能够生存的环境条件上限和下限是种的耐受限度。

9 Niche: the ecological niche of an organism is the position it fills in its environment, comprising the conditions under which it is found, the resources it utilizes and the time it occurs there.
生态位:有机体的生态位是在环境中所处的位置,包括它存在的条件,它所利用的资源和时间。

10 Habitat: the habitat of an organism is the physical environment it is found in, for example, temperate broad-leaved woodland. Each habitat provides numerous niches.
栖息地:有机体的栖息地(或生境)是它所处的物理环境,例如,温带阔叶林。每一个栖息地可提供许多生态位。

11 Fundamental niche: the niche space an organism can fill in the absence of competition or predation is known as the fundamental niche.
基础生态位:在没有竞争和捕食条件下,有机体的生态位空间叫做基础生态位。

12 Global wind patterns: the major wind systems of the earth result from the upward movement of warm air around the equator, which is replaced by cooler air coming from the north and south, forming the trade winds. The Coriolus effect deflects moving air to the right in the Northern Hemisphere and to the left in the Southern Hemisphere. The trade winds meet just to the north of the equator in the intertropical convergence zone. At 40° north and southwesterly, upper atmosphere 'jet stream' winds occur. Ice packs at the poles increase the surface reflectance which reduces the heating of the air, resulting in zones of dense cold air.
地球的风型:地球上主要的风系是由赤道的热空气的上升运动而引起的,从南北而来的较冷空气替代了热空气,形成了信风。科里奥利效应使空气在北半球向右偏转,在南半球向左偏转。南北两半球的信风恰好在热带辐合区中的赤道北部相遇。在南北纬40°地区的高层大气中,发生从西而来的"急流"风。地球两极的冰群增加了地球表面的反射系数,降低了空气的升温,产生了一些密集的冷空气带。

13 The circulation of oceans: the world's oceans are stirred by winds. Trade winds pile up water against continents, causing an imbalance in sea levels. For example, in North America sea levels are 1 or 2 m higher on the Atlantic side ran the Pacific side. This difference drives oceanic currents. Thus, for instance, warm water piled up in the Caribbean moves northwards along the American coast as the Gulf Stream and then veers towards northern Europe, where it provides a strong warming influence.
洋流:风搅动着地球上的海洋,信风将海水掀起海浪冲向大陆,造成海平面的不平衡。例如,北美洲的海平面在大西洋一侧比太平洋的一侧高一两米,这种差异形成了洋流。因此,在加勒比海堆聚的温暖海水,沿着美洲海岸向北移动,即墨西哥湾流,然后转向欧洲北部,对欧洲具有显著的增温作用。

14 Rain: rain falls when moist air cools. Warm air can hold more water than cool air, so cooling causes water droplets to condense and fall as rain.
降雨:潮湿的空气遇冷形成降雨。温暖的空气比冷空气携带更多的水,所以冷却使热空气凝结成小水滴,并以雨的形式降落。

15 Local variations in climate: global maps of temperature and rain obscure local variation. However, on an even smaller scale there is a great deal of microclimatic variation. The microclimate is the climate in which plants and animals live.
气候的局部变化:全球的温度与降雨图掩盖了局部地区的变化。然而,即使在很小的尺度范围内,也有大量的微气候变化。微气候是动植物生活在其中的气候。

16 Heat: a major storage location for heat in temperate ecosystems is the soil, which acts as a sink during the day and a source at night. Thermal conductivity (a measure of the standardized rate of heat flow) of soils determines the rate of heat transfer. The temperature change that a body experiences as a result of the heat transfer will vary with its heat capacity. The heat capacity of a substance is the amount of heat required to raise 1 cm^3 by 1℃.
热:在温带生态系统中,热的主要储存位置是土壤,白天土壤储存热,夜间作为热量的来源。土壤的热传导(热流标准率测量)决定了热转换速率。物体经历的温度变化是热传导的结果,将随着它的热容量而改变。物质的热容量是 1cm^3 物质上升 1℃ 所需要的热量。

17 Temperature profiles: temperature profiles develop in the air above the ground. During the light ground and crop surfaces cool rapidly so that surface becomes the coldest location in the profile. Thus, a temperature inversion develops in which air temperature increases with elevation. During the day the surface is the warmest location in the profile and under these conditions air temperature decreases with elevation.

温度剖面:温度剖面是在地面上的空气中形成的。夜间,地面和庄稼表面迅速冷却,以至于表面变成温度剖面中最冷的部位。因此,形成温度倒置,空气的温度随着高度而增加。白天,地表面是温度剖面中最温暖的部位,在这种状态下,空气的温度随高度而下降。

18 Relative humidity: relative humidity is the air water vapor content expressed as a ratio to the saturated water content at that water temperature. Water vapor is derived from the evaporating surfaces of land, water and vegetation.

相对湿度:相对湿度是指空气的水蒸气含量,用该温度下饱和水含量比率来表示。水蒸气来自陆地表面、水和植被的蒸发。

19 Soil water: for terrestrial plants the main source of water is the soil, which serves as a reservoir. Water enters the reservoir as rain or melting snow and passes into the soil pores. The upper limit of the water-holding capacity of a soil is called the field capacity. This is the amount of water which can be held by soil pores against the force of gravity. Plants cannot extract all the water held in the soil, as they cannot exert sufficient suction force to extract water from the narrower soil pores. The lower limit of water availability is thus determined by the physiology of the plant species and is known as the permanent wilting point- the soil water content at which plants wilt and are unable to recover.

土壤水:对于陆地植物,水的主要来源是土壤,它起了蓄水池的作用。当下雨或雪融化时,水进入蓄水池,并流进土壤孔隙。土壤的水容量上限称为田间持水量。这是土壤孔隙抗地心引力所储蓄的水量。植物不能吸取土壤中储蓄的全部水,因为它们不能产生足够的吸力从更细的土壤孔隙中吸水。因此,可利用水的下限是由植物物种的生理特性所决定,被称为永久萎蔫点 —— 土壤水含量在这个点以下,植物枯死,不能恢复。

20 Aquatic plants and water: water is apparently freely available in aquatic environments. However, the osmotic regulation of internal fluids can be energetically expensive, especially in saline environments. The salinity of an aquatic environment and of terrestrial habitats bordering the sea has an important influence on plant distribution and abundance. Plants which grow in high salinity, halophytes, accumulate electrolytes in their vacuoles, but the concentration in the cytoplasm and organelles is kept low.

水生植物和水:在水环境中,水显然是随意可获得的。然而,内部体液的渗透压调节可能消耗能量,特别是在盐水环境中。水环境的盐度和与沿海栖息地的盐度,对植物分布和多度有重要的影响。生长在盐度中的植物,即盐生植物,它们的液泡中累积电解质,但在细胞质和细胞器中保持低浓度。

21 Water availability and plant productivity: precipitation is a key determinant of plant productivity in forests, whilst in arid regions there is an approximately linear increase in primary productivity with increasing precipitation. The amount of water that would be transpired from a site, assuming no soil water limitation and complete vegetation over, is the potential evapotranspiration rate. The difference between this index and the precipitation rate defines whether the environment is moist or arid.

水的可利用性与植物生产力:降雨量是森林植物生产量的关键决定因子,而在干旱地区,初级生产量随降雨量的增加呈线性增长。假设一个地区没有土壤的水限制,并完全地被植物覆盖,水从这个地区的蒸发量就是潜在蒸发速率。这个指标和降雨量之间的差异决定了潮湿环境,还是干旱环境。

22 Water conservation by terrestrial animals: a major problem faced by terrestrial organisms is the loss of a continuous supply of water necessary to keep tissue surfaces moist. When air is inhaled, it passes along the respiratory tract into the lungs, where it is in contact with the moist respiratory tissues. If the moist air was exhaled, water would be lost. The recovery of respiratory moisture by most

terrestrial animals involves countercurrent exchange. Exhaled air from the lungs encounters a countercurrent-like gradient on the way out. This interaction between the departing air and the respiratory surfaces results in an efficient return of moisture to the tissues.

陆生动物的水保持:陆生动物面对的主要问题是连续失水,这是供应到组织表面维持湿度所必需的水。当空气被吸入时,它沿着呼吸道进入肺和潮湿的呼吸组织相接触。如果潮湿的空气被呼出,水就会丢失。大多数陆生动物呼吸湿度的维持包括了逆流交换。从肺呼出的气体,在呼出的通道上有一个像逆流的梯度。呼出的气体与呼吸表面间的相互作用,导致水分有效地返回组织。

23 Homeotherms and poikilotherms: it is possible to categorize organisms according to their temperature regulation processes. One possible division is between homeotherms and poikilotherms. As environment temperature rises, homeotherms maintain an approximately constant body temperature, while the body temperature of poikilotherms varies with environmental temperature.

恒温动物和变温动物:按照有机体的体温调节过程对有机体进行分类。一种划分是恒温动物和变温动物。当环境温度升高时,恒温动物保持着大致恒温的体温,而变温动物的体温随环境温度而改变。

24 Ectotherms and endotherms: ectotherms are organisms such as plants, reptiles and protista which are largely reliant on external sources of heat to raise their body temperature. Endotherms are organisms capable of generating heat internally in order to raise their body temperature. Birds and mammals make up this group. The thermoneutral zone is the range of environmental temperatures in which an endotherm has only to exert a minimum metabolic effort in order to maintain a constant body temperature. The further away from the thermoneutral zone that the environment temperature moves, the more energy the endotherm has to expend to maintain body temperature.

外温动物和内温动物:像植物、爬行动物和原生生物等有机体都属外温动物,它们很大程度地依赖于外来热来提高自己的体温。内温动物是能够在体内产热以升温的有机体。鸟和哺乳动物属于这一类。热中性区是环境的温度范围,在这个温度范围内,内温动物仅使用一个最小的代谢率以维持恒定的体温。环境温度离热中性区越远,内温动物维持恒定体温消耗的能量越多。

25 Temperature and rates of enzyme reaction: The rate of an enzyme catalyzing reaction increases with temperature. In ectotherms this means that metabolic activity will be faster at higher environmental temperatures. The temperature coefficient (Q10) is an index of the effect of a 10℃ temperature rise on metabolic rate, and is often near 2.0.

温度和酶反应速度:酶催化反应的速度随温度而增加。在外温动物中这意味着在较高的温度下动物代谢活力更快。温度系数(Q_{10})是温度升高 10℃ 对代谢速度影响的指数,经常接近于 2.0。

26 Rates of development and growth: within the nonlethal temperature range the most important effect on ectotherms of temperature is likely to be its effect on rate of development and growth. When rate of development is plotted against body temperature where exists an extended range of temperatures over which the relationship is linear. 'Physiological time' is a measurement combining temperature and time and applied to ectothermic organisms is dependent on environmental temperature as well as time.

发育和生长速度:在非致死温度范围内,温度对外温动物最大的影响很可能是影响了发育和生长的速度。当用发育速度对体温作图时,存在着一个直线相关的温度范围。"生理时间"是将温度和时间结合的一个测量,应用于外温动物,它反映了这些有机体的生长和发育依赖于环境温度及时间。

27 Acclimation and acclimatization: Temperature may also act as a stimulus, determining whether the organisms will begin development. Vernalization is the induction of flowering by low temperatures. Exposure of an organism to higher (or lower) temperatures in the laboratory can alter the organism's temperatures response. The habituation of an organism's response to changes in laboratory environmental conditions is termed acclimation. Acclimation is the habituation of an organism's physiological response to changes in natural environmental conditions.

驯化和气候驯化:温度能够作为一种刺激物起作用,决定有机体是否将开始发育。春化法是通过低温诱导开花。有机体在实验室里暴露到较高(或较低)的温度能够改变有机体的温度反应。有机体对实验环境条件变化产生的适应性反应称为驯化。有机体对自然环境条件变化产生的生理适应性反应称为气候驯化。

28 Radiant energy and photosynthesis: radiant energy is the sole energy source that can be used by green plants. When a leaf intercepts radiant energy it may be absorbed, reflected or transmitted. Part of the fraction absorbed reaches the chloroplast, fuelling photosynthesis, the process where radiant energy is used to convert water and CO_2 into sugars. Solar radiation contains a spectrum of different wavelengths. However, only a restricted bang of this spectrum is effective for photosynthesis. This is the band of photosynthetically active radiation and for green plants lies between 380 and 710 mm.

辐射能和光合作用:绿色植物能够利用的唯一能源是辐射能。当叶子截获辐射能时,它能被吸收、反射或者透射。吸收的部分能量到达叶绿体,引发了光合作用,在这个过程中,辐射能被用于转化水和二氧化碳生成糖。太阳辐射包含了不同波长的光。然而,仅有一个有限的光谱带对光合作用是有效的。这就是光合活性辐射带,对绿色植物位于380mm到700mm之间。

29 C3 and C4 plants: a major difference in the photosynthetic capacity of plants is that between C3 and C4 plants. C4 plants are able to capture CO_2 with greater water use efficiency than C3 plants, but this advantage comes at an energy cost. In C4 plants the rate of photosynthesis increases with light intensity, whilst photosynthesis tails off with increasing light intensity in C3 plants.

C3和C4植物:植物光合能力中的主要差别是在C3和C4植物之间。C4植物能捕获CO_2,伴随着水的利用效率比C3植物更大,而这需要消耗能量。在C4植物中,光合作用率随光强度而增加,而C3植物随光强度增加光合作用渐渐减小。

30 Strategic and tactical response of plants to radiation: A major strategic difference between plant species in their response to the intensity of radiation is exhibited by 'sun species' and 'shade species', which possess a range of adaptations to high and low light levels, respectively. Also, plants may grow leaves which develop differently under different light conditions as part of a tactical response to the light environment. This is most clearly seen in the formation of sun leaves and shade leaves within a leaf canopy of a single plant.

植物对辐射的战略和战术响应:植物种间对辐射强度反应的主要战略差异显示为"阳地种"和"阴地种",它们分别具有适用于高的和低的光辐射范围。同样,植物能够在不同光条件下生长不同的叶子,作为对光环境的部分战术反应。这一点最清楚地在反映在单株植物叶冠内的阳叶和阴叶的结构上。

31 The soil profile: At the surface of soil is a litter of dead or rotting plant parts, with one or more distinctly different layers underneath. These layers have come to be called 'horizons'. Soil horizons form as rotting plant parts mixing with the upper layers of the mineral soil. The subsoil underneath is the earth from which the soil was made and is called the parent material. This layered appearance is called the soil profile. Soil scientists recognize three kinds of soil horizon above the parent material; the A, B and C horizons. Under these three groups of horizons is the unaltered parent material. These are known as R horizons. The soil profile is an instant indicator of important ecosystem processes.

土壤剖面:土壤表面是死亡或腐烂植物部分的枯枝落叶层,下面由完全不同的物质组成一层或多层。层的形成是由于腐烂植物与矿物质土壤的上层相混合。下面的亚土壤是土壤起始形成的地方,称为母质。这种分层的外貌叫做土壤剖面。土壤科学家认为母质以上的土层有三类:A、B和C层。这三层之下是不变的母质。这些以R层著称。土壤剖面是重要生态系统过程中一个即时指标。

32 Photosynthesis: the process of using the sun light energy by chlorophyll-containing plants to covert carbon dioxide and water into complex chemical bonds forming simple carbohydrates such as glucose and fructose.

光合作用:是指含有叶绿素的植物利用太阳能,把二氧化碳和水转化成碳水化合物的过程,比如生成葡萄糖和果糖。

第三节　种群生态学
Section 3　Population Ecology

种群动态　population dynamics
密度　density
迁入　immigration
迁出　emigration
性比　sex ratio
增长率　growth rate
构件　modular
生命表　life table
存活曲线　survivorship curve
年龄结构　age structure
环境容纳量　carrying capacity
集群　aggregation
他感作用　allelopathy
竞争　competition
种内竞争　intraspecific competition
种间关系　interspecific interaction
信息　communication
领域　territory
社会等级　social hierarchy
利他行为　altruism
分布格局　distribution pattern
扩散　dispersal
生态位　niche
竞争排斥原理　competition exclusive principle
专化　specialization
泛化　generalization
性状替换　character displacement
捕食　predation
最佳摄食理论　optimal foraging theory
互利　mutualism
生活史对策　life-history strategy
能量分配原则　principle of energy allocation
协同进化　coevolution

1 Population：a population is a group of organisms of the same species which occupies a given area. The boundaries between populations can be arbitrary. Populations may be categorized as consisting of either unitary or modular organisms. In unitary populations，each zygote gives rise to a single individual. In modular organisms，the zygote develops into a unit of construction which gives rise to further modules and a branching structure. The structure may then fragment producing many individual ramets.

种群：种群是一定区域内同种生物个体的集合。种群间的边界可以是任意的。种群可以根据组成种群的生物是单体生物还是构件生物进行分类。在单体生物种群中，每一受精卵发育成一单个个体。在构件生物种群中，受精卵发育成一个结构单位，这一结构单位再形成更多的构件和分支结构。然后这些结构可能分裂，形成许多无性系分株。

2 Population size：the population size for unitary organisms，such as mammals，is simply the number of individuals in a given area. For modular organisms，such as plants and corals，the situation is more complex.

种群大小：对于单体生物种群如哺乳类，其种群大小就是一定区域内个体的数量，非常简单。对于构件生物，如植物和珊瑚，情况就较复杂。

3 Survivorship：survivorship is the converse of mortality. Survivorship data are often shown as a survivorship curve for a particular population；a graph showing the proportion of survivors on a logarithmic scale through each phase of life. There are three generalized patterns of age-specific survivorship depending on whether the probability of dying is highest later in life (Type Ⅰ)，constant through life (Type Ⅱ) or highest for young stages (Type Ⅲ).

存活率：存活率是死亡率的倒数。对于一个特定种群，存活率的数据通常以存活曲线的形式来表示；存活曲线表示的是在每一个生活期存活个体所占的比率的对数值。根据各生活期死亡率的高低，特定年龄存活曲线一般有三种模式：后期死亡率最高(类型 Ⅰ)，各期死亡率相等(类型 Ⅱ)，早期死亡率最高(类型 Ⅲ)。

4 Life tables: life tables summarize the fate of a group of individuals born at approximately the same time from birth to the end of the life cycle. Such a group is known as a cohort and investigation of this kind is termed cohort analysis. Life tables show the number of individuals present at different life stages or ages together with age-specific survival rates and age-specific mortality rates calculated for each stage. Mortality at each stage is expressed by k-values which are derived from logarithms and can be summed to give total mortality.

生命表:生命表总结了一组出生时间大体相同的个体从出生到死亡的命运,这样的一组个体称为同生群,这样的调查称为同生群分析。生命表表示存在于不同生命阶段或年龄个体的数量以及每一阶段的年龄特定存活率和年龄特定死亡率。每一阶段的死亡率用 k 值表示,k 是通过对数函数推导出来的,并且可以相加得出总死亡率。

5 k-factor analysis: This technique allows the identification of key factors contributing to mortality. Stage-specific k-values obtained over successive years are compared to the values for total mortality (k_{total}). k-factor analysis highlights those stages suffering the greatest mortality which are responsible for fluctuations in loss rate and hence population size.

k 因子分析:这一方法可以辨明关键因子对死亡率的作用。连续几年获得的特定阶段 k 值与总死亡率($k_{总}$)相比。k 因子分析强调那些死亡率最高的阶段,这些阶段是种群丧失率和种群大小波动的关键。

6 Population growth: the changes in population size over time can be calculated by adding birth (B) and the number of immigrants (I) to the original population at time t (N_t), and subtracting the number of deaths (D) and emigrants (E) to give a new population size at the time $t+1$ (N_{t+1}). This is represented by the equation:

$$N_{t+1} = N_t + B + I - D - E$$

For a particular set of conditions, an individual has a maximum potential for reproduction which is its intrinsic natural rate of increase, r. This is the theoretical maximum that may be reached in a given environment if the population is not resource-limited.

种群增长:种群大小随时间的变化可以按如下方法计算,t 时间种群初始数量(N_t),加上新出生的个体数(B) 和迁入个体数(I),减去死亡个体数(D) 和迁出的个体数(E),就可得到 $t+1$ 时间种群的数量(N_{t+1}),这可用以下方程表示:$N_{t+1} = N_t + B + I - D - E$。在一组特定条件下,个体具有最大的繁殖潜力,称为内禀自然增长率 r。这是种群在不受资源限制的情况下,于一定环境中可达到的理论最大值。

7 Density-dependent growth — the logistic equation: The logistic equation describes the growth of a simple population in a confined space, where resources are not unlimited. In the early stages resources are abundant, the death rate is minimal and reproduction can take place as fast as possible allowing the individuals to attain their intrinsic rate of increase. The population increases geometrically until the maximum number of individuals the environment can sustainably support is approached. This maximum number is called the carrying capacity (K). The population growth rate declines to zero as the population becomes more crowded and the population size stabilizes. This can be described as the logistic equation where rate of change of population size at time t = Intrinsic rate of increase × Population size × Density dependent factor

$$dN/dt = rN(1 - N/K)$$

the density-dependent factor, $(1 - N/K)$ approaches zero as the population approaches the carrying capacity and intraspecific competition becomes more intense. This equation predicts growth of a population over time to be sigmoidal, as is commonly observed in real populations.

8 密度制约性种群增长 —— 逻辑斯谛方程:逻辑斯谛方程描述的是一个在有限资源空间中的简单种群的增长。在早期,资源丰富,死亡率最小,繁殖快,种群内个体可达到内禀增长率。种群呈几何式增长，直到种群数量达到环境可持续支持的最大程度，这个最大数量称为环境容纳量（K）。当种群更加拥挤时，种群增长率减少到零，种群大小处于稳定状态。这可用逻辑斯谛方程来表示：t 时间种群大小变化率 = 内禀增长率乘种群大小乘密度制约因子，$dN/dt = rN(1-(N/K)$。当种群达到环境容纳量，种间竞争变得更激烈时，密度制约因子（$1-N/K$）接近零。该方程预测种群增长随时间变化呈现"S"形，与在真实种群中观察结果一致。

9 Equilibrium population density: the equilibrium population density occurs when the per capita death rate exactly balances the per capita birth rate so that the density is neither increasing nor decreasing. The equilibrium population density is equivalent to the carrying capacity K.

平衡种群密度：当单位个体出生率正好平衡单位个体死亡率，种群密度既不增加也不减少时，此时的种群密度为平衡种群密度。平衡种群密度与环境容纳量 K 值相等。

10 Chaos: the mathematical definition of chaos is quite distinct from the colloquial use of this term. A chaotic system is driven by a deterministic process; it is not random. The outcome of chaos depends on the precise value of the starting conditions. As perfect accuracy is impossible to attain, chaotic systems are effectively unpredictable. It is currently unclear to what extent observed fluctuations in population sizes are influenced by chaotic processes. Some analyses suggest that measles outbreak and some insect populations may display chaotic dynamics.

混沌：数学上有关混沌的定义与俗语中的用法是截然不同的。一个混沌系统是受确定性作用控制的，而不是随机的。混沌的结果依赖于初始状态的精确值。由于完全精确是不可能达到的，因此，对混沌系统不能进行有效的预测。目前，对于观察到有关种群大小在何种程度上是受混沌过程影响的还不清楚。一些分析认为，麻疹的爆发和一些昆虫种群可能表现出混沌的动态变化。

11 Competition: competition is an interaction among individuals utilizing a limited resource, resulting in reduced fitness in the competing individuals. Competition occurs both between species utilizing a shared resource (interspecific competition) and among members of a species (intraspecific competition). The niche of an individual or species (the conditions under which it is found, the resources it utilizes and the time it occurs there) is critical in determining the degree of competition with other species or individuals. Large niche overlap generally results in intense competition.

竞争：竞争是共同利用有限资源的个体间的相互作用，会降低竞争个体间的适合度。竞争既可在利用共同资源的物种间发生（种间竞争），也可在同种个体间发生（种内竞争）。个体或物种的生态位（它所处的环境，利用的资源和发生的时间）是决定该个体或物种与其他个体或物种竞争程度的关键。大范围的生态位重叠一般导致激烈竞争。

12 Intraspecific competition: as individuals are quite similar in their resource requirements, such competition may be particularly intense. Intraspecific competition is a major force in ecology and is responsible for phenomena such as dispersal and territoriality, as well as being the primary cause of population regulation via density-dependent processes.

种内竞争：当个体对资源的需要非常相似时，竞争会特别强烈。种内竞争是生态学的一种主要影响力，是扩散和领域现象的原因，并且是种群通过密度制约过程进行调节的主要原因。

13 Interspecific competition: interspecific competition occurs between two species using the same limited resource. Very few species can escape from the effects of other species competing for the same resource.

种间竞争：种间竞争发生在利用同样有限资源的两物种之间。极少物种能够逃脱其他种与之竞争共同资源的影响。

14 Interference competition: in interference competition, individuals interact directly, most obviously, in the case of some animal species, by fighting, but also by producing toxins (e.g. plant allelopathy). Fitness reduction in the 'loser' in such interactions may be due to the interference (e.g. injuries or death) as well as the lack of resource access.

干扰性竞争：在干扰性竞争方式下，个体直接相互作用，在一些动物种类中，最明显的通过打斗，也通过产生毒物（如植物异株克生）进行竞争。在这种相互作用中，“败者”可能由于干扰（如受伤或死亡），也或许由于缺乏可用资源造成适合度下降。

15 Dispersal: organisms can respond to high levels of intraspecific competition by dispersing away from the area of high population density. Even in species that are sessile for most of the lifecycle, there is a mobile dispersal stage. Dispersal is often undertaken by the younger members of a population, whilst in many mammal species males disperse more than females do.

扩散：生物可通过扩散离开种群密度高的地区应对高水平的种内竞争。即使在生活周期内大部分时间营固着生活的种类，也存在一个可运动的扩散期。扩散通常在种群内较年轻的个体进行，而在许多哺乳动物中雄性比雌性更多。

16 Territoriality: in many animal species (including insects, birds and mammals) individuals or groups compete for areas of space. There is active interference between individuals to maintain the territory boundaries.

领域性：许多动物（包括昆虫、鸟类和哺乳动物）个体或种群为争夺空间而竞争。保持领域边界的个体间有积极的相互干扰行为。

17 Self-thinning: sessile organisms, including plants, cannot escape competition by movement, and therefore the losers in the competitive battle die. In a group of plants of the same age, this results in fewer individuals of larger size surviving. This process is described as 'self-thinning'. Self-thinning results in a relationship between density and individual plant mass, which typically has a slope of $-3/2$ on a log-log plot.

自疏：包括植物在内的固着生长的生物不能通过运动逃避竞争，因此，竞争中的失败者死去。在同样年龄大小的植物群中，这种竞争结果使较少量的较大个体存活下来。这一过程叫做“自疏”。自疏导致密度与植物个体大小之间的关系，这在双对数作图时呈现典型的 $-3/2$ 斜率。

18 Competitive exclusion: If two species compete in a stable environment, there are two possible outcomes: (i) one species is excluded, or (ii) both species coexist. The competitive exclusion principle states that coexistence can only occur in a stable environment if the species niches are differentiated.

竞争排斥：如果两个物种在稳定环境中竞争，则有两种可能的结果：(i) 一种被排斥，或(ii) 两种共存。竞争排斥原理认为，共存只能发生在两物种生态位分化的稳定环境中。

19 Limiting similarity: how much niche differentiation is needed for species to coexist? This critical threshold of differentiation in resource utilization is termed the limiting similarity.

极限相似性：共存种需要多少生态位分化？在资源利用分化上的临界值叫做极限相似性。

20 Competitive release: in the absence of a competitor, a species may expand its niche. Examples of such competitive release include ground doves in New Guinea and gerbils in Israel.

竞争释放：在缺乏竞争者时，物种会扩张其实际生态位。这种竞争释放的例子包括新几内亚岛的地鸽子和以色列的沙鼠。

21 Apparent competition: if a predator attacks two prey species, then each prey species may adversely affect the other, by increasing the local predator population. Therefore, the interaction between the two prey species is exactly as if they were competing, yet they may utilize entirely different resources. This phenomenon is known as apparent competition.

表观竞争：如果捕食者进攻两种猎物，每一种猎物会因为增加了捕食者种群的数量而影响另一种猎物。因此，两个猎物种群之间的相互作用看上去似乎它们确实是在竞争，然而它们或许利用完全不同的资源。这一现象叫做表观竞争。

22 Spatial and temporal heterogeneity: environments are patchworks of habitats of varying qualities and differing resource levels. Some patches are temporary and/or unpredictable in occurrence. In time and space, the environment is heterogeneous. This means that competitive 'battles' may not be completed before the environment changes and the balance of play alters. Therefore, species that coexist in the real world may do so because of environmental heterogeneity rather than niche differentiation.

空间和时间异质性:环境是许多不同质量、不同资源水平的生境的拼缀。一些斑块的发生是暂时的和不可预测的。在时间和空间上,环境是异质性的。这意味着竞争"战斗"在环境变化和竞争平衡变化之前不能完成。因此,在现实世界中共存的种类,其共存可能是因为环境的异质性,而不是因为生态位分化。

23 Predation: predation can be defined as the consumption of all or part of another individual (the prey). This wide definition thus encompasses: (i) "true predators", which kill their prey soon after attacking them; (ii) grazers, which consume only part of a prey individual; (iii) parasites, which live in very close association with a single prey individual (the host), often inside the host's tissues.

捕食:捕食可定义为摄取其他个体(猎物)或部分个体为食。这一广泛的定义包括(i) 典型的捕食者,在袭击猎物后迅速杀死而食之;(ii) 草食者,只消费对象个体的一部分;(iii) 寄生者,与单一个体(寄生)有密切关系,通常生活在寄主的组织中。

24 Generalists and specialists: predators vary in the number of species of prey they will feed on, with some species being specialists, whilst others are more generalist. Generally, parasites tend to be more specialist than true predators and herbivores tend to be more specialist than carnivores.

泛化种和特化种:捕食者随其摄取猎物种的数量多少而变化,某些捕食者是特化种,而另一些是泛化种。一般来说,寄生者比典型捕食者更为特化,而草食者比肉食者更为特化。

25 The diversity of parasites: parasites are a subgroup of predators (which consume the issue of other living organism) that live in close association with their host. Parasites can be classified into two broad groups: (i) microparasites, which multiply within, or on the surface of, the host, and (ii) macroparasites, which grow in or on the host, but do not multiply. The main microparasites are viruses, bacteria, fungi and protozoans. Helminth worms and insects are important macroparasites. A large group of insect macroparasites are parasitoids, which lay eggs in or on the body of their insect host, and usually cause the death of the host.

寄生多样性:寄生物是摄取其他生物的组织、与其寄主紧密相连而生的捕食者的子群。寄生物可以分为两大类(i) 微寄生物,在寄主体内或表面繁殖,(ii) 大寄生物,在寄主体内或表面生长,但不繁殖。主要的微寄生物有病毒、细菌、真菌和原生动物。寄生蠕虫和昆虫是重要的大寄生物。一大类昆虫大寄生物是拟寄生生物,它们在昆虫寄主身上或体内产卵,通常导致寄主死亡。

26 Parasitoids: parasitoids (also known as hyperparasites) comprise a large group of insect macroparasites (mainly wasps and flies) which lay eggs in or on the body of their insect host and usually cause the death of the host.

拟寄生物:拟寄生物(也称作超寄生物),包含一大类昆虫大寄生物(主要是寄生蜂和蝇),它们在昆虫寄主身上或体内产卵,通常导致寄主死亡。

27 Parasite-host dynamics: vertebrate hosts commonly acquire immunity to parasites. These immune hosts are no longer available to the parasites, and the size of the susceptible population is reduced. This results in a fall in the incidence of disease. However, as new susceptible hosts enter the population (for example, by birth), the disease will increase in incidence again. Thus, there is a tendency of diseases to cycle, rising as the number of new susceptible increases and falling as the level of immunity rises.

寄生物-寄主动态:脊椎动物寄主一般对寄生产生免疫。这些免疫寄主不能再被寄生物利用,因而易感种群大小减小。这种疾病的传染力降低。然而,随着新的易感寄主加入种群(如新个体出生),疾病的传染力会再次增加。因此，传染病有循环的趋势，在新的易感个体增加时上升，在免疫水平上升时下降。

28 Host-parasite evolution: the close association between parasites and their hosts often results in evolutionary interactions, or coevolution. Coevolution may give rise to defense mechanisms in the host and routes to overcome these defenses in the parasite-so-called 'arms races'. Not all coevolution between host and parasite is escalatory, however. For example, coevolution may lead to the reduction in the virulence of a parasite.

寄主-寄生物进化：寄生物与其寄主间的关联经常导致二者在进化上的相互作用，或协同进化从而使寄主的防御机制提高，而寄生物克服这些防御的通路——叫做“军备竞赛”。但是，寄主与寄生物之间协同进化并不都是提高性的。例如，协同进化会降低寄生物毒性。

29 Mutualism and symbiosis: mutualism is a positive reciprocal relationship between two individuals of different species which results in increased fitness for both parties. Mutualism may be symbiotic, in which the organisms live together in close physical association.

互利共生和共生现象：互利共生是不同种体间一种正的互惠关系，可增加双方的适合度。互利共生可以是共生性的，生物体以一种紧密的物理关系生活在一起。

30 *r*- and *K*- selection: Two distinct life history strategies are described by *r*- and *K*- selection theory, which states that *r*- selected species are adapted to be competitive. Thus, *r*-selected species have rapid development, small adults, many and small offspring and a short generation time. In contrast, *K*-selected species have slow development, large adult size, few, large offspring and long generation times. Although some species fit this theoretical dichotomy, many do not, and this theory is now out-dated as better models have wider predictive powers.

r-选择和*K*-选择：*r*-选择和*K*-选择理论描述了两种明显的生活史对策，该理论中，*r*-选择种类适于竞争，这样，*r*-选择种类发育快，成体小，后代数量多但体积小，时代时间短。与此相反，*K*-选择种类发育慢，成体大，后代数量少但体积大，时代时间长。尽管有些种类符合这种理论上的两分法，但也有许多种不符合。由于具有更广预测性的更好的模型出现，该理论已过时。

31 Migration: organisms may also avoid local harsh conditions by moving to another location. Migration is directional movement, such as the autumn flight of swallows from Europe to Africa. Dispersal, in contrast, is a nondirectional movement away from the birth or breeding site. There are three categories of migration, depending on whether an individual makes either (i) repeated return trips, (ii) a single return trip, or (iii) a one-way trip.

迁移：生物可通过移动到别处来躲避当地恶劣的环境。迁徙是方向性运动，如燕子从欧洲到非洲的秋季飞行。相反，扩散是离开出生或繁殖地的非方向性运动。依据生物个体行为，迁移有三种，(i) 反复的往返旅行，(ii) 单次往返旅行，或(iii) 单程旅行。

32 Altruism: altruism occurs when an individual causes an increase in the fitness of another individual of the same species at a cost to its own survival or offspring production. An example is the sterile worker in a eusocial ant society, which sacrifices its opportunity of reproducing and instead supports its mother's reproduction. This appears paradoxical, as it implies that the altruist is actively reducing its fitness. However, if the altruist is related to the beneficiary, this strategy may result in more of the altruist's genes passing to the next generation.

利他：当一个个体行为导致另一个个体适合度增加，同时以其牺牲自身存活或产生后代为代价时，利他行为出现。例如，真社会性的蚂蚁群体中不育的工蚁，它们牺牲自身生殖的机会，代之以对其母后生殖的支持。这意味着，利他主义者积极地减少其适合度。利他行为看起来似乎是矛盾的。然而，如果利他行为者与受惠者具有亲戚关系，此对策可使利他行为者有更多的基因遗传给下一代。

33 Sensitivity analysis: in general, population dynamics is not useful for making accurate quantitative predictions. It's useful for making qualitative predictions comparing different scenarios.

灵敏度分析：一般来说，种群动态研究并不适于进行精确的定量预测，它适用于对不同情境的定性预测。

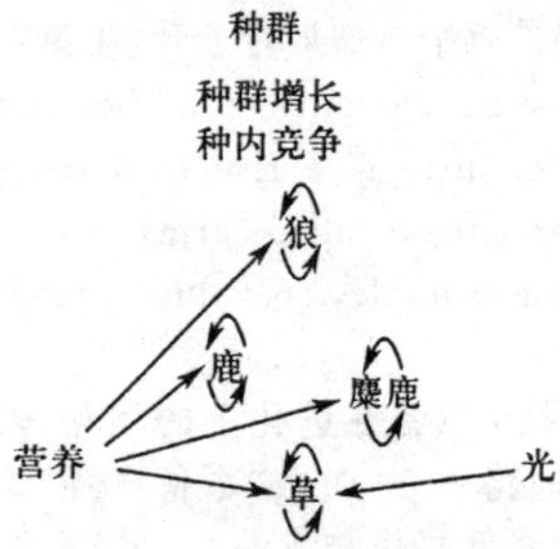

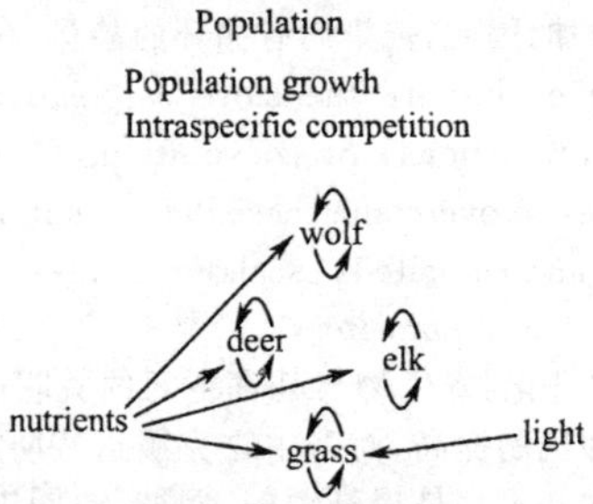

第四节 群落生态学

Section 4 Community Ecology

群落 community
外貌 physiognomy
结构 structure
垂直结构 vertical structure
水平结构 level structure
物种组成 species composition
亚优种 subdominant species
相对多度 relative density
多度 abundance
相对盖度 relative coverage
相对显著度 relative dominance
相对频度 relative frequency
生物量 biomass
重要值 important value，IV
多样性指数 diversity index
均匀度 evenness
物种多样性梯度 gradient of species diversity
稳定性 stability
复杂性 complexity
顶级群落 climax community
季相 seasonal aspect
季节动态 seasonal dynamic
年际动态 annual dynamic
演化 evolvement
演替 succession
中生演替系列 mesosere
沙丘演替 dune succession
湖泊演替 lake succession
先锋物种 pioneer species
气候顶级 climax
单元顶级 monoclimax
多元顶级 polyclimax
亚顶级 subclimax
偏途顶级 disclimax
预顶级 preclimax
超顶级 postclimax
干扰 disturbance
热带雨林 tropical
老茎生花 cauliflory
绞杀植物 strangle
藤本植物 vines
附生植物 epiphytes
季雨林 seasonal rain forest
红树林 mangroves
亚热带常绿阔叶林 subtropical evergreen broad-leaved forest
温带落叶阔叶林 temperate deciduous broad-leaved forest
针叶林 coniferous forest
泰加林 taiga
草原 grassland
干草原 steppe
草甸草原 meadow
稀树干草原 savannah
沼泽 marsh or swamp
水生植被 aquatic vegetation
荒漠 desert
干荒漠 hot desert
冻荒漠 cool desert
冻原 tundra

1 Community: the community is an assemblage of species population that occurs together in the same place at the same time. It has properties determined by the interactions among individuals such as competition and parasitism. The community can also be viewed from the broader perspective in terms of species diversity, food-webs, energy flow and the interactions among guilds of species.

群落：群落是在同一时空聚集在一起的许多物种种群的集合。它的特性是由个体之间的相互作用，如竞争和寄生决定的。群落也能够从更广阔的角度，如物种多样性、食物网、能量流和同资源种种间相互作用，来观察它的特性。

2 Community structure: the species diversity of a community depends on the number of different species it contains (the species richness) and the evenness of species abundance across species. Diversity indices can be calculated to take into account both of these factors. Another way of representing species richness and evenness together is to plot the relative abundance of species against rank order of species abundance. Diversity can be measured over the three different spatial scales of the local community, the region and the broadest geographic scale (e. g. the continent). This yields alpha (α)-, beta (β)- and gamma (γ)- diversity.

群落结构：一个群落的物种多样性取决于群落含有的不同种的数量，即种丰富度和种多度的均匀性。多样性指数可以通过这两个因子计算。综合表示种丰富度和种均匀度的另一种方式是通过种的相对多度对种的顺序作图。多样性能够在局域群落、地区和最广阔的地理尺度(如大陆)三个不同的空间尺度上测算。这样就产生了 α、β 和 γ 多样性。

3 Keystone species: a keystone species has a significant and disproportionate effect on the community. Keystone species can be top predators such as the northern sea otter; however, the term can be usefully applied to any species whose removal would have a significant effect on community structure.

关键物种：关键物种在群落中有一个重要的和不成比例的作用。关键种可能是顶端的捕食者，如北方的海獭；然而，这个词可用在任何一个会对群落结构造成重大影响的物种。

4 Succession: ecological succession is defined as a continuous, unidirectional, sequential change in the species composition of a natural community. This sequence of community is termed a sere, and culminates in the climax community.

演替：生态演替是指在一个自然群落中，物种的组成连续地、单方向地、有顺序地变化。这一顺序被称为一个演替序列，最后达到的阶段称为顶级。

5 Disturbance: a disturbance is an interruption or interference that occurs sufficiently often for it to have exerted some selection pressure on the species experiencing it. Disturbance includes trees falling in a forest, grazing by herbivores, tidal action, fire, unusual climatic conditions or human activity. Plants of environment prone to disturbance such as fire have morphological and life history adaptations to ensure the survival of themselves or their offspring.

干扰：干扰是指林中倒树、食草动物的啃食、潮汐活动、火灾、反常的气候变化或人类活动等经常发生的扰乱或干涉，它们迫使物种经历某些选择压力。比如，生长在易发生火灾的环境中的植物，在形态和生活史方面有其独特的适应，以保证它们自己和后代在这样的环境中生存下去。

6 Patch dynamics: localized disturbance of a community creates gaps which can then be colonized by individuals of the same or another species. Potentially, competing species may not encounter on another because of the stochastic nature of gap colonization, allowing species-rich communities to persist. In a late successional community such as a forest, gaps undergo a mini-succession resulting in a mosaic of patches at different stages of succession.

斑块动态：群落的局部性干扰产生了许多断层，相同的或不同的物种个体就乘机迁移到这些断层上来。由于物种迁移到哪个断层是随机的，因此，相互竞争的物种可能不会在同一断层上相遇。因此，物种丰富的群落可以生存下来。在晚期演替群落中，如森林群落，断层经历了一个微型演替的过程，形成了一个由处于不同演替阶段的各类斑块组成的镶嵌体。

7 Spatial heterogeneity: spatial heterogeneity refers to the heterogeneity and complexity of ecological processes and patterns in the spatial distribution. The concept is widely used in the field of ecology, but there is also a variety of its meaning and usage. Specifically, the spatial heterogeneity is generally understood as a sum of the patchiness and the gradient.

空间异质性：是指生态学过程和格局在空间分布上的不均匀性及其复杂性。这一名词在生态学领域应用广泛，其含义和用法亦有多种。具体地讲，空间异质性一般可以理解为是空间斑块性和梯度的总和。

8 Degradative succession: degradative succession is a type of autogenic succession involving colonization and subsequent decomposition of dead organic matter. Different species invade and disappear in turn, as the degradation of the organic matter uses up some resources and makes others available. This process leads to the production of humus and is important in soil formation.

退行性演替：退行性演替是涉及建群和死亡个体分解的自发演替的一种类型。当有机物质降解耗尽了一些资源和制造了其他可以利用的资源的时候，不同的物种就交替地出现和消失。这个过程导致了腐殖质的生产，并且对土壤形成过程起着重要作用。

9 Allogenic succession: allogenic succession results from external environmental factors, such as long term climatic change (e.g. ice ages) or environmental change over a short time (e.g. sediment accretion). Changes in community structure over time are apparent from pollen analysis of sediment cores. Successional progress depends on species tolerance to environmental conditions such as salinity.

异发演替：异发演替是由外界环境因素引起的，像长期的气候变化（如冰河时期）和环境在短时间内的变化（如沉积物的增加）。群落结构随时间的变化从沉积物中心的花粉分析来看是十分明显的。演替依赖于物种对环境条件（如盐分）的忍耐性。

10 Successional processes: succession is strongly influenced by three processes.

(i) Facilitation: changes in the abiotic environment that are imposed by the developing community and allow other species to invade.

(ii) Inhibition: species of one stage resist invasion by later successional species such that invasion is only possible following disturbance or death.

(iii) Tolerance: late successional species invade because they are able to tolerate lower resource levels and can outcompete early successional species. Highly competitive species, which are tolerant of low resource levels will replace opportunistic good colonizers and come to dominate the climax community.

演替过程：演替受三个过程的强烈影响。

(1) 促进：正在发展的群落所产生的非生物环境变化，允许其他物种的入侵。

(2) 抑制：一个阶段的物种抵制后来演替的物种的入侵，以至于入侵仅仅在干扰或死亡后才有可能。

(3) 耐受：后期演替的物种的入侵，因为它仍能够忍受较低的资源水平和排斥早期演替的物种。能忍受低资源水平的强竞争物种，将统治顶级群。

11 Fluctuations and the climax community: There is no single climax community for a geographic area, but a continuum of climax types varying along environmental gradients. Climax communities are not stable, but are in a state of continual flux. Unidirectional succession to the climax does not always happen as succession can be arrested at an earlier stage. In general, biomass and species diversity increase with succession but often peak at an intermediate stage and not at the climax community.

波动和顶级群落：作为一个地理区域，没有一个单一的顶级群落，但都有一个连续的顶级类型系列沿着环境梯度变化。顶级群落不稳定，它处于一个连续涨落之中。当演替在早期阶段被阻的时候，朝向顶级的单项演替并不总发生。总之，生物量和物种多样性随着演替的进行而增加，但常常在演替中期阶段达到最大，而不是在顶极群落时。

12 Two species, A and B

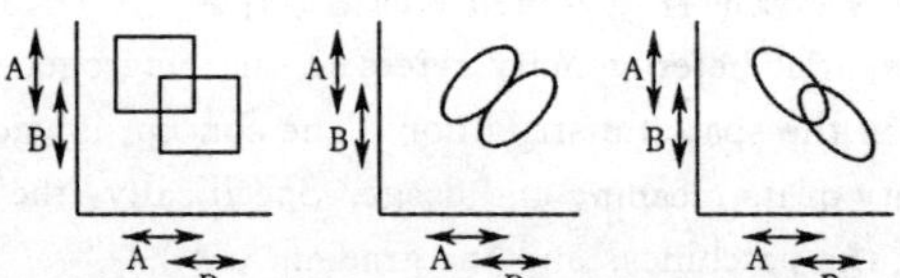

(a) (b) (c)

(a) Niches are independent.
(b) and (c) Niches are partially dependent.
两个物种,A 和 B
(a) 各自独立
(b) 和(c) 生态位部分重叠

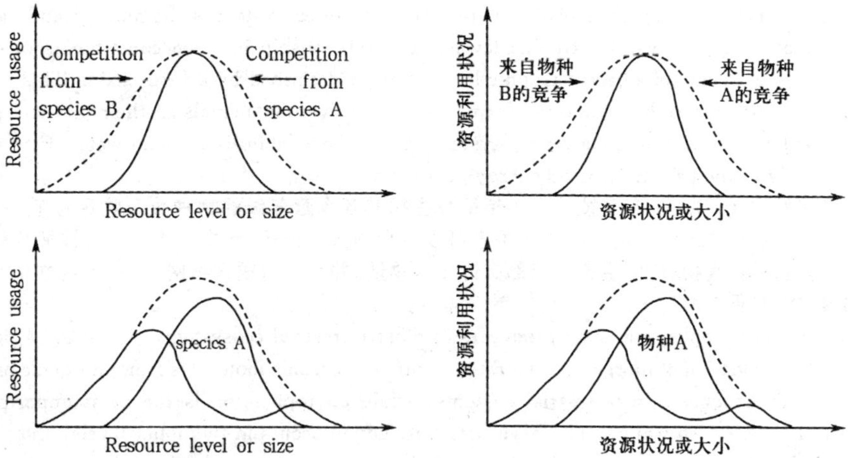

13 Independent vs. inter-dependent niches
独立生态位和相互依赖的生态位

第五节　生态系统
Section 5　Ecosystem

生态系统　ecosystem
初级消费者　primary consumer
次级消费者　secondary consumer
草食动物　herbivore
肉食动物　carnivore
杂食动物　omnivore
分解者　decomposer
真菌　fungus
蘑菇　mushroom
细菌　bacteria
菌丝体　mycelia
酶　enzyme
食物链　food chain
捕食动物　predator
猎物　prey
寄生生物　parasite
寄主　host
碎屑　detritus
食碎屑生物　detritus feeder
腐生植物　saprophyte
食物网　food web
数量金字塔　pyramid of number
生物量　biomass
第一性生产　primary production
初级生产力　primary productivity
能量流动　energy flow
物质循环　nutrient cycle
气体循环　gaseous cycle
沉积循环　sedimentary cycle
生物地球化学循环　biogeochemical cycle
微量元素　micronutrient
大量元素　macronutrient
行为信息　behavior signal
化学信息　chemical signal
营养信息　trophic signal
物理信息　physical signal
自然生态系统　natural ecosystem
人工生态系统　artificial ecosystem
进化　evolution
演替　succession
反馈　feedback

1 Ecosystem: the ecosystem concept was proposed by Tansley in 1935, and was originally defined to include all the animals, plants and physical interactions of a defined space. Modern ecologists tend to think of ecosystems in terms of energy flow, carbon flow or nutrient cycles.

生态系统:1935 年 Tansley 提出,最初定义为在一定空间中所有的动物、植物和物理因子的相互作用构成生态系统。近代生态学家更倾向于能量流、碳流或营养物流来考虑生态系统。

2 Interactions between trophic levels: a characteristic of an ecosystem is the number and nature of the species that occupy its various trophic levels. The relationship between constituents of one trophic level and constituents of adjacent trophic levels may be described by a food chain. This is a chain of eating and being eaten that connects, for example, carnivorous animals to their ultimate plant food. Many food chains exist in any given ecosystem and can be combined into food webs. Ecosystems vary considerably in the pattern of their energy-nutrient webs.

营养级间的相互作用:生态系统的一个特征是占据其各个营养级的物种数及物种性质。一个营养级的成员与邻近营养级的成员之间的相关系可以用食物链来描述。食物链是捕食与被捕食的链,例如,食肉动物到植物食物。任何生态系统都有许多食物链,并可以组成食物网。生态系统在能量 - 营养物网的模式上有很大的变化。

2 Producers: autotrophs, the main producers in most terrestrial ecosystems, use energy from the sun or from the oxidation of inorganic substrates to produce organic molecules from inorganic ones. Plants are the main producers in terrestrial systems, while phytoplankton is the most important in open oceans. In these deep waters, photosynthetic production occurs in the upper levels, the photic zone, where light can penetrate. In deeper waters, light does not penetrate and no photosynthesis occurs in this aphotic zone.

生产者:大多数陆地生态系统的主要生产者是自养生物,它们利用太阳能或无机物质氧化作用的能量,使无机分子成为有机分子。在陆地系统中植物是主要的生产者,而浮游植物在开放海洋中最重要。在深水域中,光合产物只在上层部分(透光区)产生,因为有光的透入。在更深水域中,光不能透入,故这无光区没有光合作用发生。

3 Consumers: plants are eaten by primary consumers, such as grazing mammals and insects. In oceans, phytoplankton is consumed by zooplankton. These primary consumers are in turn eaten by secondary consumers, carnivores such as mammals and spiders. In oceans, many fish are secondary consumers as they feed on zooplankton.

消费者:植物被初级消费者所食,如草食哺乳动物和昆虫。海洋里,浮游动物以浮游植物为食。这些初级消费者依次又被次级消费者 —— 肉食动物所食,如哺乳动物和蜘蛛。在海洋中,很多鱼类以浮游动物为食成为次级消费者。

4 Decomposers: the organic material that composes living organisms in an ecosystem is eventually recycled, broken down and returned to the abiotic environment in forms that can be used by plants. Decomposers, which feed on nonliving organic material, are the organisms in this recycling. The most important decomposers are bacteria and fungi.

分解者:在生态系统中组成生命有机体的有机物质,最终经分解以可被植物利用的形式回到非生物环境中被重新利用。分解者以无生命的有机物质为食,是这个再循环中的有机体。最重要的分解者是细菌和真菌。

第六节 生态系统中的能量流动
Section 6 Energy Flow in the Ecosystem

动能 kinetic energy
势能 potential energy
热力学第一定律 the first law of thermo-dynamics
热力学第二定律 the second law of thermo-dynamics
生态效率 ecological efficiency
转化效率 transfer efficiency

同化效率	assimilation efficiency	消费效率	consumption efficiency
生长效率	growth efficiency	太阳辐射	solar radiation

1 Ecosystems and the laws of thermodynamics: the first law of thermodynamics states that energy can neither be created nor destroyed. The second law of thermodynamics states that every transformation results in a reduction of the free energy of the system. Because energy transformation cannot be 100% efficient (form the second law), heterotrophs must have less energy, and must therefore be rarer than the plants they feed on. The complexity of ecological interaction means that it is not possible to construct predictive mathematical models of living systems based on these laws of thermodynamics.

生态系统与热力学定律:热力学第一定律指出,能量既不能被创造,也不能被破坏。热力学第二定律指出,每一次转换都导致系统的自由能的减少。因为能量转换不可能 100% 有效,因此,异养生物能量必然比所食的植物少。生态相互作用的复杂性意味着以热力学定律为基础去构建用于预测的数学模型几乎不可能。

2 Solar energy drives climatic processes; energy from the sun strikes the earth where high-energy wavelengths of light are absorbed and re-emitted in the form of radiant heat, warmed air rises, and expands as it does so. This expansion takes energy from the air, resulting in temperature reduction, a process known as adiabatic cooling.

太阳能使地球上的气候产生变化。太阳能到达地球上时,高能量的光被地球吸收,并以热辐射的形式再释放出来,被加热的空气上升并膨胀。此膨胀吸收了空气中的能量,导致空气温度下降,这个过程称为绝热冷却。

3 Primary production: global net primary productivity is approximately 120×10^9 tons dry weight per year on land, and 50×10^9 tons per year in the sea. This productivity is very unevenly distributed across the earth. The most productive systems are found amongst swamp and marshland, estuaries, reefs and cultivated land. Productivity decreases moving away from the equator, indicating the importance of temperature and radiation.

初级生产力:全球陆地净初级生产力大约为年 120×10^9 吨干物质,而海洋大约为年 50×10^9 吨干物质。生产力在地球上分布很不均匀。最富有生产力的系统出现在沼泽、湿地、河口湾、珊瑚礁和耕田。生产力随离赤道距离增大而降低,它表明温度与辐射的重要性。

4 Relationship of productivity to biomass: community productivity can be related to the standing crop that produces it by comparing the ratio of productivity (P) to standing crop biomass (B). The resulting P: B ratios (i. e. kg produced $year^{-1}$ standing crop) average 0.042 for forests, 0.29 for other terrestrial systems, and 17.0 for aquatic communities. An alternative way of looking at P: B ratios would reduce these large differences between communities. However, accurate measurement of the proportion of biomass alive is difficult.

生产力对生物量的关系:通过生产力(P)与现存生物量(B)之比,可以把群落生产力与产生它的现存量联系起来。对于森林,所得到的 P:B 比率(即每年每公斤现存量所生产出的公斤数)平均为 0.042,其他陆地系统是 0.29,而水体群落是 17.0。考虑 P:B 比率的另一种选择时以活组织的质量确定的生物量,它将缩小群落之间的这些区别。当然,精确测定活的生物量比例很困难。

5 Secondary production: secondary production is defined as the rate of production of new biomass by heterotrophic organisms. Heterotrophs are organisms, such as animals and fungi, with a requirement for energy-rich organic molecules. Secondary production by heterotrophs is inevitably dependent on primary productivity.

次级生产力:次级生产力是异养生物生产新生物量的速率。异养生物如动物、真菌是需要高能量有机分子的生物。异养生物的次级生产力必然地依存于初级生产力。

6 Transfer efficiencies: the proportion of net primary production that flows through trophic levels depends on transfer efficiencies in the way energy is used and passed from one step to the next.

Knowledge of just three categories of transfer efficiency is all that is required to understand the pattern of energy flow. These are consumption efficiency (CE), assimilation efficiency (AE) and production efficiency (PE).

转换效率：净初级生产力通过营养级流动的比例取决于能量从一级到下一级的利用和通过途中的转化效率。理解能量流动格局只需要掌握以下三类转换效率：消费效率(CE)、同化效率(AE)和生产效率(PE)。

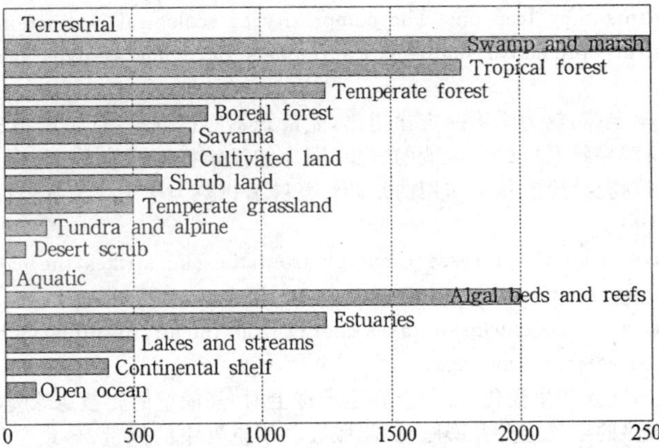

The distribution of the primary production on the earth

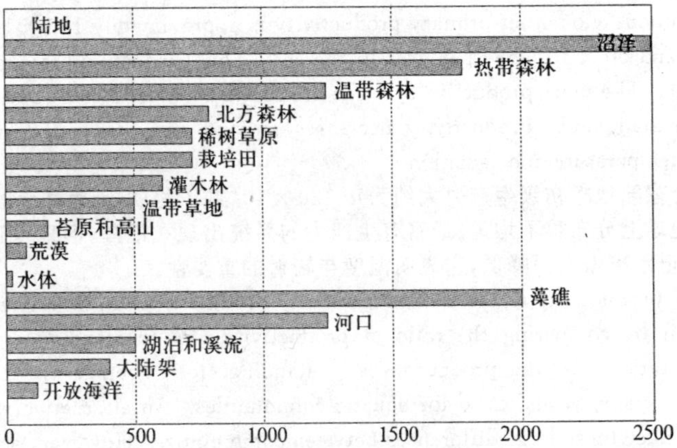

地球上初级生产力的分布

The relationship between matter and energy: once energy is transformed into heat, it can no longer be used by living organisms to do work or to fuel the synthesis of biomass. The heat is lost to the atmosphere and can never be recycled. Life on earth is possible because a fresh supply of solar energy is made available every day. In contrast, nutrients such as carbon can be used. Chemical nutrients, the building blocks of biomass, can be used again, and recycling is a critical feature. Unlike the energy in solar radiation, nutrients are not in unalterable supply. If plants and their consumers were not eventually decomposed, the supply of nutrients would become exhausted and life on earth would cease.

物质与能量的关系：能量一旦转化为热，它就不能再被有机体用于做功或合成生物量。热损失到大气中不能进入再循环。地球上生命的存在是由于太阳辐射每天提供可用能量。相反，像碳一样的营养物质能被再利用。构成生物量基本单位的化学营养物也可以被再利用，这是再循环的特征。与太阳辐射能不同，营养物的供应是可变的。如果植物、消费者没有被最终分解掉，营养物将会耗尽，地球上的生命也将终止。

第七节 生态系统的物质循环

Section 7 Cycles of Material in the Ecosystem

大量营养元素 macronutrient
贮存库 reservoir pool
交换库 exchange pool
流通率 flow rate
周转率 turnover rate
水循环 water cycle
气体型循环 gaseous cycle
沉积型循环 sedimentary cycle
生物泵 biological pump
固氮作用 nitrogen fixation
氨化作用 ammonification
硝化作用 nitrification
反硝化作用 denitrification
磷循环 phosphorus cycle

1 Carbon cycle: the carbon cycle in the biosphere mainly in green plants absorbs carbon dioxide from the air, transforms into glucose through photosynthesis and release oxygen. Photosynthesis and respiration are the two opposing processes that drive the global carbon cycle, with CO_2 as the main vehicle of flux between atmosphere hydrosphere and biota.

碳循环:生物圈中的碳循环主要是绿色植物从空气中吸收二氧化碳,经光合作用转化为葡萄糖,并放出氧气。光合作用和呼吸作用是驱动全球碳循环的两个相反过程,CO_2 是碳在大气圈、水圈和生物群之间流动的主要贮存形式。

2 Nitrogen deposition: nitrogen deposition is the process of atmosphere nitrogen compounds (including natural source and human activities source) through the biological way into the ecological system.

氮沉降:氮沉降是指大气中的含氮化合物(包括自然和人类活动来源)通过生物途径进入生态系统的过程。

3 Nutrient budgets in terrestrial communities: a particular nutrient taken up by an organism may cycle continuously until eventually the nutrient will be lost through any one of a number of processes that remove nutrients from the system. Release may be direct to the atmosphere or via bacterial action (as in the case of methane). For many elements, the most substantial pathway of loss is in streamflow. Other pathways of nutrient loss include fire, the harvesting of crops and deforestation.

陆地群落营养物预算:生物体吸收的特定营养物可连续地循环,直到营养物最终通过许多过程中的任何一个都可从系统中去除。营养物可直接释放到空气或通过细菌作用完成(如甲烷)。对于许多元素来说,最重要的流动途径是水流。营养物损失的其他途径包括火灾、庄稼的收获和森林砍伐。

4 Nutrient budgets in aquatic communities: aquatic systems obtain the bulk of their supply of nutrients from streamflow. In streams, rivers and lakes with a stream outflow, export in outgoing stream water is a major factor. Commonly, phases of fast inorganic nutrient displacement alternate with periods when the nutrient is locked in biomass. In lakes, plankton plays a key role in nutrient cycling. Oceans contain warm surface waters, where most plant life is found and cold deep waters (which make up 90% of the total water volume). Nutrients in the surface waters come from two sources: (i) upwelling from deep water (which comprise over 95% of the nutrient budget), and (ii) river input.

水生群落营养物质预算:水生系统从水流中得到它们所需的大量营养物质。在具有出口的溪流、河流和湖泊中,水流出口是个重要因素。通常,无机营养物质转移与营养物质在生物量中的固定交替进行。浮游生物在湖泊中的营养物质循环中起关键作用。海洋含有温暖的表层水,生活着大多数的植物,还有冷的深层水(占总水容量的 90%)。表层水的营养物来源于两方面:(i) 深层水的上涌流(组成营养物预算的 95% 以上),(ii) 河流的输入。

5 Global biogeochemical cycles: terrestrial plants utilize atmospheric CO_2 as their carbon source for photosynthesis, whereas aquatic plants use dissolved carbonates (hydrosphere carbon). Respiration releases the carbon locked in photosynthetic products back to the atmospheric and hydrospheric carbon compartments. The atmospheric phase is predominant in the global nitrogen cycle, in which nitrogen fixation and denitrification by microorganisms are of particular importance. The main stocks of phosphorus occur in soil water, rivers, lakes, rocks and ocean sediments, whilst sulfur has both atmospheric and lithospheric components.

全球生物地球化学循环：陆地植物利用空气中CO_2作为光合作用的碳源，而水生植物使用溶解的碳酸化合物（水圈的碳）。呼吸作用把固定在光合产物中的碳，再释放到气圈和水圈的碳圈层中。在全球氮循环中，气相是占优势的，其中，氮的固定和微生物的脱氮作用尤其重要。磷主要储存在土壤水、河流、湖泊、岩石和海洋沉淀物中，而硫储存在大气和岩石的组分中。

第八节　生态系统的信息流
Section 8　Information Flow in the Ecosystem

生物信息学　bioinformatics
信息传递　information transfer
信源　source
信道　channel
接收器　receiver
报警信息素　alarm pheromone
聚集信息素　aggregation pheromone
再生　regeneration
信息素　pheromone
异种化感物　allelochemics
追踪信息素　trail pheromone

第九节　陆地生态系统
Section 9　Terrestrial Ecosystem

热带雨林　tropical rain forest
热带落叶林　tropical deciduous forest
热带旱生林　tropical thorn forest
温带草原　temperate steppe
温带落叶林　temperate deciduous forest
北方针叶林　boreal coniferous forest
冻原　tundra
乔木层　tree stratum
灌木层　fruticose stratum
草本层　herb layer
地被层　field stratum/ground stratum
生物圈　biosphere
森林生态系统　forest ecological system
调节气候　regulate weather
涵养水源　conserve water
沙漠化　desertification

1 Grassland: grasslands occur where rainfall is intermediate between that of deserts and forests. There are two major types of grasslands depending on the temperature: tropical grassland, and temperate grassland.

草原：草原存在于降雨量介于荒漠和森林之间的区域。依据温度的不同，草原有两种主要类型：热带草原和温带草原。

2 Tundra: the Arctic tundra forms a circumpolar band between the Arctic Ocean and the polar ice caps to the north and the coniferous forests to the south.

冻原：北极冻原在北部的北冰洋和极地冰盖与南部的针叶树森林之间，形成了一个环绕极地带。

3 Forest: forest type is dependent on rainfall and temperature (latitude or altitude). Coniferous boreal forest is characteristic of cold climates and high elevation. Temperate forests occur at lower latitudes where there is sufficient rainfall, giving way to tropical rain forest toward the equator.

森林：森林的类型依赖于降雨和温度（纬度和高度）。北方针叶森林是寒冷气候与高海拔的象征。温带森林出现在低纬度地区，近赤道有充足降雨的地方为热带雨林。

4 Vegetation and climate: the vegetation of the earth is distinct blocks or formations which reflect climatic conditions; this pattern arises as a result of the adaptation of form to temperature and water availability.

植被与气候：地球植被被明显地分为许多区域或形式，这些区域或形式在很大程度上反映了气候条件。这种分布是植物对温度和水分适应的结果。

5 Climate and soils: boreal forests have long, cold winters with light precipitation in the winter and

more rain in the summer. The climate of temperate forests is also seasonal, with higher rainfall. Soils are well developed and rich. In the tropics, the forest is non-seasonal and warm with frequent heavy rainfall. Soils are leached, acid and nutrient poor.

气候与土壤：北方针叶林具有漫长而寒冷的冬季，同时冬季降水少，夏季降雨多。温带森林的气候是季节性的，具有很高的降雨量。土壤发育好并且肥沃。热带森林气候是非季节性且温暖的，降雨量大且降雨频度高。土壤是淋溶的、酸性的和养分贫瘠的。

6 Forest animals: temperate and boreal forests support herbivorous mammals and predatory species such as wolves. Forests are important habitats for birds and, in temperate forests, small mammals which inhabit the dense understory. Forest ecosystems also contain a wide range of specialist and generalist insect herbivores. Tropical rain forests support tremendous animal diversity, particularly of insects, amphibians, reptiles, birds and small mammals. Many species ate tree dwelling, feeding on fruit and/or seeds.

森林动物：温带和北方森林养育了食草的哺乳动物和像狼一样的食肉种类。森林对鸟类来说是重要的栖息地，而且在温带森林里，小型的哺乳动物居住在稠密的森林下层。森林生态系统也含有范围很广的专食性和泛食性的食草昆虫。热带雨林具有巨大的动物多样性，尤其是昆虫、两栖类、爬行类、鸟类和小型的哺乳动物。许多物种居住在树上，以吃果实和/或种子为生。

第十节 城市生态系统
Section 10 Urban Ecosystem

城市化 urbanization
大城市连绵区 megalopolitan region
社会—经济—自然复合体 social-economic-natural complex
同心圆结构 concentric zone structure
扇形结构 sector structure
多核心式结构 multiple nuclei structure

1 The aim of pest control: although in some instances the aim of pest control is to eradicate the pest species, generally the aim of pest control is to reduce the pest population to a level at which no further reductions are profitable. This is known as the economic injury level (EIL) for the pest, or if social and amenity benefits are included, the aesthetic injury level (AIL).

有害生物防治目标：尽管在有些情况下，有害生物防治的目标是根除有害物种，然而通常有害生物防治目的是降低有害生物种群到一个水平上，在这个水平上进一步降低是无利的。这被称为有害生物的经济损害水平，如果考虑到社会的和环境的舒适利益，也可以称为美学损害水平。

2 In the case of disease, total eradication can be justified on the basis that the saving of one single life far exceeds any economic costs. In practical pest control, the EIL is not as important as the control action threshold (CAT) — the pest density at which action should be taken in order to prevent an impending pest outbreak.

就疾病而言，根除全部有害生物被认为是有理的，这是基于救活一个人远超过任何经济价值。在有害生物的防治实践中，EIL 不像防治活动阈值(CAT) 那样重要，在经济活动阈值这个有害生物密度值上，其作用应该被理解是阻止即将来临的有害生物的爆发。

第十一节 水域生态系统
Section 11 Water Ecosystem

湿地 wetland
外洋流域 exorheism region
盐沼 salt marsh
淡水 freshwater
静水 standing water
流水 running water
海洋生态系统 marine ecosystem
赤潮 red tide
海滨区 seashore region
沿岸区 coastal region

1 Water and temperature: two features of water make it particularly suitable as a medium for life-forms to inhabit. Water has (i) a high heat capacity and (ii) maximal density at 4℃. A high heat capacity means that it can absorb heat energy with only a small increase in temperature. As a result, aquatic life-forms are buffered from temperature fluctuations. Water becomes increasingly dense and therefore heavier at lower temperatures, with the maximum density occurring at 4℃. Thus, ice floats on water and a body of water freezes form the top down. This phenomenon protects aquatic organisms because ice acts as insulator to prevent further decreases in temperature in the underlying water.

水和温度：水的两个特性使它特别适合于作为许多生物栖息的一种介质。水有高的热容量，4℃时水的密度最大。高热容量意味着它能吸收热能，而只增加很少的温度。其结果是减少了温度波动对水生生物的影响。在较低温度时，水的密度增加，因而也更重，最大的密度在4℃。因此，冰浮在水上，水体从上向下冻结。冰作为一种绝热体，阻止下层水的温度进一步降低，这种现象保护了水生生物。

2 Saltwater regions: the saltwater biomes are open oceans, continental shelves, the intertidal zone and coral reefs, salt marsh, mudflats and mangroves. Oceans cover 70% of the world's surface. Physical factors, namely tides, currents, temperature, pressure (depth) and high intensity, determine the makeup of biological salt water communities.

盐水区域：盐水生物群系包括开阔的海洋、大陆架、潮间带和珊瑚礁、盐水沼泽、潮泥滩和红树林。海洋覆盖了70%的地球表面。物理因素即潮汐、环流、温度、压力(深处的)和光强度决定着生物盐水群落的组成。

3 Freshwater and wetland regions: freshwater biomes include lakes, rivers, bogs, marshes and swamps. These systems are fed by water and nutrients leaching from the surrounding catchments area.

淡水和湿地区域：淡水生物群系包括湖泊、河流、湿地、草本沼泽和木本沼泽。该系统靠附近排水区沥滤的水和营养物生存。

4 Fisheries: the seas have historically provided a valuable food source to human populations. The current global fish catch is about 100 million tons, although a substantial minority of this is not used for direct human consumption, but as animal feed source.

渔业：海洋一直为我们人类提供丰富的食物资源，现在全球渔获量约1亿吨，其中一小部分并不直接被人类消费，而是用做动物饲料。

5 Ocean productivity: the open oceans are unproductive habitats compared to terrestrial systems. Productivity is higher on the continental shelves, but the overall global fish catch represents a very low yield per hectare. This is partly because of the very low productivity of most of the ocean area, and partly because the harvest consists on animals high up the food chain, which have low production efficiencies.

海洋生产力：与陆地系统相比，海洋是生产力低的生境。大陆架生产力较高，但全球单位公顷渔获量很低，部分是因为海洋区极低的生产力，部分是由于渔获物处于食物链高营养级，而生产效率低。

第十二节 景观生态学

Section 12 Landscape Ecology

景观生态学 landscape ecology
生态规划 ecological planning
多样性指数 diversity index
镶嵌度指数 patchiness index
距离指数 distance index
生境破碎化指数 fragmentation index
尺度 scale
景观异质性 heterogeneity
景观格局 landscape pattern
斑块 patch
廊道 corridor
基质 matrix
景观 landscape
景观生态学 landscape ecology
空间自相关 spatial autocorrelation
地统计学 goestatistics

谱分析 spectral analysis
聚块样方方差分析 blocked quadrat variance analysis
分形几何学 fractal geometry
小波分析 wavelet analysis
细胞自动机 cellular automata
趋面分析 trend surface analysis
亲和度分析 affinity analysis
生态交错带 ecotone
干扰 disturbance
生物区系 biotopes
景观破碎化 landscape fragmentation
岛屿生物地理学 island biogeography
异质种群理论 metapopulation theory
大陆和岛屿型 mainland and island
斑块型种群 patchy population
卫星型种群 satellite population
非平衡型 non-equilibrium
生态恢复 ecological restoration
等级系统理论 hierarchy theory
景观要素 landscape element
尺度推绎 scaling
尺度上推 scaling up
尺度下推 scaling down
粗粒化 coarse-graining
细粒化 fine- graining
外推 extrapolation
空间内插值 spatial interpolation
简单聚合法 lumping
直接外推法 direct extrapolation
期望值外推法 extrapolation by expected value
复合种群 metapopulation
亚种群 subpopulation
局部种群 localpopulation
景观尺度 landscape scale
景观连接度 landscape connectivity
渗透理论 percolation theory
临界阈现象 critical threshold characteristic
中性模型 neutral model
等级理论 hierarchy theory
整体元 holon
范式 paradigm
范式变迁 paradigm shift
空间异质性 spatial heterogeneity
空间尺度 spatial scale
时间尺度 temporal scale
地理区域尺度 geographical region scale
组织尺度 organizational scale
生态流 ecological flow
流动镶嵌体 shifting mosaic
干扰斑块 disturbance patch
环境资源斑块 environmental patch
引进斑块 introduced patch
种植斑块 planted patch
固有边缘 inherent edge
诱导边缘 induced edge
边缘种 edge species
内部种 interior species
弯曲度 curvilinearity
连通性 connectivity
狭点 narrow
节点 node
孔隙度 porosity
生态风险 ecological risk
生态风险评价 ecological risk assessment
区域生态风险 regional ecological risk
生态风险指数 ecological risk index
景观生态规划 landscape ecological planning
城市生态规划 urban ecological planning
生态环境规划 eco-environment planning
水域生态规划 waterfront ecological plan
生态建设规划 ecological construction planning
遥感 remote sensing
遥感图像 Remote-sensing image
地理信息系统 geographical information system (GIS)

1 Landscape ecology is the science of studying and improving relationships between ecological processes in the environment and particular ecosystems. This is done within a variety of landscape scales, development spatial patterns, and organizational levels of research and policy.
景观生态学是研究与改善环境与特定生态系统中各种生态过程之间相互关系的科学，主要致力于开展各种各样的景观尺度、空间格局发展和组织水平上的研究和政策制定。

2 Ecological planning is the process of understanding, evaluating, and providing options for the use of landscape to ensure a better fit with economic development and human habitation.
生态规划是了解、评价并提供景观利用的选择方案以确保其更适合经济发展与人类居住。

第十三节 环境保护与可持续发展

Section 13 Environmental Protection and Sustainable Development

全球变化 global change
全球变暖 global warming
土地覆盖 land cover
土地利用 land use
生物地球化学循环 biogeochemical cycle
国际地圈生物圈计划 International Geosphere-Biosphere Program (IGBP)
生物多样性 biodiversity
生物群落 biome
脆弱性 vulnerability
适应性 adaptability
生态过渡区 ecotone
荒漠化 desertification
变异性 variability
多样性 diversity
生态系统多样性 ecosystem diversity
景观多样性 landscape diversity
丰富度 richness
均匀度 evenness
国际自然保护协会 The Nature Conservancy (TNC)
国际联盟 International Conservation Union (IUCN)
生态恢复 ecological restoration
恢复生态学 restoration ecology
位移 displacement
受损生态系统 damaged ecosystem
恢复 restoration
生态工程 ecological engineering
自生原理 self-resiliency
自我组织 self-organization
自我优化 self-optimum
自我调节 self-regulation
可持续发展 sustainable development
温室效应 greenhouse effect
温室气体 greenhouse gases
臭氧洞 ozone depletion
酸雨 acid rain
水土流失 soil erosion
富营养化 eutrophication
环境保护 environmental protection
环境污染 environmental pollution
自然资源 natural resources
可更新的 renewable
非更新的 non-renewable
可回收的 recyclable
生态资源 ecological resources
生物能 life energy
水土保持 water and soil conservation

例1 Greenhouse effect and global warming: The 'greenhouse affect' is a theory which proposes that pollution by common anthropogenic (i. e. created by humans) pollutants such as CO_2 and methane may lead to an increased global temperature. Over the last century, CO_2 concentrations have risen and global air temperatures have increased by 0.4-0.7℃, which supports this theory. The atmosphere formed around the earth insulates the planet from the full effects of heat loss by trapping heat in the atmosphere using greenhouse gases, which include water vapor as well as CO_2 and other anthropogenic pollutants. Doubling of the atmospheric CO_2 concentration from its present level is predicted to lead to a further warming of around 3.5℃.

Global warming is the earliest awareness of climate change. Global warming is the value of the surface earth's atmosphere, the soil, the water and the vegetation temperature rising slowly between years. Global warming may cause organisms to migrate in order to seek optimal temperatures (organism tracking), or to adapt to the changed conditions or face extinction.

温室效应和全球变暖："温室效应"是这样一种理论，它假设二氧化碳和甲烷造成的污染会导致全球气温的上升。在过去一个世纪中，二氧化碳浓度升高，而全球气温也上升了0.4～0.7℃。这个事实支持了上述理论。因为大气中的温室气体(包括水蒸气、二氧化碳和其他人为污染物)能够吸收热量，所以围绕地球的大气层防止了地球热量的全部丧失。如果将大气中的二氧化碳浓度在已有水平上翻倍，预计气温将上升3.5℃左右。

全球变暖最早引起人们对气候变化的广泛关注。全球变暖是地球表层大气、土壤、水体及植被温度年际间缓慢上升。全球变暖会引起生物的迁移，这种迁移或者是为寻求适宜的温度，或者是为适应变化的环境，或者是面临灭绝的反应。

2 Biodiversity: biodiversity is a term encompassing the variety of organisms at all levels, from genetic variants belonging to the same species, to species diversity and including the variety of ecosystems.
生物多样性：生物多样性是包含生物在所有水平上的多样性的术语，从属于同一物种的遗传变型，到物种多样性和生态系统变异。

3 Sustainable development: sustainable development is not only mecting the contemporary people's needs, but also satisfying the future generations' reguirements without harm. Sustainable development is a correct concept of development; it contains the fairness of the sustainable development principle, sustainable principle, and common principle.
可持续发展：可持续发展是既满足当代人的需要，又不对后代满足其需要的能力构成危害的发展。这个定义里包含了可持续发展的公平性原则、持续性原则、共同性原则，它是一种正确的发展观。

4 Restoration ecology: restoration ecology is a sub-discipline of ecology. Restoration ecology is application of ecological principles and methods to restore the community or ecosystem structure function changes caused by human disturbance.
恢复生态学：恢复生态学是应用生态学原理、方法，对人为干扰引起的群落或生态系统的结构和功能的改变进行恢复的科学，它是生态学的一门分支学科。恢复生态学总是和生态恢复联系在一起，生态恢复作为一种研究方法和手段，同时又是一种目的，恢复实践为恢复生态学提供了发展理论的基础，反过来，恢复生态学又为生态恢复提供理论基础。

5 Ecological restoration: ecological restoration is mainly for damage which on ecological system structure, function and relationship. Therefore, ecological restoration is to restore the ecological system's reasonable structure, function and coordination.
生态恢复：生态恢复是针对受损而言的，受损就是生态系统结构、功能和关系的破坏。因而，生态恢复就是恢复生态系统合理的结构、高效的功能和协调的关系。

6 Ecoengineering: Ecological engineering is designed to promote the production of material layer multistage process system, that combined the analysis of the system optimization method with the ecological system's principles, which includes species coexistence and species recycling principle, structure and function coordination principle.
生态工程：生态工程是应用生态系统中物种共生与物种循环再生原理、结构与功能协调原则，结合系统分析的最优化方法，设计的促进分层且多级利用物质的生产工艺系统。

7 Desertification: the process of improper management of arid or semiarid land to the extent that it is no longer agriculturally productive is desertification. High stock densities may over-consume the vegetation and expose fragile soils. Another problem is irrigation of dry areas where evaporation is high, which results in soils accumulating salt (salinization). It is a serious and growing problem that results from climate and human activities. In addition, desertification may occur naturally. Approximately 30% of the earth's land suffers from some degree of desertification. Actual global losses to desertification are estimated to be around six million hectares annually.
荒漠化：干旱地区和半干旱土地因管理不当而使其不再适于农业生产，这个过程称为荒漠化。牲畜密度过高会过度消耗植被，暴露出脆弱的土壤。另外，干旱地区蒸发量大，灌溉会引起土壤盐分的积累即盐渍化。这是气候和人类活动造成的一个日益严重的问题。此外，荒漠化可能自然发生。全球约有30% 的土地受到不同程度的荒漠化。全世界每年因荒漠化损失的土壤约六百万公顷。

第十四节　应用生态学
Section 14　Applied Ecology

石油污染　oil pollution
放射污染　radioactive pollution
农药污染　pesticide pollution
水体污染　water pollution

水体自净 self-purification of water body
污水处理 sewage treatment
噪声 noise trouble
空气污染 air pollution
氟化物 fluoride
沉淀 sedimentation
稀释 dilution
混合 mixture
水质标准 water quality standard
活性污泥法 activated membrane method
生物膜法 biological membrane method
废水厌氧生物处理法 anaerobic treatment of sewage
生物膜 biomembrane
本底值 background level
无机污染物 inorganic pollutant
有机污染物 organic pollutant
土壤净化 soil purification
降解作用 degradation function
固体废物 solid waste
堆肥 compost
煤灰渣 coal ash
焚烧炉 incinerator
放射性废物 radioactive waste
市政管理 municipal administration
生物修复 bioremediation
植物吸收 phytoextraction
植物挥发 phytovalatilization
植物固定 phytostabilization
植物过滤 phytofiltration
生物氧化塘 oxidation pond
替代技术 alternative technology
环境容量 environmental capacity
不枯竭的 inexhaustible

1 Photochemical smog: an atmospheric haze that occurs above industrial sites and urban areas resulting from reactions, which take place in the presence of sunlight, between pollutants produced in high temperature and pressurized combustion processes(such as the combustion of fuel in a motor vehicle). The primary component of smog is ozone.

光化学烟雾：发生在工业区或城区上空的大气薄雾，通常由高温、高压燃烧产生的污染物在强阳光辐射的作用下反应生成(比如汽车尾气)。光化学烟雾的主要成分是臭氧。

2 Environmental assessment: environmental assessment is the means by which the conservation value of a site may be assessed without detailed and time-consuming surveys of its entire biodiversity.

环境评估：环境评估是在没有进行某地整个生物多样性的详细和耗时调查以前，确定该地保护价值的一种方法。

第十三章　生物化学
Chapter 13　Biochemistry

第一节　绪论
Section 1　Introduction

大分子　macromolecule
进化　evolution
多细胞的　multicultural
前景　perspective
最新的　updated
说明　elucidation

第二节　糖类
Section 2　Carbohydrates

糖　carbohydrate/saccharide/sugar
单糖　monosaccharide
葡萄糖　glucose
果糖　fructose
醛糖　aldose
酮糖　ketose
丁糖　tetrose
戊糖　pentose
己糖　hexose
庚糖　heptose
斐林试剂　Fehling's reagent
吐伦试剂　Tollen's reagent
甘油醛　glyceraldehyde
旋光度　optical rotation
比旋光度　specific rotation
变旋　mutarotation
哈沃斯投影式　Haworth projection
差向异构体　epimer
糖苷　glycoside
二糖　disaccharide
低聚糖,寡糖　oligosaccharide
多糖,聚糖　polysaccharide
麦芽糖　maltose
乳糖　lactose
纤维二糖　cellobiose
同多糖　homopolysaccharide
杂多糖　heteropolysaccharide
淀粉　starch
淀粉酶　amylase
支链淀　amylopectin
糖元　glycogen
复合糖　glycoconjugate

1 Glycogen: A highly branched polymer of glucose units linked by α-1,4-glucosiduc bond and α-1,6-glucosiduc bond, present in animal tissues.
糖元:结构与支链淀粉相似,主要是α-D-葡萄糖,按α(1-4)糖苷键缩合失水而成,另有一部分支链通过α(1-6)糖苷键连接,存在于动物体内。

第三节　脂类
Section 3　Lipids

脂肪　fat
脂肪酸　fatty acid
硬脂酸　stearic acid
油酸　oleic acid
单纯脂　simple lipid
油　oil
蜡　wax
复合脂　compound lipids/complex lipids
脂蛋白　lipoprotein
饱和的　saturated
不饱和的　unsaturated
皂　soap
甘油　glycerol
甘油三酸酯　triglyceride (TAG)
三酰甘油　triacylglycerol
两性分子的　amphiphilic
单分子层　monolayer
去垢剂　detergent

表面活性剂 surfactant
表面张力 surface tension
胶束 micelle
双层结构 bilayer structure
碘值 iodine number
皂化值 saponification number
鞘氨醇 sphingosine
鞘磷脂 sphingomyelin
髓磷脂鞘 myelin sheath
乙醇胺 ethanolamine
胆碱 choline
丝氨酸 serine
萜 terpene
类固醇 steroid
胆固醇 cholesterol
外周膜蛋白 peripheral membrane protein
整合膜蛋白 integral membrane protein
液态镶嵌模型 fluid-mosaic model
脂质体 liposome

1 Lipoprotein: A class of protein complexes by which lipids are transported in the blood.
脂蛋白:是由脂质和蛋白质以非共价键结合的复合体,广泛存在于血液当中。

2 Triacylglycerol:A naturally occurring ester of three fatty acids and glycerol that is the chief constituent of fats and oils.
三酰甘油:是一个甘油分子和三个脂肪酸分子形成的三酯,是脂肪和油的主要成分。

3 Liposome: A microscopic vesicles of lipid bilayer consisting of an aqueous core enclosed in one or more phospholipid layers made spontaneously or artificially.
脂质体:当两性分子如磷脂和鞘脂分散于水相时,分子的疏水尾部倾向于聚集在一起,避开水相,而亲水头部暴露在水相,形成具有双分子层结构的微小囊泡,称为脂质体。

第四节 蛋白质
Section 4 Proteins

肽 peptide
氨基酸 amino acid
抗体 antibody
疫苗 vaccine
球蛋白 globulin/globular protein
催化 catalyze
调节 regulate
保护 protect
清蛋白 albumin
色蛋白 chromoprotein
抗生素 antibiotic
纤维蛋白 fibrous protein
单纯蛋白质 simple protein
结合蛋白质 conjugated protein
蛋白质缺乏症 protein deficiency
必需的 essential
代谢过程 metabolic process
α-氨基酸 α-amino acid
甘氨酸 glycine
丙氨酸 alanine
缬氨酸 valine
亮氨酸 leucine
异亮氨酸 isoleucine
苯丙氨酸 phenylalanine
脯氨酸 proline
苏氨酸 threonine
酪氨酸 tyrosine
半胱氨酸 cysteine
蛋氨酸 methionine
赖氨酸 lysine
精氨酸 arginine
组氨酸 histidine
色氨酸 tryptophan
天冬氨酸 aspartic acid
谷氨酸 glutamic acid
天冬酰胺 asparagine
谷氨酰胺 glutamine
含氮杂环 nitrogen heterocyclic ring
两性离子 zwitterion
两性电解质 ampholyte
偶极离子 dipolar ion
解离常数 dissociation constant
丙氨酸的滴定曲线 the titration curve for alanine
甲醛滴定 formol titration
等电点 isoelectric point (pI)
完全质子化 fully protonated state
丹磺酰氯 dansyl chloride

茚三酮 ninhydrin
巯基乙醇 mercaptoethanol
过甲酸 performic acid
中和 neutralization
电泳 electrophoresis
缓冲液 buffer solution
阴极 cathode
阳极 anode
电场 electric field
静电荷 static charge
带正电荷基团 positively charged group
带负正电荷基团 negatively charged group
有助消化的 digestive
胃蛋白酶 pepsin
纸色谱 paper chromatography
矩形的 rectangular
层析柱 chromatographic column
肽键 peptide bond
天然肽 natural peptide
生物活性肽 active peptide
胰岛素 insulin
二硫键 disulfide link
C-端 the C-terminus
N-端 the N-terminus
构像 conformation
多肽的一级结构 the primary structure of peptide
Edman 降解 Edman degradation
蛋白质测序 protein sequencing
氨肽酶 amino peptidase
羧肽酶 carboxypeptidase
二级结构 secondary structure
三级结构 tertiary structure
四级结构 quarternary structure
肽键的平面化 planarity of peptide bond
酰胺平面 protein amide plane
共振相互作用 resonance interaction
二面角 dihedral angle
氨基酸残基 amino acid residue
α-螺旋 α-helix
β-折叠 β-pleated sheet
β-转角 β-bend/β-turn
无规卷曲 random coil
发夹结构 hairpin structure
反平行 antiparallel
变性 denaturation
复性 renaturation
自我装配 selfassembly
沉降 sedimentation
沉淀 precipitate
等离子点 isoionic point
构建进化树 construction of evolutionary tree
极端条件 extreme condition
紫外辐射 ultraviolet radiation
免疫球蛋白 immunoglobulin
可变区 variable region/variable domain
不变区 constant region/constant domain
多克隆抗体生产 production of polyclonal antibodies
单克隆抗体生产 production of monoclonal antibodies
酰胺羰基与 N-H 供体间的氢键 hydrogen bonding of amide carbonyl groups to N-H donors
邻位基团的立体位阻 steric crowding of neighboring groups
带电基团间的排斥与吸引 repulsion and attraction of charged groups
绿色荧光蛋白标记 green fluorescent protein as a marker
自动多肽固相合成 automated solid-phase peptide synthesis
凝胶过滤色谱 gel-filtration chromatography
离子交换色谱 ion-exchange chromatography
亲和色谱 affinity chromatography
高压液相色谱 high-pressure liquid chromatography
凝胶电泳 gel elctrophoresis
等电聚焦 isoelectric focusing
二维电泳 two-dimensional electrophoresis
超速离心 ultracentrifugation
多肽质量指纹识别 peptide mass fingerprinting
蛋白质纯化 protein purification
盐析 salting out
透析 dialysis

1 Electrophoresis: Movement of charged molecules in response to an electrical field, used to separate mixtures of charged proteins, nucleic acids, etc..
电泳:在外加直流电源的作用下,胶体微粒在分散介质里向阴极或阳极做定向移动,这种现象叫做电泳。

2 Peptide bond: An amide linkage between the α-amino group of one amino acid and the α-car-boxyl group of another by the elimination of an element of water.
肽键:一分子氨基酸的 α－羧基和一分子氨基酸的 α－氨基脱水缩合形成的酰胺键。

3 Conformation: The spatial arrangement of substituent groups with different positions in space, based on the freedom of bond rotation.
构象:由于分子中单键自由旋转以及键角有一定的柔性,具有同一结构式和同一构型的分子在空间中可有多种形态,叫构像。

4 Denaturation: Disordering of the specific native conformation of a polypeptide proteins, or nucleic acid.
变性:是指蛋白质和核酸在某些物理和化学因素作用下其特定的空间构象被改变,从而导致其理化性质的改变和生物活性的丧失。

5 Isoelectric focusing: An electrophoretic procedure for separating proteins on the basis of their isoelectric point.
等电聚焦:是指依据等电点来分离蛋白质的电泳过程。

第五节 酶
Section 5 Enzymes

催化剂 catalyst
生物催化 biocatalyst
专一性 specificity
绝对专一性 absolute specificity
立体异构专一性 stereospecificity
脱辅基酶蛋白 apoenzyme
辅助因子 cofactor
辅酶 coenzyme
辅基 prosthetic group
全酶 holoenzyme
单体酶 monomeric enzyme
寡聚酶 oligomeric enzyme
多酶体系 multienzyme complex
氧化还原酶类 oxido-reductase
转移酶类 transferase
水解酶类 hydrolase
裂合酶类 lyase
异构酶类 isomerase
连接酶类 ligase
底物 substrate
活性中心 active center
活化能 activation energy
热力学 thermodynamics
酶-底物复合物 the enzyme-substrate complex
诱导契合 induced fit
动力学的 kinetic
转换数 turnover number
酶活力 enzyme activity
比活力 specific activity
自我剪接 self-splicing
抗体酶 abzyme
酶-底物复合物 enzyme-substrate (ES) complexes
三维结构 three-dimensional structure
稳态 steady state
米氏方程 Michaelis-Menten equation
米氏常数 the Michaelis constant (*km*)
最大反应速度 the maximal rate (*Vmax*)
序列反应 sequential reaction
有序反应 ordered reaction
随机反应 random reaction
乒乓反应 ping-pong reaction
别构酶 allosteric enzyme
协同作用 concerted action
亚基 subunit
代谢途径 metabolic pathway
抑制 inhibition
不可逆抑制 irreversible inhibition
可逆抑制 reversible inhibition
竞争性抑制 competitive inhibition
非竞争性抑制 noncompetitive inhibition
反竞争性抑制 uncompetitive inhibition
最适温度 optimum temperature
最适 pH optimum pH
激活剂 activator
失活作用 inactivation
电荷中继网 charge relay network
修饰 modifying
磷酸化 phosphorylation

共价催化　covalent catalysis
酸碱催化　general acid-base catalysis
靠近　approximation
定向效应　orientation
形变　distortion
协同性　cooperativity
松弛型　relaxed state
紧张型　tensed state
同工酶　isoenzymes
激酶　kinase
固定化酶　immobilized enzyme
人工酶　artificial enzyme
酶联免疫测定　enzyme-linked immunosorbent assay (ELISA)

1 Coenzyme: Small organic molecules bound noncovalently to the enzyme, and essential for activity of enzymes.
辅酶:是一类可以将化学基团从一个酶转移到另一个酶上的有机小分子,与酶非共价地结合,对于特定酶的活性发挥是必要的。

2 Multienzyme complex: An enzyme complex assembled by several different functional enzymes.
多酶体系:在完整细胞内的某一代谢过程中,由几种不同的酶联合组成的一个结构和功能的整体,催化一组连续的密切相关的反应

3 Allosteric enzyme: An enzyme which activity is adjusted by reversible, changing its conformations to response its specific modulators.
别构酶:活性受别构调节物调控的酶。

4 Michaelis-Menten equation: A equation expressing the relationship between the reactive rate and the concentrations of substrate.
米氏方程:表示酶促动力学基本原理的数学表达式,此方程式表明了底物浓度与酶反应速度间的定量关系。

5 the Michaelis constant, KM: A constant equal to the substrate concentration when the reaction rate is half its maximal velocity.
米氏常数:其值是当酶促反应速度达到最大反应速度一半时的底物浓度。

6 Isoenzyme: A group of enzymes catalyzing the same reaction with different primary structure.
同工酶:是指生物体内催化相同反应而分子结构不同的酶。

7 Uncompetitive inhibition: A reversible inhibition of an enzyme by binding of uncompetitive inhibitors only to the enzyme-substrate complex.
反竞争性抑制:抑制剂只与酶-底物复合物结合,而不与游离酶结合的一种酶促反应抑制作用。

第六节　核酸
Section 6　Nucleic Acids

基因　gene
常染色体　autosome
性染色体　sex chromosome
遗传的　hereditary
反义 RNA　antisense RNA
核糖核酸　ribonucleic acid (RNA)
嘌呤衍生物　derivatives of purine
嘧啶衍生物　derivatives of pyrimidine
脱氧核糖核酸　deoxyribonucleic acid (DNA)
脱氧核　deoxyribose
脱氧腺苷　deoxyadenosine
脱氧鸟苷　deoxyguanosine
脱氧胞　deoxycytidine
胸苷　thymidine
磷酸二酯键　phosphodiester bridge
骨架　backbone
核苷　nucleoside
N-糖苷键　N-glycosidic linkage
碱基对　base pairing
双螺旋　double helix
进化的　evolutionary
转录　transcription
翻译　translation
基因表达　gene expression
DNA 聚合酶　DNA polymerase
模板　template
遗传密码　genetic code
真核的　eukaryotic

初生的 nascent
成熟的 mature
内含子与外显子 intron and exon
互补 complementary
疏水相互作用 hydrophobic interaction
螺旋直径 the diameter of the helix
大沟 major groove
小沟 minor groove
沃森-克里克模型 Watson-Crick model
假设 hypothesis
平行碱基对 parallel base pair
碱基堆垛 base stacking
螺旋桨状扭曲 propeller twisting
子代 DNA daughter DNA molecule
父代 DNA parent DNA molecule
回文结构 palindrome sequence
超螺旋 superhelical form
拓扑学 topology
标记的 labeled
^{15}N 氮的重同位素 ^{15}N heavy isotope of nitrogen
密度梯度平衡沉降 density-gradient equilibrium sedimentation
离心机 centrifuge
氯化铯 cesium chloride
探测 detect
紫外吸收 absorption of ultraviolet light
增色效应 hyperchromic effect
减色效应 hypochromic effect
一代 generation
杂交 DNA hybrid DNA
复制 duplication
退火 annealing
逆转录酶病 retrovirus
逆转录酶 reverse transcriptase
突变 mutation
编码 encode
切除 excise
错配 mismatched
发夹 hairpin
自身互补序列 self-complementary sequence
起始 initiation
终止 termination
Western 印迹法 Western blotting
限制性酶分析 restriction-enzyme analysis
Southern 和 Northern 印迹分析 Southern and Northern blotting techniques
X-射线结晶学 X-ray crystallography
Sanger 双脱氧 DNA 序列分析法 Sanger dideoxy method of DNA sequencing
核酸固相分析 solid-phase analysis of nucleic acids
聚合酶链反应 polymerase chain reaction (PCR)
转基因动物 transgenic animal
位点特异性突变 site-specific mutagenesis
琼脂糖 agarose
聚丙烯酰胺 polyacrylamide

1 Transcription: The process whereby the genetic information contained in a region of one strand of DNA flows to a complementary sequence of bases in an RNA molecule.
转录:是通过碱基互补的原则来生成一条带有互补碱基的 mRNA,通过它携有的密码子到核糖体中可以实现蛋白质的合成。

2 Translation: The process whereby the genetic information present in an mRNA molecule specifies the sequence of amino acids during protein synthesis.
翻译:是根据遗传密码的中心法则,将成熟的 mRNA 分子中核苷酸序列解码,并生成对应的特定氨基酸序列的过程。

3 Gene expression: A process by which inheritable information from a gene (DNA sequence) is made into a functional gene product, such as protein or DNA.
基因表达:是指细胞在生命过程中,把储存在 DNA 顺序中遗传信息经过转录和翻译,转变成具有生物活性的蛋白质分子。

4 Intron: a sequence of nucleotides in gene that is transcribed but excised before the gene is translated.
内含子:在转录后的加工中,从最初的转录产物除去的内部的核苷酸序列。

5 Exon: A segment of a eukaryotic gene that encodes a portion of the final product of the gene, such as a part of a protein or a part of a mature RNA.
外显子:是真核生物基因的一部分,它在剪接后仍会被保存下来,并可在蛋白质生物合成过程中被表达为蛋白质。

6 Reverse transcriptase: An RNA-directed DNA polymerase in retroviruses that is capable of making DNA complementary to an RNA.
逆转录酶:是以 RNA 为模板指导三磷酸脱氧核苷酸合成互补 DNA(cDNA) 的酶。
7 Mutation: An inheritable change in the nucleotide sequence of a chromosome.
突变:就是指染色体上某一基因位点内部发生了化学性质的变化。
8 Nucleoside: A unit consisting of a base bonded to a sugar is referred to as a nucleoside.
核苷:一个碱基与一分子糖结合组成的个体。

第七节 维生素
Section 7 Vitamins

维生素 vitamin
维生素缺乏症 avitaminosis
脚气病 beriberi
视黄醇 retinol
视黄醛 retinal
β-胡萝卜素 β-carotene
视紫红质 rhodopin
生育酚 tocopherol
凝血酶原 prothrombin
硫胺素 thiamine
烟酸 nicotinic acid
烟酰胺 nicotinamide
核黄素 riboflavin
泛酸 pantothenic acid
磷酸吡哆醛 pyridoxal phosphate
四氢叶酸 tetrahydrofolate
钴胺素 cyanocobalamin
血卟啉环 heme porphyrin ring
生物素 biotin
活动羧基载体 mobile carboxy group carrier
硫辛酸 lipoic acid
抗坏血酸 ascorbic acid

1 Vitamin: Organic substances required in small quantities in the diet of some species, generally functions as a component of coenzymes.
维生素:是人和动物为维持正常的生理功能而必须从食物中获得的一类微量有机物质。
2 Biotin: A vitamin as a prosthetic group of carboxylase enzymes.
生物素:是一种作为多种羧化酶的辅酶的维生素。

第八节 代谢概论
Section 8 Metabolism: Basic Concepts and Design

食物 foodstuff
代谢中间物 metabolic intermediate
主要代谢途径 central metabolic pathway
两用代谢途径 amphibolic pathway
能量代谢 energetic metabolism
基团转移 group transfer
供体 donor
肌收缩 muscle contraction
被动运输 active transport
单向运输 uniport transport
光合生物 photosynthetic organism
开尔文循环 Calvin cycle (dark reactions)
自养生物 autotroph
异养生物 heterotroph
细胞色素氧化酶体系 cytochrome oxidase system
电子载体 electron carrier
铁硫蛋白 iron-sulfur protein
最终电子受体 ultimate electron acceptor
高势能电子 high potential electron
电化学势能 electrochemical potential
氧化还原电势测量 measurement of redox potential
低能分子 energy-poor molecule
高能分子 energy-rich molecule
高能键 high-energy bond
氧化磷酸化 oxidative phosphorylation
化学渗透假说 chemiosmotic hypothesis
离子梯度 ion gradient
中间代谢 intermediary metabolism
途径 pathway
相互依赖的 interdependent
分解代谢的 catabolic
合成代谢的 anabolic

自发地 spontaneously
负的 negative
正的 positive
三磷酸腺苷 adenosine triphosphate(ATP)
磷酸酐键 phosphoanhydride bond
正磷酸 orthophosphate (Pi)
焦磷酸 pyrophosphate (PPi)
生物合成 biosynthetic
不利的 unfavorable
有营养的 nutrient
通用货币 universal currency
钠钾泵 sodium-potassium pump
水解反应 hydrolytic reaction
去磷酸化 dephosphorylate
水合作用 hydration
周转 turnover
再生 regenerating
放能的 exergonic
吸能的 endergonic
捕获的 trapped
三羧酸循环 tricarboxylic acid (TCA) cycle/Krebs cycle/citric acid cycle
吡啶核苷 pyridine nucleotide
黄素 flavin
肌氨酸 creatine
磷酸肌酸 phosphocreatine
降级 degrade
乙酰辅酶 A acetyl CoA
前体 precursor
还原剂 reductant
标准自由能 standard free energy
能荷 energy charge

1 Oxidative phosphorylation: The phosphorylation of ADP to ATP that occurs in conjunction with the transit of electrons down the electron transport chain in the inner mitochondrial membrane.

氧化磷酸化：是物质在体内氧化时释放的能量供给 ADP 与无机磷合成 ATP 的偶联反应。主要在线粒体中进行。

2 Chemiosmotic hypothesis: A theory proposing that a proton concentration gradient established during the electrons transport which provides the energy to drive processes of the formation of ATP and P.

化学渗透理论：主要论点是电子传递过程中建立的质子浓度梯度提供了形成 ATP 和 P 的能量。

3 Citric acid cycle: A cycle system of enzymatic reaction for the oxidation of acetyl moiety of acetyl CoA to carbon dioxide and reducing power for passing the respiratory chain to produce H_2O, and releasing energy, the first step is the formation of citrate; this cycle is also known as Krebs cycle or tricarboxylic acid cycle.

柠檬酸循环：是一个由一系列酶促反应构成的循环反应系统，在该反应过程中，首先由乙酰辅酶 A 与草酰乙酸缩合生成含有 3 个羧基的柠檬酸，经过 4 次脱氢，2 次脱羧，生成四分子还原当量和 2 分子 CO_2，重新生成草酰乙酸的这一循环反应过程成为三羧酸循环。

第九节 糖代谢

Section 9 Glycolysis and Gluconeogenesis

糖酵解 glycolysis
厌氧的 anaerobic
乳酸发酵 lactic acid fermentation
乙醇发酵 alcoholic fermentation
磷酸烯醇丙酮酸 phosphoenolpyruvate
2-磷酸甘油酸 2-phosphoglycerate
3-磷酸甘油醛 glyceraldehyde 3-phosphate
1,3-二磷酸甘油酸 1,3-diphosphoglycerate
脱水 dehydration
基团转移反应 group-transfer reaction
反馈抑制 feedback inhibition
肾上腺素 epinephrine
触发信号转换 trigger signal transduction
梯瀑 cascade
高度放大变化 highly amplified change
细胞间的 intercellular
细胞内的 intracellular
使不稳定 destabilize
己糖激酶 hexokinase
磷酰基 phosphoryl group
二价金属离子 divalent metal ion
构象变化 conformational change
果糖的直链状形式 the open-chain form of fructose
酮基 keto group
裂开 splitting

醛缩酶 aldolase
羟醛缩合 aldol condensation
磷酸二羟丙酮 dihydroxyacetone phosphate
限速步 committed step
载体 transporter
调整 mediate
跨膜 transmembrane
分泌 secretion
饥饿 starvation
骨骼肌 skeletal muscle
羧化作用 carboxylation
参与 participate
动态平衡 homeostasis
激发 stimulate
生存能力 viability
双功能酶 bifunctional enzyme
硫酯中间体 thioester intermediate
柠檬酸合成酶 citrate synthase
α-酮戊二酸脱氢酶 α-ketoglutarate dehydrogenase
蜕变中间物质 metabolon
乙醛酸循环 glyoxylate cycle
异柠檬酸裂解酶 isocitrate lyase
苹果酸合成酶 malate synthase
乙醛酸循环体 glyoxysome
多功能性 versatility
构件 building block
浓缩 condense
琥珀酸盐 succinate
硫氧还蛋白 thioredoxi
质子化作用 protonation
磷酸戊糖途径 pentose phosphate pathway
转酮酶 transketolase
转醛酶 transaldolase
谷胱甘肽 glutathione

1 Glycolysis: The catalytic pathway for one molecule of glucose being broken down into two molecule of lactate.

糖酵解:催化 1 个分子的葡萄糖分解成 2 个分子乳酸的过程。

2 Pentose phosphate pathway: A pathway for providing reducing power NADPH and pentoses for biosynthetic processes.

磷酸戊糖途径:磷酸已糖先氧化脱羧形成磷酸戊糖及 NADPH,磷酸戊糖又可重排转变为多种磷酸糖脂的一种途径。

第十节 脂代谢
Section 10 Lipid Metabolism

肠 intestine
食物脂 dietary lipid
乳化作用 emulsification
胆酸盐 bile salt
胆囊 gallbladder
摄取 ingestion
酯酶 esterase
胰脂酶 pancreatic lipase
磷脂酶 phospholipase
二酯酰甘油 diacylglycerol
2-酯酰甘油 2-acylglycerol
溶血磷脂 lysophospholipid
上皮细胞 epithelial cell
脂-蛋白复合体 lipid-protein complex
乳糜微粒 chylomicron
黏膜 mucosa
淋巴系统 lymphatic system
血流 bloodstream
血浆 plasma
毛细管 capillary
脂肪组织 adipose tissue
残余的 remnant
肝的 hepatic
利用 utilize
丙二酸单酰辅酶 A amalonyl-CoA
乙酰辅酶 A 羧化酶 acetyl-CoA carboxylase
延伸 elongating
酰基载体蛋白 acyl carrier protein (ACP)
乙酰辅酶 A 转移 acetyl-CoA transacylase
丙二酸单酰辅酶 A 转移酶 malonyl-CoA transacylase
β-酮脂酰 ACP 合成酶 β-keto-ACP synthase
β- ACP 还原酶 β-keto-ACP reductase
3-羟脂酰 ACP 脱水酶 3-OH acyl-ACP dehydratase
烯脂酰辅酶 A 还原酶 enoyl-CoA reductase

不饱和化 unsaturation
三羧酸转运系统 tricarboxylate transport system
精巧调控 exquisitely regulate
激素敏感型脂酶 hormone-sensitive lipase
酮体 ketone body
脂酰辅酶A去饱和酶 fatty acyl-CoA desaturase
细胞色素 cytochrome
亚油酸盐 linoleate
亚麻酸盐 linolenate
必需脂肪酸 essential fatty acid
β-氧化途径 β-oxidation pathway
渗透性屏障 permeability barrier
肉碱穿梭 carnitine
磷脂合成 phospholipid synthesis
卵磷脂 lecithin
肺表面活性剂 pulmonary surfactant
乙醇胺 ethanolamine
心磷脂 cardiolipin
神经酰胺 ceramide
神经细胞 nerve cell
紊乱 disorder
缺陷 defect

1 Chylomicron: A largest plasma lipoprotein particle that carries dietary lipids (primarily triacylglycerols) from the intestine to other tissues.
乳糜微粒:主要含有外源性甘油三酯,是运输外源性甘油三酯及胆固醇的主要形式。

2 Carnitine shuttle: A carnitine mediated transport system in mitochondria by which fatty acyl-CoA is transferred from cytosol to matrix for β-oxidation.
肉碱穿梭:线粒体中肉碱介导的脂肪酰CoA转移到细胞质进行β氧化的转运系统。

3 Ketone body: Three water-soluble compounds including acetone, that are produced in the liver when fatty acids are broken down and used as energy source in extrahepatic tissues.
酮体:酮体是指包括丙酮在内的三种水溶性化合物,脂肪酸分解时作为肝外组织的能量源,由肝脏产生。

4 Essential fatty acid: The fatty acids including linoleic acid, linolenic acid and arachidonic acid, that cannot be synthesized within an organism and must be obtained from the diet.
必需脂肪酸:脂肪酸包括亚油酸、亚麻酸和花生四烯酸,机体无法合成,必须从食物中摄取。

第十一节 蛋白质代谢
Section 11 Protein Metabolism

固氮 nitrogen fixation
固氮酶复合体 nitrogenase complex
细菌固氮酶 bacterial nitrogenase
生物圈 biosphere
谷氨酸脱氢酶 glutamate dehydrogenase
氨 ammonia
酰胺基 amide group
转氨作用 transamination
谷草转氨酶 glutamate-oxaloacetate-aminotransferase (GOT)
谷丙转氨酶 glutamate-pyruvate aminotransferase (GPT)
外围组织 peripheral tissues
尿素 urea
尿素循环 urea cycle
尿 urine
乌氨酸 ornithine
氨甲酰磷酸 carbamoyl phosphate
瓜氨酸 citrulline
延胡索酸盐 fumarate
精氨琥珀酸 argininosuccinate
精氨酸酶 arginase
毒害神经的 neurotoxic
碱毒症 alkalosis
酸毒症 acidosis
酸性碳酸盐 bicarbonate
生酮氨基酸 ketogenic amino acid
生糖氨基酸 glucogenic amino acid
胃蛋白酶原 pepsinogen
自身催化地 autocatalytically
蛋白质水解 proteolysis
十二指肠 duodenum
胰腺 pancreas
肽链内切酶 endopeptidase
肽链外切酶 exopeptidase
二肽 dipeptide

三肽　tripeptide
迂回的　circuitous
循环入口　portal circulation
易化扩散　facilitated diffusion
肾脏的　renal
氮平衡　nitrogen balance
L-氨基酸氧化酶　L-amino acid oxidase
死亡率　mortality rate
成人期　adulthood
反密码子　anticodon
氨酰-tRNA 合成酶　aminoacyl-tRNA synthetase
多核糖体　polysome
肽酰基转移酶　peptidyl transferase
起始因子　initiation factor
延伸因子　elongation factor
释放因子　release factor
操纵子模型　operon model

1 Transamination: The enzymatic process whereby the amino group is transferred from an α-amino acid to an α-keto acid to form a new α-keto acid and a new α-amino acid, respectively.
转氨作用：氨基从一个 α- 氨基酸转移到另一个 α- 酮酸，形成一个新的 α- 酮酸和新的 α- 氨基酸的催化过程。

2 Nitrogen balance: The balance between the dietary intake of nitrogen and the excretion of nitrogen.
氮平衡：氮平衡是指氮的摄入量与排出量之间的平衡状态。

3 Anticodon: Three-base sequence in the anticodon loop of tRNA that base pairs with the codon in mRNA during protein synthesis.
反密码子：每个 tRNA 的 3 个碱基可以与 mRNA 上的密码子互补配对，因而叫反密码子。

4 Elongation factor: Proteins responsible for the elongation phase of transcription.
延伸因子：是在 mRNA 翻译时促进多肽链延伸的蛋白质因子。

5 Operon model: A functioning unit of the prokaryotic gene expression, consisting of a regulatory region (an operator, a common promoter, and upstream genes encoding repressor proteins) and one or more structural genes.
操纵子模型：指启动基因、操纵基因和一系列紧密连锁的结构基因的总称。转录的功能单位。很多功能上相关的基因前后相连成串，由一个共同的控制区进行转录的控制，包括结构基因以及调节基因的整个 DNA 序列。

第十二节　核酸代谢
Section 12　Nucleic Acid Metabolism

同源碱基　cognate base
核酸内切酶　endonuclease
核苷磷酸化酶　nucleoside phosphorylase
寡核苷酸　oligonucleotide
向内　inward
核糖-1-磷酸　ribose-1-phosphate
脲酸　uric acid
重新合成　denovosynthesis
补救合成　salvage synthesis
磷酸核糖焦磷酸　5-phosphoribosyl-1-pyrophosphate (PRPP)
消耗　consume
次黄苷，肌苷　inosine
次黄嘌呤　hypoxanthine
甲酸盐　formate
S-腺苷蛋氨酸　S-adenosylmethionine (SAM)
二分之一，半族　moiety
积聚　accumulation
超额　excess
嘌呤核苷酸循环　purine nucleotide cycle
分解代谢　catabolism
抗代谢物　antimetabolite
免疫的　immune
自杀性底物　suicide substrate
5-氟尿嘧啶　5-fluorouracil
叶酸　folic acid
胸腺嘧啶核苷酸　thymidylate synthase
重组　recombination
修复　repair
直接修复　direct repair
碱基切除修复　base-excision repair
核苷切除修复　nucleotide-excision repair
端粒　telomere
端粒酶　telomerase

重组酶 recombinase
DNA 连接酶 DNA ligase
复制的起点 origin of replication
引物酶 primase
复制叉 replication fork
岗崎片段 Okazaki fragment
滞后链 lagging strand
前导链 leading strand
引物 primer
核酸外切酶 exonuclease
螺旋酶 helicase
连环数 linking number (Lk)
拓扑异构体 topoisomer
5'帽子 5'cap
多聚腺苷酸尾巴 poly(A) tail

1 Endonuclease: An enzyme that hydrolyzes that interior phosphodiester bond other than the terminal bonds of a nucleic acid.
核酸内切酶:核糖核酸酶和脱氧核糖核酸酶中能够水解核酸分子内磷酸二酯键的酶。

1 Lagging strand: The DNA strand that is synthesized in the direction opposite to the direction of replication fork movement during replication.
滞后链:在 DNA 复制中复制方向与复制叉方向相反的那条链。

2 Leading strand: The DNA strand that is synthesized in the same direction of the replication fork movement during replication.
前导链:在 DNA 复制中复制方向与复制叉方向相同的那条链。

3 Replication fork: A moved Y-shaped structure found at the point where the double strands of DNA are separating and DNA is being synthesized.
复制叉:DNA 复制时,双链 DNA 分开并在生长点形成的移动的 Y 字型结构称为复制叉。

第十三节 细胞代谢调控
Section 13 Metabolic Regulation in Cell

单向反应 opposing unidirectional reaction
前馈 feedforward
反馈调节 feedback regulation
单价反馈 monovalent feedback
二价或多价反馈 divalent or multivalent feedback
质量作用比 mass-action ratio
级联反应 cascade
腺苷酸化 adenylylation
脱腺苷酸化 deadenylation
腺苷酰转移酶 adenylytransferase
光面内质网 smooth endoplasmic reticulum
运输泡 transport vesicle
质膜体 mesosome
双关酶 ambiguous enzyme
信号假说 signal hypothesis
凋亡 apoptosis
适应条件 adaptive regulation
动作电位 action potential
神经脉冲 nerve impulse
电位门控离子通道 voltage-gated ionchannel
配体门控离子通道 ligand-gated channel
神经递质 neurotransmitter
旁分泌 paracrine

1 Feedback regulation: A regulatory effect on an enzyme by its product.
反馈调节:酶被它的产物所调节的调控机制。

2 Cascade reaction: A chain reaction, when an enzyme is activated other enzymes are also excited in turn to bring primal signal amplifying.
级联反应:是指酶启动的连续激活的反应链。

第十四章 普通生物学
Chapter 14 General Biology

第一节 绪论
Section 1 Introduction

生物圈 biosphere
稳态 homeostasis
应激性 irritability
适应 adaptation
生物多层次结构 hierarchy of biological organization
五界系统 five kingdom system
双名法 binomial nomenclature

1 Five kingdom system: R. H. Whittaker in 1969 proposed a five kingdom system, which classified living organisms into five kingdoms according to the cellular structure and nutrition type, including monera, protista, fungi, plantae and animalia.
五界系统:惠特克在 1969 年提出了五界系统,他根据细胞结构和营养类型将生物分为五界,即原核生物界、原生生物界、真菌界、植物界和动物界。

2 Hierarchy of biological organization: Atom — molecule — biomacromolecule — organelle — cell — tissue — organ — organ system — organism — population — community — ecosystem — biosphere.
生物多层次结构:原子 — 分子 — 生物大分子 — 细胞器 — 细胞 — 组织 — 器官 — 系统 — 个体 — 种群 — 群落 — 生态系统 — 生物圈。

3 Homeostasis: The maintenance of a relatively constant internal environment by an organism in a variable external environment is called homeostasis.
稳态:有机体在可变动的外部环境中维持一个相对恒定的内部环境,称为稳态。

第二节 生命的化学基础
Section 2 Chemical Context of Life

同位素示踪 isotopic tracing
必需元素 essential element
生物大分子 macromolecule
多聚体 polymer
糖类 carbohydrate
氨基酸 amino acid
必需氨基酸 essential amino acid
肽键 peptide bond
肽和多肽 peptide and polypeptide
蛋白质一级结构 protein primary structure
蛋白质二级结构 protein secondary structure
蛋白质三级结构 protein tertiary structure
蛋白质四级结构 protein quaternary structure
蛋白质变性 protein denaturation
核苷酸 nucleotide
DNA 双螺旋 double helix of DNA
三磷酸腺苷 adenosine triphosphate (ATP)

1 Biological macromolecules: Some organic molecules in organisms are large complex assemblies called macromolecules. Biological macromolecules are traditionally grouped into four major categories: proteins, nucleic acids, lipids, and carbohydrates.
生物大分子:在生物体中的一些有机分子是大的复合结构,称为生物大分子,传统上可分为 4 个主要类型:蛋白质、核酸、脂质和碳水化合物。

2 Polymers: are large molecules consisting of long chains of similar subunits joined by dehydration reactions. In a dehydration reaction, a hydroxyl (-OH) group is removed from one subunit and a hydrogen atom (H) is removed from the other. Protein, nucleic acid, polysaccharide are polymers.
多聚体:由相似的亚基通过脱水反应连接而成的长链组成的大分子。在脱水反应中,羟基(-OH) 被从一个亚基上移走,氢原子(H) 被从另一个亚基上移走。蛋白质、核酸、多糖都是多聚体。

3 Amino acids: An amino acid is a molecule containing an amino group (—NH_2), a carboxyl group (-COOH). Amino acids are the monomers of proteins.
氨基酸:含氨基(-NH_2)和羧基(-COOH)的分子,是蛋白质的结构单体。

4 Peptide bond: A covalent bond (-NH-CO-) that links two amino acids is called a peptide bond, a peptide bond forms when the -NH_2 end of one amino acid joins to the -COOH end of another by a condensation reaction.
肽键:连接两个氨基酸的共价键(-NH-CO-)被称为肽键,是由一个氨基酸分子中的氨基端与另一氨基酸分子中的羧基端脱水缩合形成。

5 Protein denaturation: If a protein's environment is altered, the protein may change its shape or even unfold, this process is called denaturation. Proteins can be denatured when the pH, temperature, or ionic concentration of the surrounding solution is changed. When proteins are denatured, they are usually rendered biologically inactive.
蛋白质的变性:当蛋白质的环境改变的,蛋白质可能改变它的构型,甚至展开,这个过程叫做变性。当周围溶液的pH、温度或离子浓度发生变化时,蛋白质可能变性。当蛋白质变性时,通常会失去生物活性。

6 Nucleotides: are monomers of nucleic acid(DNA and RNA). Each nucleotide consists of three components: a five-carbon sugar (ribose in RNA and deoxyribose in DNA), a phosphate (—PO_4) group, and an organic nitrogen containing base.
核苷酸:核酸(DNA 和 RNA)的结构单体。每一核苷酸分子包括3个组分:一个戊糖(在 RNA 中为核糖,在 DNA 中为脱氧核糖)分子,一个磷酸基团和一个含氮的有机碱。

第三节 细胞结构与细胞通讯
Section 3 Cell Structure and Cell Communication

细胞学说 cell theory	染色体 chromosome
细胞质 cytoplasm	组蛋白 histone
生物膜 biomembrane	核仁组织者 nucleolus organizer
核膜 nuclear envelope	高尔基体 Golgi complex/Golgi apparatus
核纤维 nuclear lamina	质体 plastid
染色质 chromatin	液泡 vacuole
常染色质 euchromatin	细胞连接 cell junction
异染色质 heterochromatin	细胞通讯 cell communication

1 Cell theory: Schleiden and Schwann put forward cell theory, pointed out that: all plant kingdom and animal kingdom are composed of cells; new cells arise only by division of a previously existing cell.
细胞学说:施莱登和施万提出细胞学说,指出所有的植物界和动物界都是由细胞构成的,新细胞只能由原来的细胞分裂而产生。

2 The cytoplasm: the rest of the cell's interior exclusive of the nucleus lying within it, plasma membrane is the most outer layer of the cytoplasm.
细胞质:指细胞内除细胞核以外的所有部分,质膜是细胞质的最外层。

3 Chromosome: The structure within cells that contains the genes. In eukaryotes, it consists of a single linear DNA molecule associated with proteins. The DNA is replicated during S phase, and the replicas separated during M phase.
染色体:包含基因的细胞结构。在真核细胞内,由与蛋白质相连的单条线性DNA分子构成,DNA在S期被复制,在M期复制后的DNA被分开。

4 Golgi complex: an organelle that receives newly synthesized proteins from the endoplasmic reticulum and processes them for subsequent delivery to other destination. It is composed of several flattened membrane disks arranged in a stack.

高尔基体：一种多个膜囊堆积而成的细胞器，主要功能是接收内质网转运来的新合成的蛋白质并对其进行加工后送到其他目的地。

5 Cell junctions: Cells attach themselves to one another with long lasting bonds called cell junctions. Cell junctions are divided into three categories, based upon the functions they serve: tight junctions, anchoring junctions, and communicating junctions.

细胞连接：细胞与相邻细胞之间分化而成的长期性特定连接，即细胞连接。细胞连接根据它们的功能分成三种类型：紧密连接、锚定连接和通讯连接。

第四节 细胞代谢

Section 4 Cell Metabolism

代谢 metabolism
势能 potential energy
热力学 thermodynamics
熵 entropy
自由能 free energy
活化能 activation energy
酶 enzyme
核酶 ribozyme
活性部位 active site
竞争性抑制 competitive inhibition
非竞争性抑制 noncompetitive inhibition
扩散 diffusion
渗透 osmosis
膨压 turgor
质壁分离 plasmolysis
单纯扩散 simple diffusion
易化扩散 facilitated diffusion
被动转运 passive transport
主动转运 active transport
胞吐 exocytosis
胞吞 endocytosis
吞噬作用和胞饮作用 phagocytosis and pinocytosis
受体介导的胞吞 receptor-mediated endocytosis
细胞呼吸 cellular respiration
电子传递链 respiration chain
氧化磷酸化 oxidative phosphorylation
化学渗透偶联假说 chemiosmotic coupling hypothesis
光合作用 photosynthesis
光反应 light reaction
希尔反应 Hill reaction
光合膜 photosynthetic membrane
色素 pigment
捕光色素 light harvesting pigment
反应中心色素 reaction center pigment
荧光和磷光 fluorescence and phosphorescence
光系统 photosystem
原初电子供体 primary electron donor
原初电子受体 primary electron acceptor
光合电子传递链 photosynthetic electron transfer chain
光合磷酸化 photophosphorylation
解偶联剂 uncoupler
碳反应 carbon reaction
C_4 途径和 C_4 植物 C_4 pathway and C_4 plant
景天酸代谢途径 Crassulacean acid metabolism pathway
光呼吸 photorespiration
光合速率 photosynthetic rate
光饱和点 light saturation point
CO_2 饱和点 CO_2 saturation point

1 Metabolism: The sum of all chemical processes occurring within a living cell or organism.

代谢：活细胞或生物体内发生的所有化学反应的总称。

2 Enzymes: A kind of biological catalysts that facilitate specific chemical reactions. Enzymes are globular proteins, with a three dimensional shape that fits snugly around the chemicals they work on, facilitating chemical reactions by stressing particular chemical bonds

酶：一种促进特定化学反应的生物催化剂，为球状蛋白，具有三维构型，可与其所作用的化合物紧密结合，通过加压于特定化学键而促进化学反应。

3 Active transport: The cells can move substances across the membrane up their concentration gradients. Like facilitated diffusion, active transport involves highly selective protein carriers within the membrane; this process requires the expenditure of energy, typically ATP.

主动转运：细胞能够逆浓度梯度移动物质穿过膜。类似于易化扩散，主动转运需要膜内具有高度选择性的蛋白载体，这个过程也需要消耗能量，通常是 ATP。

4 Passive transport: is a movement of biochemicals and other atomic or molecular substances across membranes. Unlike active transport, it does not require an input of chemical energy, being driven by the growth of entropy of the system. The rate of passive transport depends on the permeability of the cell membrane, which, in turn, depends on the organization and characteristics of the membrane lipids and proteins. While diffusion, facilitated diffusion, and osmosis are passive transport processes that move materials down their concentration gradients.

被动转运：指生化物质和其他原子或分子物质穿过膜的移动。不同于主动转运，它不需要化学能的消耗，被系统熵的增加所驱动。被动转运的效率依赖于细胞膜的渗透性，其次，依赖于膜脂和膜蛋白的结构和特性。扩散、易化扩散和渗透是顺浓度梯度移动物质的被动转运过程。

5 Cellular respiration: is the set of the metabolic reactions and processes that take place in the cells of organisms to convert biochemical energy from nutrients into adenosine triphosphate (ATP), and then release waste products. It includes aerobic and anaerobic respiration.

细胞呼吸：是一系列发生在生物体细胞内的代谢反应和过程，将来自营养物质的化学能转化成三磷酸腺苷(ATP)并释放废物，包括有氧呼吸和无氧呼吸。

6 Light reaction: The energy of light is used to energize electrons in the reaction center chlorophyll molecules of a photosystem. The electrons are then passed on to an electron transport chain. The missing electrons in the reaction center are replenished when $2H_2O$ is split to O_2, $4H^+$ and $4e^-$. The electron moves through the electron transport chain until it energizes the cytochrome b/f complex to pump protons into the thylakoid lumen and is then passed to photosystem I (PS-I) where it is re-energized and passed to NADP reductase, generating NADPH. PS-I can also function alone to transport H^+. Adenosine triphosphate(ATP) is generated by H^+ passing through the enzyme ATP synthase located in the thylakoid membrane. The two products of the light reactions are therefore NADPH and ATP.

光反应：光能可激活光系统中反应中心叶绿素分子中的电子，随后电子沿着电子传递链传递。当发生反应时，反应中心失去的电子由水解产生的电子补充，即 $2H_2O \rightarrow O_2 + 4H^+ + 4e^-$。电子沿着电子传递链将能量传至细胞色素 b/f 复合体，把质子泵入类囊体腔。然后电子又传递给光系统 I(PS-I) 并补充能量，再传给 NADP 还原酶，生成 NADPH。PS-I 也可单独运送 H^+。当 H^+ 穿过类囊体膜上的 ATP 合成酶时产生 ATP。因此，光反应的两个产物是 NADPH 和 ATP。

第五节 细胞的分裂和分化
Section 5 Division and Differentiation of Cells

细胞周期 cell cycle
细胞增殖 cell proliferation
染色单体 chromatid
着丝粒 centromere
动粒 kinetrochore
星体 cytaster
赤道面 equatorial plane
主缢痕 primary constriction
次缢痕 secondary constriction
性染色体 sex chromosome
染色体组型 karyotype
G_0 期细胞 G_0 phase cell
无丝分裂 amitosis
有丝分裂 mitosis
减数分裂 meiosis
联会 synapsis
联会复合体 synaptonemal complex
四分体 tetrad
端粒 telomere
细胞分化 cell differentiation
干细胞 stem cell
细胞全能性 totipotency
复制性细胞衰老 replicative senescence
Hayflick 界线 Hayflick boundary
细胞凋亡 cell apoptosis
细胞坏死 cell necrosis

1 Mitosis: is the process of chromosome segregation and nuclear division that follows replication of the genetic material in eukaryotic cells. This process assures that each daughter nucleus receives a complete copy of the organism's genome.
有丝分裂：是真核生物细胞内遗传物质复制之后的染色体分离和核分裂的过程，这个过程确保了每个子细胞核获得母体基因的完整复制。

2 Meiosis: Meiosis occurs to produce haploid cells. It involves an extra round of cell division. The first phase, prophase to anaphase I, results in exchange of DNA between the pairs of chromosomes followed by their separation with both chromatids present. The second phase (metaphase II to telophase II) is a mitosis resulting in separation of the chromatids and formation of four haploid cells.
减数分裂：减数分裂产生单倍体细胞，包含增加的一次细胞分裂的过程。第一阶段，即分裂前期到后期I，配对染色体发生DNA交换，随后具有两个染色单体的染色体分离；第二阶段（中期II和末期II）是有丝分裂，染色单体分离并形成4个单倍体细胞。

3 The cell cycle: Cell division occurs in stages: interphase G_1—cell enlarges, nucleus migrates to center, protein synthesis; interphase S—DNA replication occurs; interphase G_2—preprophase band and structures of mitosis form, chromosomes condense. In M-phase mitosis, chromosomes replicate and divide to form two daughter nuclei. In M-phase cytokinesis, the cytoplasm divides and a cell plate and new cell walls form.
细胞周期：细胞分裂经历以下几个阶段：间期 G_1—— 细胞增大，细胞核移向细胞中心，合成蛋白质；间期S——DNA复制；间期 G_2——前中期带和有丝分裂结构形成，染色体紧密。在有丝分裂M期，染色体进行复制并且细胞核分裂形成两个姐妹子核。在M期胞质分裂时，细胞质分裂，细胞板和新细胞壁形成。

第六节 动物的形态与功能
Section 6 Animal Morphology and Function

组织 tissue
器官 organ
系统 system
自养 autotrophic nutrition
异养 heterotrophic nutrition
营养素 nutrient
消化 digestion
吸收 absorption
胞内消化 intracellular digestion
胞外消化 extracellular digestion
内环境 internal environment
体液 body fluid
细胞内液 intracellular fluid
细胞外液 extracellular fluid
心动周期 cardiac cycle
高血压 hypertension
粥样动脉变化 atherosclerosis
呼吸 respiration
变温动物 poikiothermic animal
恒温动物 homeothermic animal
异温动物 heterothermic animal
淋巴 lymph
免疫 immunity
特异性免疫 specific immunity
体液免疫 humoral immunity
细胞免疫 cellular immunity
补体系统 complement system
干扰素 interferon
抗原 antigen
抗体 antibody
单克隆抗体 monoclonal antibody
激素 hormone
内分泌 endocrine
体液调节 humoral regulation
神经元 neuron
突触 synapse
反射 reflex
反射板 reflex arc
自发脑电活动 spontaneous brain electrical activity
适宜刺激 adequate stimulus
不适宜刺激 not adequate stimulus
感觉刺激 sensory stimulation
感受器 receptor

本体感受器 proprioceptor	受精 fertilization
效应器 effector	分娩 parturition
肌丝滑行学说 muscle silk taxi theory	绝育 sterilization
有性生殖 sexual reproduction	人工流产 abortion
无性生殖 asexual reproduction	生物重演率 biological repeat rate

1 Water balance in amphibians: Amphibian kidneys function much like those of freshwater fishes. However, on land, where dehydration is the most important problem in terms of osmoregulation, frogs conserve body fluid by reabsorbing water across the epithelium of the urinary bladder.
两栖类的水平衡:两栖类的肾功能很像淡水鱼的肾功能。然而,在陆地上,脱水对渗透调节是最重要的问题,蛙保存体液是通过膀胱上皮细胞的重吸收水。

2 Water conservation by terrestrial animals: A major problem faced by terrestrial organisms is the loss of a continuous supply of water necessary to keep tissue surfaces moist. When air is inhaled, it passes along the respiratory tract into the lungs, where it is in contact with the moist respiratory tissues. If the moist air was exhaled, water would be lost. The recovery of respiratory moisture by most terrestrial animals involves countercurrent exchange. Exhaled air from the lungs encounters a countercurrent-like gradient on the way out. This interaction between the departing air and the respiratory surfaces results in an efficient return of moisture to the tissues.
陆生动物的水保持:陆生动物面对的主要问题是连续的水供应,这是供应到组织表面维持潮湿所必需的水。当空气被吸入时,它沿着呼吸道进入肺,和潮湿的呼吸组织相接触。如果潮湿的空气被呼出,水就会丢失。大多数陆生动物呼吸湿度的维持包括了逆流交换。从肺呼出的气体,在呼出的通道上有一个像逆流的梯度。呼出的气体和呼吸表面的相互作用,导致水分有效地返回组织。

3 Tissue: Groups of cells similar in structure and function are organized into tissues. Vertebrates have epithelial tissue, connective tissue, muscle tissue and nerve tissue, the four basic tissues.
组织:结构与功能相同的细胞组合成的细胞群体。脊椎动物有上皮组织、结缔组织、肌肉组织和神经组织四种基本组织。

4 Organ: is a collection of tissues joined in a structural unit to serve a common function.
器官:由多种组织组成的结构单元,以完成某一共同的功能。

5 Organ systems: Two or more organs working together in the execution of a specific body function form an organ system, also called a biological system or body system. The functions of organ systems often share significant overlap. For instance, the nervous and endocrine system both operate via a shared organ, the hypothalamus. For this reason, the two systems are combined and studied as the neuroendocrine system.
器官系统:生物体内 2-3 个器官一起工作执行特定的功能形成器官系统,也叫作生物系统或身体系统。器官系统的功能经常有效地重叠,例如,神经系统和内分泌系统都会通过一个共同器官起作用,下丘脑。由于这个原因,这两个系统被组合在一起作为神经内分泌系统研究。

6 Extracellular fluid: usually denotes all body fluid outside of cells. In some animals, including mammals, the extracellular fluid can be divided into two major subcompartments, interstitial fluid and blood plasma. The extracellular fluid also includes the transcellular fluid, making up only about 2.5 percent of the ECF.
细胞外液:通常指细胞外的所有体液。在一些动物中,包括哺乳动物,细胞外液可分成两个主要组分,组织液和血浆。细胞外液也包括细胞透过液,仅占细胞外液的 2.5%。

7 Immunity: resistance of an organism to infection or disease.
免疫:指生物体对感染或疾病的抵抗力。

8 Antigen: In immunology, an antigen is a substance that evokes the production of one or more antibodies. Each antibody binds to a specific antigen by way of an interaction similar to the fit between a lock and a key. The substance may be from the external environment or formed within the

body. The immune system will try to destroy or neutralize any antigen that is recognized as a foreign and potentially harmful invader.

抗原：在免疫学中，抗原指能够引起一个或更多个抗体产生的物质。每个抗体通过一种类似于锁和钥匙之间的交互作用与特定抗原结合。抗原可能来自于外部环境或在生物体内形成。免疫系统将尽力破坏或抵消被认作外来物或潜在有害入侵者的任何抗原。

9 Antibody: also known as an immunoglobulin, is a large Y-shaped protein produced by B-cells that is used by the immune system to identify and neutralize foreign objects such as bacteria and viruses. The antibody recognizes a unique part of the foreign target, called an antigen.

抗体：也被称为免疫球蛋白，是一种由 B 细胞产生的大的 Y 形蛋白，被免疫系统用于识别和抵消细菌和病毒等外来物质。抗体识别被叫作抗原的外来物的特定部分。

10 Neuron: also known as a neurone or nerve cell, is an electrically excitable cell that processes and transmits information through electrical and chemical signals. A chemical signal occurs via a synapse, a specialized connection with other cells. Neurons connect to each other to form neural networks. Neurons are the core components of the nervous system. A typical neuron possesses a cell body (often called the soma), dendrites, and an axon.

神经元：也被称为神经细胞，是一种电传导的可兴奋细胞，通过电和化学信号处理和传输信息。化学信号通过突触(一种与其他细胞的特定连接）产生。神经元彼此连接形成神经网。神经元是神经系统的核心成分。一个典型的神经元具有胞体、树突和轴突三部分。

11 Reflection: is a series of regular responses to the certain stimulation through the nervous system activity of the animal, is the most basic actitity pattern of the nervous system by the reflex arc.

反射：是动物通过神经系统的活动对一定刺激的规律性反应，是神经系统最基本的活动方式，通过反射弧进行。

12 Sexual reproduction: is a process that creates a new organism by combining the genetic material of two organisms, it occurs only in eukaryotes. There are two main processes during sexual reproduction, they are: meiosis, involving the halving of the number of chromosomes; and fertilization, involving the fusion of two gametes and the restoration of the original number of chromosomes. During meiosis, the chromosomes of each pair usually cross over to achieve homologous recombination.

有性生殖：通过组合两个生物体的遗传物质创造新生物体的过程，仅发生在真核生物中。包括两个主要过程：减数分裂 —— 涉及染色体数目的减半 —— 和受精 —— 涉及两个配子的融合和原来染色体数目的恢复。减数分裂期间，每对染色体通过交叉联会完成同源重组。

第七节 植物的形态与功能
Section 7 Plant Morphology and Function

分生组织 meristem
平周分裂和垂周分裂 periclinal division and anticlinal division
定根和不定根 fixed root and adventitious roots
直根系和须根系 straight roots and fibrous root system
初生生长 primary growth
初生结构 primary structure
形成层 cambium
维管组织 vascular tissue
维管系统 vascular system
维管束 vascular bundle
次生生长 secondary growth
次生结构 secondary structure
根瘤 nodule
凯氏带 casparian strip
树皮 bark
年轮 annual ring
髓射线 pith ray
维管射线 vascular ray
边材 sapwood
心材 heartwood
春材 spring wood
秋材 autumn wood

完全花 perfect flower
不完全花 incomplete flower
花序 inflorescence
心皮 carpel
胎座 placenta
单雌蕊 single pistil
传粉 pollination
双受精 double fertilization
花粉败育 pollen abortion
雄性不育 male sterility
单性结实 parthenocarpy
上位子房 superior ovary
下位子房 inferior ovary
真果 true fruit
假果 pseudocarp
假种皮 aril
聚花果 syncarp
聚合果 aggregate fruit
世代交替 alternation of generations
生活史 life history
蒸腾作用 transpiration
共质体 symplast
根压 root pressure
压流模型 pressure flow model
必需元素 essential element
水培法 hydroponics
顶端优势 apical dominance
向光性 phototropism
光周期现象 photoperiodism
春化作用 jarovization/vernalization
光敏色素 phytochrome
光周期 photoperiod
长日植物 long-day plant
短日植物 short-day plant

1 Common features of plants: Plants derive from a green algal ancestor. Plants are photosynthetic and autotrophic (with very few exceptions), have chlorophyll *a* and *b*, have a cellulose cell wall and a cell vacuole and have an alternation of diploid and haploid generations. Vegetative structure is similar across most vascular plants, but reproductive structures differ.
植物的共同特征：植物源自古老的绿藻。植物是能进行光合作用的自养生物(极少数除外)。植物具有叶绿素 *a* 和叶绿素 *b*；具有含纤维素的细胞壁和液泡，有二倍体和单倍体的世代交替现象。大多数维管植物的营养体结构相似，而生殖结构则不同。

2 Meristems: Meristems form the new cells of a plant. Apical meristems occur near the tips of roots and shoots and give primary growth. They give rise to the epidermal tissue, vascular tissue and ground meristems that form the parenchyma. Intercalary meristems are formed at the nodes of grass stems. Lateral meristems generate secondary growth. The vascular cambium is a cylinder that forms new phloem and xylem. The cork cambium is also a cylinder, that forms the bark.
分生组织：分生组织可产生植物体的各种新细胞。顶端分生组织位于茎和根的近顶端，进行初生生长，由它们产生表皮组织、维管组织和形成薄壁组织的基本分生组织。居间分生组织分布在草本植物茎的节处。侧生分生组织进行次生生长。维管形成层为圆筒状，由它形成新的韧皮部和木质部。木栓形成层也呈圆筒状，形成树皮。

3 Vascular tissue: Vascular tissue is made up of xylem and phloem. Xylem carries water and dissolved minerals, and phloem carries the solutes from sites of synthesis to sites of storage or use. Xylem cells have thickened secondary cell walls and no cell contents. Xylem vessel elements join to form vessels. Tracheids are tapered cells that interconnect via pits. Phloem has two cell types: sieve tube elements and companion cells. Sieve tube elements have no nucleus, but contain organelles and link through sieve plates. Companion cells are narrower than sieve tubes and function with them.
维管组织：维管组织由木质部和韧皮部组成。木质部运输水分和溶解的矿物质，韧皮部将溶质从合成部位运输到储藏或利用部位。木质部导管分子有加厚的次生壁，但无细胞内容物，它们连接形成导管。管胞是两端尖的细胞，细胞间通过纹孔相互联系。韧皮部有两种类型的细胞，即筛管分子和伴胞。筛管分子无核，但有细胞器，通过筛板相连。伴胞比筛管狭窄，与筛管分子共同行使功能。

4 Water transport in roots: Water predominantly enters roots via root hairs and mycorrhizae. Water transport may be apoplastic or symplastic (transcellular via plasmodesmata or transmembrane via membrane transporters). At the endodermis, water movement must be cytoplasmic. Water enters xylem as a result of the low water potential generated by transpiration stream.

根部水分运输：水分通过根毛和菌根进入根。水分的运输可能通过质外体途径，也可能通过共质体途径（通过胞间连丝的跨细胞运输和通过膜运载蛋白的跨膜运输）。在内皮层，水分运动必须通过细胞质。水分进入木质部是蒸腾流形成低水势的结果。

5 Plant types: Plants are categorized as mesophytes adapted to an environment in which water is available, xerophytes adapted for areas of low water availability and hydrophytes adapted to the presence of large amounts of water or to growth entirely in water.

植物类型：植物可分为：中生植物 —— 适应生存于易获得水分的环境，旱生植物 —— 适应生存于可利用水少的地方，水生植物 —— 适应生存于有大量水的环境中或完全生活在水中。

6 Photoperiodism: is the response of a plant to the length of a day. It governs processes such as dormancy and flowering. Species may be either long-day, short-day or day-neutral in their response to day length.

光周期：光周期是植物对日长的反应，它控制着诸如休眠和开花过程。根据植物对日长的反应，可以将它们划分为长日植物、短日植物和日中性植物。

7 Vernalization and acclimation: temperature may also act as a stimulus, determining whether the organisms will begin development. Vernalization is the induction of flowering by low temperatures. Exposure of an organism to higher (or lower) temperature in the laboratory can alter the organisms' temperature response. The habituation of an organism's response to changes in laboratory environmental conditions is termed acclimation. Acclimatization is the habituation of an organism's physiological response to changes in natural environmental conditions.

春化和驯化：温度能够作为一种刺激物起作用，决定有机体是否将开始发育。春化法是通过低温诱导开花。有机体在实验室里暴露到较高（或较低）的温度能够改变有机体的温度反应。有机体对实验环境条件变化产生的适应性反应称为驯化。有机体对自然环境条件变化产生的生理适应性反应称为气候驯化。

第八节 遗传与变异

Section 8 Heredity and Variation

遗传变异 genetic variation
性状 character
相对性状 relative character
显性性状 dominant character
性状分离 segregation
显性基因 dominant gene
隐性基因 recessive gene
等位基因 allele
表型 phenotype
基因型 genotype
纯合体 homozygote
杂合体 hybrid
基因的分离定律 law of segregation
测交 test cross
基因的自由组合定律 independent assortment of genes
基因重组 gene recombination
性别决定 sex determination
伴性遗传 sex-linked inheritance
连锁 linkage
细胞核遗传与细胞质遗传 nuclear inheritance and cytoplasmic inheritance
DNA 的复制 DNA replication
半保留复制 semi-conservative replication
基因 gene
遗传信息 genetic information
转录 transcription
密码子 codon
翻译 translation
中心法则 central dogma
染色体畸变 chromosome aberration
单体性 monosomy
三体性 trisomy
基因突变 gene mutation
自然突变 natural mutation
诱发突变 induced mutation
诱发育种 mutation breeding
基因组 genome

多倍体　polyploid
单倍体　haploid
人工诱导多倍体　synthesized polyploid
多倍体育种　polyploid breeding
DNA 损伤修复　DNA damage repair
基因表达　gene expression
基因调控　gene regulation
断裂基因　interrupted gene
基因工程　genetic engineering
聚合酶链式反应　polymerase chain reaction
限制性内切核酸酶　restriction endonuclease
DNA 连接酶　DNA ligase
反转录酶　reverse transcriptase
载体　vector
质粒　plasmid
基因文库　gene library
基因治疗　gene therapy
基因组学　genomics
结构基因组学　structural genomics
功能基因组学　functional genomics
人类遗传性疾病　human hereditary disease
单基因病　monogenic disorder
多基因遗传病　polygenic disease

1 Genotype and phenotype: The genotype is the genetic composition of an individual. The phenotype is the individual organism, a product of the interaction between its genotype and environmental influences. The ability of the phenotype to vary due to environmental influences on its genotype is known as phenotypic plasticity (e. g. human suntan, wind-shaped plants, locust morph solitary or migratory).

基因型和表型：基因型是个体的遗传组成。表型是各个有机体，它是基因型与环境的相互作用的产物。由于环境对基因型的影响，表型由于环境对基因型的影响而发生改变的能力称为表型可塑性(如人的晒黑、风造形的植物、蝗虫的单生或群居型)。

2 Genes and alleles: Genes are pieces of DNA which contain the genetic code necessary to produce a specific protein. A gene can have many alleles and individuals can have a homozygous or heterozygous genotype. The phenotype expressed depends on whether alleles are dominant, recessive or codominant. The total set of genes and alleles present in a population is known as the gene pool.

基因和等位基因：基因是带有可产生特定蛋白质的遗传密码的DNA片段。一个基因可有许多等位基因，而个体可有纯合或杂合的基因型。表达的表现型取决于等位基因是显性的、隐性的还是共显性的。种群中存在的所有基因组和等位基因叫做基因库。

3 Genetic drift: Genetic drift is a random change in gene frequency arising through chance alone. It tends to be more apparent in small populations, allele frequencies drift away from their starting values, increasing, decreasing or fluctuating up and down. It occurs because there is an element of chance in which individuals and which gametes will produce offspring and which individuals will die before reproducing. Random changes in allele frequency can lead to fixation and the progressive loss of genetic variation from the population.

遗传漂变：遗传漂变是基因频率的随机变化，仅偶然出现。在小种群中，遗传漂变更明显。基因频率"漂离"起始值，增加、减少或上下波动。发生遗传漂变是因为在个体、产生后代的合子以及繁殖前死亡的个体中都有偶然要素。基因频率的随机变化导致来自种群的遗传变异的固定和逐渐丧失。

第九节　生物进化
Section 9　Biological Evolution

达尔文主义　Darwinism
古生物学　paleontology
胚胎学　embryology
比较解剖学　comparative anatomy
同源器官　homologous organs
生存斗争　struggle for existence
自然选择　natural selection
物种　species
品种　variety
亚种　subspecies
隔离　isolation
邻地种形成　parapatric speciation
同地种形成　sympatric speciation
异地种形成　allopatric speciation

现代综合论 modern synthesis
哈迪—温伯格平衡定律 Hardy-Weinberg equilibrium
中性理论 neutral theory
间断平衡论 punctuated equilibrium
大进化 macroevolution
小进化 microevolution
常规灭绝 normal extinction
集群灭绝 mass extinction
趋同进化 convergent evolution
趋异进化 divergent evolution
适应辐射 adaptive radiation
平行进化 parallel evolution
内共生学说 endosymbiotic theory
联适应 simultaneous adaptation
内聚体学说 cohesion theory
微球体 microsphere

1 Natural selection: The individuals in a species which have the highest fitness will contribute disproportionately to the subsequent generations. If fitness differences have a genetic component, then the genetic make-up of the subsequent generations will be altered. This process is known as natural selection or 'survival of the fittest'.

自然选择：种中具有最高适合度的个体将会对未来世代做出特别高的贡献。如果适合度的差别含有遗传的成分，则后代的遗传组成会有改变。这个过程称为自然选择或"最适者生存"。

2 Biological species concept: This concept views species as a group of interbreeding natural populations that are reproductively isolated from other such groups. It allows populations that are indistinguishable morphologically to be classified as separate, sibling species because they do not interchange genes. Biological species are maintained by reproductive isolating mechanisms.

生物种的概念：生物种概念认为种是一组可以相互杂交的自然种群，它们与其他种群间具有繁殖隔离。该概念将没有明显形态学差别的种群分成分离的姊妹种，因为它们不能彼此交换基因。生物种由繁殖隔离机制来保持。

3 Island biogeography: MacArthur and Wilson's theory of island biogeography states that the number of species found on an island is determined by a dynamic equilibrium between the immigration of new colonizing species and the extinction of previously established ones. As the number of colonizing species increases, the number of immigrants arriving on the island decreases over time. In contrast, as competition among species becomes more intense, the extinction rate increases. The point at which extinction and colonization rates are equal gives the number of species at equilibrium.

岛屿生物地理学：麦克阿瑟和威尔逊的岛屿生物地理学理论指出，岛屿上物种的数目是由新移植来的物种和以前存在物种的灭绝之间的动态平衡决定的。当移植种的数目增加时，到达岛屿的移植来的物种的数目会随着时间的推移而减少。相反，当物种之间的竞争变得强烈时，灭绝的速率就会增加。当灭绝和移植的速率达到相等时，物种的数目就处于平衡稳定状态。

4 Allopatric speciation: Allopatric speciation occurs when the new species evolves in geographic isolation from the parent species. It may take place most readily in peripheral isolates, small populations at the extreme edge of a species range. The combined effect of a small atypical population and extreme environmental conditions can cause rapid and extensive genetic reorganization (a genetic revolution) leading to speciation.

异地种形成：与原来种由于地理隔离而进化形成新种，从而形成异地性物种。异地性物种形成最易发生在边缘隔离、处在种分布区的极端边缘的小种群。小的非典型种群与极端环境条件的混合作用可产生迅速而广泛的遗传重组(遗传革命)，从而导致物种形成。

5 Parapatric speciation: This form of speciation occurs where the speciating populations are contiguous but subject to different environmental (e.g. climatic) conditions in different parts of the range. Intermediate hybrids are found, as in ring species, but the large distances involved prevent the two types from merging completely.

邻地种形成：邻地性物种形成发生在分布区相邻，但分布区内不同地点环境(如气候)条件不同的种群。如在环形种中那样，可发现中间杂种，但其间很大的距离使两种类型不能完全混合。

6 Sympatric speciation: Sympatric speciation may occur where there is no geographical separation between the speciating populations of habitat preference. Whether sympatric speciation happens at all is a contentious issue, although rapid changes in the host preference of phytophagous insects have been observed. Plants can undergo sympatric speciation through polyploidy.

同地种形成：同地性物种形成可能发生在没有地理隔离，但具有宿主选择差异、食物选择差异或生境选择差异的种群。同地性物种形成是否发生尚有争议，尽管已观察到植食性昆虫宿主选择的快速变化。植物可以通过多倍体进行同地性物种形成。

第十节 生物多样性的进化

Section 10 Evolution of Biological Diversity

生物多样性 biodiversity
古细菌 archaea
革兰氏染色 Gram's staining
巴斯德消毒法 pasteurization
外毒素 exotoxin
内毒素 endotoxin
溶菌周期 lytic period
溶原周期 lysogenic period
异形胞 heterocyst
藻殖段 hormogonia
外生孢子 exospore
内生孢子 endogenous spore
原生生物 protest
接合生殖 conjugation
孢子植物 spore plant
颈卵器植物 archegoniatae
子实体 fruitbody
孢子 spore
配子 gamete
孢子体 sporophyte
配子体 gametophyte
原丝体 protonema
假根 rhizoid
原叶体 prothallus
颈卵器 archegonium
精子器 antheridium
维管植物 vascular plant
世代交替 alternation of generations
辐射对称 radial symmetry
同形世代交替 isomorphic alternation of generations
异形世代交替 heteromorphic alternation of generations
同配生殖 homogamy
异配生殖 heterogamy
卵式生殖 oogamy
原体腔 primitive body cavity
同律分节 homonomous segmentation
后口动物 deuterostome
羊膜动物 amniote
脊索动物 chordata
系统树 phylogenetic tree

1 Biodiversity: is the degree of variation of life forms within a given species, ecosystem, biome, or an entire planet. Biodiversity is a measure of the health of ecosystems. Biodiversity is in part a function of climate. In terrestrial habitats, tropical regions are typically rich whereas polar regions support fewer species.

生物多样性：是特定物种、生态系统、生物群系或整个星球内部生命形式的变异程度。生物多样性是生态系统健康的量度，某种程度上，生物多样性是气候的作用结果。在陆地生境中，热带区域是典型的多样性丰富区域而极地只有较少的物种。

2 Spore plants: do not form the life cycle of seed, the main use of spore on breeding of plants, including algae plants, fungi, lichens, mosses and ferns, etc.

孢子植物：生活史中不形成种子，主要利用孢子进行繁殖的植物，包括藻类植物、菌类、地衣、苔藓植物和蕨类植物等。

3 Sporophyte: the multicellular, diploid phase of a plant life cycle; it is the generation that ultimately produces spores.

孢子体：植物生活史中的多细胞、二倍体阶段；是最终产生孢子的世代。

4 Gametophyte: the multicellular, haploid phase of a plant life cycle in which gametes are produced by mitosis.

配子体:植物生活史中的多细胞、单倍体阶段,在这个阶段通过减数分裂产生配子。

5 Species diversity: the species diversity of a community depends on the number of different species it contains (the species richness) and the evenness of species abundance across species. Diversity indices can be calculated to take into account both of these factors. Diversity can be measured over the three different spatial scales of the local community, the region and the broadest geographic scale (e. g. the continent). This yields alpha (α)-, beta (β)-and gamma (γ)-diversity.

物种多样性:一个群落的物种多样性取决于群落含有的不同种的数量,即种丰富度和物种丰度的均匀性,多样性指数可以通过这两个因子计算出来。多样性能够在局域群落(local community)、地区(region) 和最广阔的地理尺度(geographic scale)(如大陆) 三个不同的空间尺度上测算它,这样就产生了 a、β 和 γ 多样性。

6 Keystone species: a keystone species has a significant and disproportionate effect on the community. Keystone species can be top predators such as the northern sea otter. However, the term can be usefully applied to any species whose removal would have a significant effect on community structure.

关键物种:关键物种在群落中有一个重要的和不成比例的作用。关键物种可能是顶端的捕食者,像北方的海獭。然而,这个词可以运用在任何一个物种上,只要这个物种被移去时会对群落结构造成重大影响。

7 Vascular plants: (also known as tracheophytes or higher plants). Vascular plants are those plants that have lignified tissues for conducting water, minerals, and photosynthetic products through the plant. Vascular plants include the clubmosses, equisetum, ferns, gymnosperms (including conifers) and angiosperms (flowering plants).

维管植物:(也称为导管植物或高等植物) 指那些具有木质化的组织,能输导水分、矿物质和光合作用产物的植物,维管植物包括石松类、木贼属、蕨类,裸子植物(包括松柏类) 和被子植物(有花植物)。

第十一节 生态学和动物行为
Section 11 Ecology and Animal Behavior

生态学 ecology
环境 environment
生态因子 ecological factor
最小因子法则 law of minimum
耐受性法则 law of tolerance
种群 population
种群密度 population density
年龄组成 age composition
出生率 birth rate
死亡率 mortality
群落 community
优势种 dominant species
生态位 niche
生物群落的结构 biological community structure
演替群落 community succession
顶极群落 climax
生态系统 ecological system
生产者 producer
消费者 consumer
分解者 disintegrator
食物链 food chain
营养级 trophic level
生态金字塔 ecological pyramid
初级生产量 primary production
生物量 biomass
植物生物量 plant biomass
物质循环 material circulation
碳的循环 carbon cycle
能量流动 energy flow
系统的能量流 energy flow of system
生态效率 ecological efficiency
生态平衡 ecological equilibrium
生物富集作用 biological enrichment
自然因素 natural factor
人为因素 human factor
就地保护 in-situ conservation

自然保护区 nature reserve
迁地保护 ex-situ conservation
水体富营养化 water body eutrophication
生物净化 biological purification
绿色食品 green food
行为 behavior
本能 instinct
固定行为型 fixed action pattern
学习 learning
习惯性 habituation
印记 imprinting
联系学习 associative learning
顿悟学习 insight learning
利他行为 altruistic behavior
生物节律 biological rhythm

1 A definition of ecology: Ecology is the study of the interactions between organisms and their environment. The 'environment' is a combination of the physical environment (temperature, water availability, etc.) and any influence on an organism exerted by other organisms in the biotic environment.

生态学的定义:生态学是研究有机体与其环境相互作用的科学。"环境"是物理环境(温度、可利用水等)和生物环境(对有机体的、来自其他有机体的任何影响)的结合体。

2 Adaptation: Any heritable trait possessed by an organism which aids survival or reproduction is an adaptation. Such traits may be physiological, morphological or behavioral. Adaptation is the result of natural selection.

适应:有机体所具有的有助于生存和生殖的任何可遗传特征都是适应。适应性特征可以是生理的、形态的或行为的。适应是自然选择的结果。

3 Tolerance: Organisms can cope with variation in their external environment (though different species may differ markedly). The upper and lower extremes of environment condition under which members of species can survive are species limits of tolerance.

耐受性:有机体能够应付其外部环境的变化(虽然不同物种表现很不相同)。种内成员能够生存的环境条件上限和下限是种的耐受限度。

4 Negative feedback: Most biological homeostatic mechanisms act in a broadly similar way: if the current internal level of a factor (e. g. temperature or osmolarity) is too high, the mechanism will reduce it; if the level is too low the mechanism will increase it. This process, by which the response is opposite to the signal, is known as negative feedback.

负反馈:大多数生物的稳态机制以大致一样的方式起作用:如果一个因子的内部水平(如温度或渗透性)太高,该机制将减少它;如果水平太低,就提高它。这个过程叫作负反馈,负反馈反应的方向与信号反应相反。

5 Niche: The ecological niche of an organism is the position it fills in its environment, comprising the conditions under which it is found, the resources it utilizes and the time it occurs there.

生态位:有机体的生态位是它在环境中所处的位置,包括它发现的各种条件、所利用的资源和在那里的时间。

6 Ecological effects of temperature: The rate of an enzyme catalyzing reaction increases with temperature. The temperature coefficient (Q10) is an index of the effect of a 10℃ temperature rise on metabolic rate, and is often near 2.0. Within the nonlethal temperature range the most important effect on organism of temperature is likely to be its effect on growth and development.

温度的生态作用:酶催化反应的速度随温度而增加(温度系数(Q 10)是温度升高10℃对代谢速度影响的指数,经常大约为2.0)在非致死温度范围内,温度对生物最大的影响很可能是影响了生长和发育。

7 Allen's rule: The effects of temperature on individuals may be moderated by evolved differences. Allen's rule states that endothermic animals from cold climates tend to have shorter extremities (ears and legs) compared with animals from warmer climates, thus reducing their surface area: volume ratio. This rule has widespread applicability.

阿伦法则：阿伦法则陈述了来自冷气候中的内温动物与来自温暖气候的内温动物相比，趋向于具有更短的末端(耳朵和四肢)，因此，降低了它们的表面积对体积的比率。

8 Bergmann's rule: states that mammals tend to be larger in colder areas than warm climates, again to reduce their surface area: volume ratio.

贝格曼规律：讲述了寒冷地区的哺乳动物比温暖地区的哺乳动物个体趋向于更大，也减低了它们的表面积与体积的比率。

9 Population: A population is a group of organisms of the same species which occupies a given area. The boundaries between populations can be arbitrary.

种群：是一定区域内同种生物个体的集合。种群间的边界可以是任意的。

10 The community: is an assemblage of species populations that occur together in the same place at the same time.

群落：是在相同时间聚集在同一地段上的许多物种种群的集合。

11 Density-dependent growth-the logistic equation: The logistic equation describes the growth of a simple population in a confined space, where resources are not unlimited. In the early stages resources are abundant, the death rate is minimal and reproduction can take place as fast as possible allowing the individuals to attain their intrinsic rate of increase. The population increases geometrically until the maximum number of individuals the environment can sustainably support is approached. This maximum number is called the carrying capacity (K). The population growth rate declines to zero as the population becomes more crowded and the population size stabilizes.

密度 — 制约性种群增长 — 逻辑斯蒂方程：逻辑斯蒂方程描述的是一个在有限资源空间中的简单种群的增长。在早期，资源丰富，死亡率最小，繁殖尽可能的快，种群内个体可达到内禀增长率。种群呈几何式增长，直到种群数量达到环境可持续支持的最大程度，即环境容纳量(K)。当种群更加拥挤时，种群增长率减少到零，种群大小处于稳定状态。

12 Succession — the classical model: Ecological succession is defined as a continuous, unidirectional, sequential change in the species composition of a natural community. This sequence of community is termed a sere, and culminates in the climax community. Early successional stages are characterized by pioneer species, low biomass and often low nutrient levels. Community complexity increases as succession progresses, often peaking in the mid-successional stage. A mid-successional community is characterized by high biomass, high levels of organic nutrients and high species diversity.

演替 —— 经典模型：生态演替是指在一个自然群落中，物种的组成连续地、单方向地、有顺序地变化。这一顺序被称为是一个演替序列，最后达到的阶段称为顶级。早期的演替阶段具有先锋物种、低生物量和低营养水平的特征。随着演替的进行，群落的复杂性增加，通常在演替的中期阶段，复杂性达到最大。一个中期的演替群落具有高生物量、高有机营养水平和高的物种多样性。

第十五章　细胞生物学
Chapter 15　Cell Biology

第一节　绪论
Section 1　Introduction

细胞　cell
细胞学说　cell theory
细胞学　cytology
细胞生物学　cell biology
细胞外被　cell coat
细胞衰老　cell senescence/ cellular aging
细胞凋亡　apoptosis
细胞融合　cell fusion
细胞液　cell sap
细胞株　cell strain
细胞系　cell line
细胞表面受体　cell surface receptor
细胞壁　cell wall
细胞分化　cell differentiation
细胞周期　cell cycle
细胞分裂　cell division
细胞通讯　cell communication
细胞工程　cell engineering
非细胞体系　cell-free system
生物膜　biomembrane
分子生物学　molecular cell biology

1 Cell wall: A cell wall is a rigid layer that surrounds the plasma membrane of some cells (such as bacteria, yeast, and plant cells) and consists of extracellular molecules produced by the cells. Most animal cells do not have a cell wall. The cell wall of bacteria consists mainly of peptidoglycan, which forms a layer of the cell envelope that protects and supports the cell. The major component of plant cell walls is the polymer cellulose, which is organized into extremely strong fibers.

细胞壁：细胞是围绕在细胞质膜外的一层坚硬的结构(例如细菌、真菌和植物细胞)，由细胞产生的胞外分子组成。大多数动物细胞没有细胞壁。细菌的细胞壁主要由肽聚糖组成，作为一层细胞包被保护和支持细胞。植物细胞壁的主要成分是多聚糖，由它组成强韧的纤维。

2 Cell cycle: The cell cycle is an ordered set of stages that results in the accurate division of one cell into two.

细胞周期：细胞周期指在细胞一个变成两个的精确的分裂过程中一系列有规律发生的阶段。

3 Senescence: Senescence is the process by which cells lose the ability to divide, often after prolonged propagation in culture *in vitro*.

衰老：指细胞失去分裂能力的过程，在离体培养中细胞多次增殖之后经常发生。

第二节　细胞的基本知识
Section 2　Basic Knowledge about the Cell

细胞膜　cell membrane
细胞增殖　cell proliferation
细胞繁殖周期　cell reproductive cycle
细胞识别　cell recognition
衣壳、壳体　capsid
病毒　virus
类病毒　viroid
腺病毒　adenovirus
支原体　mycoplasma
古细菌　archaea
原细菌，古细菌　archaeobacteria
真细菌　eubacteria
细胞癌基因　cellular oncogene
病毒癌基因　viral-oncogene
(细胞)质膜　plasma membrane
原核细胞　prokaryotic cell
真核细胞　eukaryotic cell
常染色体　euchromatin
无丝分裂　amitosis
蓝细菌　cyanobacteria

双链 DNA 病毒 dsDNA virus
双链 RNA 病毒 dsRNA virus
鞭毛蛋白 flagellin
盐细菌 halobacteria
中膜体 mesosome
有丝分裂器 mitotic apparatus
原核生物 prokaryote
拟核,类核 nucleoid
核壳 nucleocapsid
复制子 replicon

1 Plasma membrane: Plasma membrane is the continuous membrane defining the boundary of every cell.
质膜:质膜是界定细胞边界的连续分布的膜。

2 Prokaryote: A prokaryote is a unicellular organism that does not have a membrane-bounded nucleus.
原核生物:单细胞有机体,没有膜包被的细胞核。

3 Eukaryote: A eukaryote is an organism that is composed of one or more cells, each containing a nucleus. The term eukaryote means "true nucleus".
真核生物:由一个或多个细胞组成的有机体,每个细胞都含有细胞核,这是真正意义上的细胞核。

4 Nucleoid: The nucleoid is the structure in a prokaryotic cell that contains the genome. The naked DNA is bound to proteins and is not enclosed by a membrane.
拟核:原核细胞中含有基因组的结构区域,无膜包绕,裸露的 DNA 与蛋白相连。

5 Euchromatin: Euchromatin comprises most of the genome in the interphase nucleus, is less tightly coiled than heterochromain, and contains most of the active or potentially active single copy genes.
常染色质:间期核中处于分散状态,与异染色质相比其压缩程度相对较低、着色较浅的染色质,主要包含有活性的或有潜在活性的单拷贝基因。

6 Heterochromatin: heterochromatin describes regions of the genome that are highly condensed, not transcribed, and late-replicating. Heterochromatin is divided into two types, which are called constitutive and facultative.
异染色质:在细胞间期保持高度凝聚状态、不具有转录活性、复制较晚的染色质,分为两类:组成性异染色质和兼性异染色质。

7 Mitosis: Mitosis is part of the process of eukaryotic cell division by which cells reproduce themselves, resulting in daughter cells that contain the same amount of genetic material as the parent cells. During mitosis, the replicated chromosomes separate and the nucleus divides.
有丝分裂:有丝分裂是真核细胞分裂过程的一种方式,通过复制,子细胞含有和母细胞等量的遗传物质。在这个过程中,复制后的染色体分离,核分裂。

8 Oncogenes: Oncogenes are genes whose expression products have the ability to transform eukaryotic cells so that they grow in a manner analogous to tumor cells.
癌基因:癌基因是指那些表达产物能够转入真核细胞,并以瘤细胞相类似的方式生长的基因。

第三节 细胞生物学研究方法
Section 3 The Research Method of Cell Biology

光学显微镜技术 light microscopy
荧光显微镜技术 fluorescence microscopy
电子显微镜 electron microscope
绿色荧光蛋白 green fluorescent protein(GFP)
分辨率 axial resolution
激光共焦点扫描显微镜 laser scanning confocal microscope
微分干涉显微镜 differential-interference microscope
录像增差显微镜技术 video-enhance microscopy
冷冻蚀刻 freeze-etching
低温电镜技术 cryoelectron microscopy
差速离心 differential centrifugation
电融合技术 electron fusion method
成纤维样细胞 fibroblast like cell
流式细胞仪 flow cytometry(FCM)
异核体,异核融合细胞 heterokaryon
同核体,同核融合细胞 homokaryon
传代(培养)细胞 subculture cell
融合核细胞 synkaryon

原位杂交 *in situ* hybridization
免疫荧光技术 immunofluorescence technique
单克隆抗体 monoclonal antibody
显微分光光度测定技术 microspectrophometry
微量注射 microinjection
显微操作技术 micromanipulation
突变体 mutant
蛋白质组学 proteomics
快速冷冻深度蚀刻 quick freeze deep etch
放射免疫沉淀 radioimmuno-precipitation
扫描电子显微镜 scanning electron microscope (SEM)
扫描隧道显微镜 scanning tunneling microscope (STM)
超薄切片 ultra thin section
免疫印迹法 western blotting
原生质体 protoplast
液泡 vacuole
核体 karyoplast
生物信息学 bioinformatics

1 *In situ* hybridization: *In situ* hybridization (also called cytological hybridization) is performed by denaturing the DNA of cells squashed on a microscope slide so that reaction is possible with an added single-stranded RNA or DNA probe; the added preparation is radioactively labeled and its hybridization is followed by autoradiography.

原位杂交:原位杂交也叫细胞杂交,其操作为将细胞在载物台上破碎后使DNA变性,加入单链的放射性标记的RNA或DNA探针,对细胞或组织中的基因或mRNA进行定位的技术,然后进行放射自显影。

2 Proteomics: Proteomics is the study of protein sequences and structural characteristics in a systemized (and usually computerized) way.

蛋白质组:指以系统性(通常为计算化的)的方式研究蛋白质序列和结构特点的科学。

3 Transmission electron microscope(TEM): Transmission electron microscope allows electrons to pass through a thin section of the object, resulting in a detailed view of the object's structure.

透射电子显微镜:电子穿透样品薄片后在显示屏上其结构成像的显微镜。

第四节 细胞膜与细胞表面
Section 4 Cell Membrane and Cell Surface

心磷脂 cardiolipin
透明质酸 hyaluronic acid
细菌紫膜质 bacteriorhodopsin
黏着带 adhesion belt
附着蛋白 attachment protein
带状桥粒 belt desmosome
半桥粒 hemidesmosome
钙黏素 cadherin
胶原 collagen
连接子 connexion
膜内在蛋白 integral membrane protein
整联蛋白 integrin
层粘连蛋白 laminin
凝集素 lectin
本质素 lignin
流动镶嵌模型 fluid mosaic model
脂伐模型 lipid rafts model
单位膜模型 unit membrane model
脂质体 liposome
跨膜蛋白 transmembrane protein
运动蛋白 movement protein
膜骨架 membrane associated cytoskeleton
去垢剂 detergent
电偶联 electrical coupling
电突触 electronic junction
细胞外基质 extracellular matrix
细胞外小页断裂面 ertrocytoplasmic face(EF)
细胞外表面 ertrocytoplasmic surface(ES)
纤连蛋白 fibronectin
光脱色恢复技术 fluorescence recovery after photobleaching (FRAP)
功能基因组学 functional genomics
间隙连接 gap junction
封闭连接 occluding junctions
紧密连接 tight junction
黏着斑 focal adhesion
糖萼 glycocalyx
血型糖蛋白 glycophorin
糖氨聚糖 glycosaminoglycan
成斑现象 patching
植物抗毒素 phytoalexin
胞间连丝 plasmodesmata

孔蛋白 porin
原代(培养)细胞 primary culture cell
早前胶原 preprocollagen
蛋白聚糖 proteoglycan
原生质小页断裂面 protoplasmic face(PF)
原生质表面 protoplasmic surface(PS)
红膜肽,血影蛋白 spectrin
突触 synapse
选择素 selectin

1 Selectin: Selectins are a class of cell-cell adhesion receptors expressed only on vascular cells.
选择素:一类细胞黏着的受体家族,只在血管细胞上表达。

2 Hemidesmosome: The hemidesmosome is a cell junction that links intermediate filaments in cells to the basal lamina via cell surface integrins.
半桥粒:一种细胞连接方式,通过整合素连接中间纤维和基底膜。

3 Cadherin: A cadherin is a cell surface receptor that forms dimers and participates in cell-cell adhesion. Some cadherins are found in adherens junctions.
钙黏蛋白:一种细胞表面受体,多以二聚物或多聚体形式参与细胞与细胞间的黏着。一些钙黏蛋白在黏合连接中发现。

4 Focal adhesion: A focal adhesion is a large protein complex that develops from a focal complex and has assembled a strong link with actin stress fibers. The proteins in focal adhesions initiate intracellular signaling pathways in response to cell-matrix interactions.
黏合斑:一个黏合斑是一个巨大的蛋白复合体,起始于黏合复合物,与肌动蛋白纤维紧密连接。黏合斑中的蛋白质启动应答细胞-基质互作的胞内信号传递。

5 Gap junction: A gap junction is a cell-cell junction that allows the direct transfer of small cytosolic components between adjacent cells.
间隙连接:间隙连接是细胞与细胞间连接的一种方式,允许小的溶质分子在相邻的细胞间直接转移。

6 Tight junction: Tight junctions are cell-cell junctions that prevent most molecules from diffusing in the spaces between the cells.
紧密连接:紧密连接是细胞与细胞间连接的一种方式,阻止大多数分子扩散到细胞间隙中。

7 Plasmodesmata: Plasmodesmata are narrow channels that connect the cytoplasm of adjacent plant cells and allow material to pass between them. Each plasmodesma is formed by a tube of plasma membrane extending between the cells and has a thin tubule of endoplasmic reticulum running down its center.
胞间连丝:相邻植物细胞之间细胞质联系的狭窄的通道,允许物质流通。胞间连丝是由相邻细胞的质膜延伸而成的管状结构,中央是由内质网形成的细管。

第五节 物质的跨膜运输与信号传递
Section 5 Transmembrane Transport of the Material and Signals Transmission

动作电位 action potential
协助扩散 facilitated diffusion
被动运输 passive transport
主动运输 active transport
通讯连接 communicating junction
接触性依赖通讯 contact-dependent signaling
接触抑制 contact inhibition
共转移 cotranslocation
协同运输 cotransport
C4 途径 C4 pathway
组成型胞吐途径 constitutive exocytosis pathway
离子通道偶联受体 ion-channel-linked receptor
接头蛋白 adaptor protein
接合素蛋白 adaptin
钙调蛋白 calmodulin(CaM)
钙调蛋白激酶 CaM-kinase
载体蛋白 carrier protein
催化性受体 catalytic receptor
有被小窝 coated pit
有被小泡 coated vesicle
网格蛋白 clathrin
网格蛋白有被小窝 clathrin coated pit
网格蛋白有被小泡 clathrin coated vesicle
自分泌 autocrine

旁分泌 paracrine
批量运输 bulk transport
胱氨酸尿 cystinuria
去极化 depolarization
二聚化 dimerization
双信使系统 double messenger system
下降调节 down regulation
早胞内体 early endosome
电化学梯度 electrochemical gradient
胞吞泡 endocytic vesicle
胞吞作用 endocytosis
表皮生长因子 epidermal growth factor (EGF)
伸展蛋白 extention protein
弹性蛋白 elastin
融合复合物 fusion complex
融合蛋白 fusion protein
鸟苷酸环化酶 G-cyclase(GC)
G蛋白偶联的受体 G-protein linked receptor
细胞内受体 intracellular receptor
配体门离子通道 ligand-gated channel
长时程增强作用 long-term potentiation (LTP)
低密度脂蛋白 low-density lipoprotein (LDL)
膜转运蛋白 membrane transport protein mitogen-activated protein kinase (MAPK)
分子开关 molecular switch
神经信号 neuronal signaling
肌球蛋白轻链磷酸酶 myosin light chain phosphatase(MLCP)
渗透压 osmotic pressure
乌本苷 ouabain
一氧化氮合酶 nitric oxide synthase(NOS)
肽信号 peptide signal
通透酶 permease
吞噬作用 phagocytosis
(激酶)磷酸化级联反应 phosphorylation cascade reaction
胞饮作用 pinocytosis
多向性效应 pleiotropic effect
受体 receptor
受体下行调节 receptor down-regulation
受体介导的胞吞作用 receptor mediated endocytosis
调节型胞吐途径 regulated exocytosis pathway
静息电位 resting potential
第二信使 second messenger
第二信使学说 second messenger theory
信号分子 signal molecule
信号网络系统 signal network system
信号转导 signal transduction
细胞信号通路 signaling pathway
简单扩散 simple diffusion
压力激活通道 stress-activated channel
电压门通道 voltage-gated channel
超极化 super polarization
开关蛋白 switch protein
共运输 symport
膜泡运输 vesicular transport
跨细胞的转运 transcytosis
递质门(离于)通道 transmitter-gated channel
酪氨酸蛋白激酶偶联受体 tyrosine kinase-linked receptor

1 Action potential: An action potential is a wave of electrical signals that is propagated rapidly and transiently along the plasma membrane of excitable cells such as neurons and some muscle and endocrine cells. It occurs when the membrane potential increases to a threshold level that triggers the opening and closing of voltage-sensitive ion channels.

动作电位：动作电位是可兴奋细胞质膜上产生的一种快速、短暂的电信号，例如神经元、一些肌肉细胞和内分泌细胞。动作电位在膜电位升高达到电位门离子通道开、关所需的阈值时发生。

2 Ion channel: A ion channel is a transmembrane protein that selectively allows the passage of one or more types of ions across the membrane. Ions pass through a central aqueous pore of the channel.

离子通道：离子通道是选择性允许一种或多种特定离子通过生物膜的跨膜蛋白。离子由其位于中央的亲水性通道通过。

3 Clathrin: Clathrin proteins interact with adaptor complexes to form the coat on some of the vesicles that bud from the cytoplasmic face of the plasma membrane and the *trans*-Golgi network. Clathrin is composed of heavy and light chains that form triskelions, which then assemble into polyhedral-curved lattices during the formation of clathrin-coated pits and vesicles.

网格蛋白：网格蛋白与接合素蛋白相互作用形成有被分泌小泡，这些小泡由细胞膜的胞质面和高尔基体反面膜囊出芽产生。网格蛋白由 3 条重链和 3 条轻链组成，在网格有被小窝、网格有被小泡形成过程中组装成多面体笼形结构。

4 Receptor: A receptor is generally a plasma membrane protein that binds a ligand in a domain on the extracellular side and, as a result, changes the activity of the cytoplasmic domain. The same term is also used for nuclear receptors, which are transcription factors that are activated by binding ligands such as steroids or other small molecules.

受体：受体一般与配体的跨膜结构域相连，从而改变胞内结构域的活性。这一术语也被用于核受体，它们作为转录因子通过连接象类固醇或其他小分子这样的配体而被激活。

第六节 细胞质基质与细胞内膜系统

Section 6 Cytoplasmic Matrix and Endomembrane System

内吞溶酶体 endolysosome
内质网 endoplasmic reticulum, ER
内膜系统 endomembrane system
嵴 cristae
细胞质基质 cytoplasmic matrix, cytomatrix
细胞质臂动蛋白 cytoplasmic dynein
协助装配 aided-assembly
氨酰 tRNA 蛋白转移酶 aminoacyl-tRNA-protein transferase
装配 assembly
自噬溶酶体 autophagic lysosome
原子力显微镜 atomic force microscope
高尔基体顺面网状结构 cis Golgi network(CGN)
复杂的 N-连接寡糖 complex N-linked oligosaccharide
胞质溶胶 cytosol
停泊蛋白 docking protein
逃逸蛋白 escaped protein
门控转运 gated transport
糖基转移酶 glycosyl transferase
己醛酸循环体 glyoxysome
高尔基体(器) Golgi body/Golgi apparatus
高尔基复合体 Golgi complex
高甘露糖 N-连接寡糖 high mannose N-linked oligosaccharide
高尔基体中间膜囊 medial Golgi
微体 microbody
微粒体 microsome
单核细胞 monocyte
黏蛋白 mucin
N-连接糖基化 N-linked glycosylation
O-连接糖基化 O-linked glycosylation
神经多肽 neuropeptide glycosylation
过氧化物酵体蛋白分选信号 peroxisomal-targeting signal
过氧化物酶体 peroxisome
吞噬细胞 phagocyte
吞噬溶酶体 phagocysosome
磷酸葡萄糖苷酶 phospho-glucosidase
磷脂转换蛋白 phospholipid exchange protein(PEP)
磷脂转位因子 phospholipid translocator
后转移 post translation
初级溶酶体 primary lysosome
蛋白二硫键异构酶 protein disulfide isomerase
蛋白酶体 protosome
蛋白质定向转运 protein targeting
蛋白质分选 protein sorting
残余小体 residual body
回收信号(序列) retrieval signal
逆行转运 retrograde transport
粗面内质网 rough endoplasmic reticulum (rER)
肌质网 sarcoplasmic reticulum
次级溶酶体 secondary lysosome
分泌溶酶体 secretory lysosome
信号假说 signal hypothesis
信号区 signal patch
信号识别颗粒 signal recognition particle (SRP)
信号序列，信号肽 signal sequence/signal peptide
依赖泛素途径 ubiquitin-dependent pathway
起(开)始转移序列 start transfer sequence
停止转移序列 stop transfer sequence
自我装配 self-assembly
高尔基反面网状结构 trans Golgi network (TGN)
易位子 translocon

1 Endoplasmic reticulum(ER): The endoplasmic reticulum is an organelle involved in the synthesis of lipids, membrane proteins, and secretory proteins. It is a single compartment extending from the outer layer of the nuclear envelope into the cytoplasm and has subdomains such as the rough ER and

the smooth ER. It is also the intracellular storage compartment for calcium. In striated muscle cells in the heart and skeletal muscle, it is commonly referred to as the sarcoplasmic reticulum, which has a specialized structure and function.

内质网：内质网是涉及脂质、膜蛋白和分泌蛋白合成的一种细胞器。由核被膜的外膜延伸至细胞质中形成，分为糙面内质网和光面内质网两类。它也是细胞内的钙库。在心脏平滑肌细胞和骨骼肌细胞中，肌质网相当于内质网，是具有特定结构和功能的内质网。

2 Signal sequence: Signal sequence is a short region of a protein that directs it to the endoplasmic reticulum for cotranslational translocation. The term signal sequence is also used more generally to refer to a sorting signal.

信号序列：信号序列是一段短的氨基酸序列，指导蛋白质进入内质网进行共翻译。这一术语也广泛用于分选信号相关的领域。

3 Signal recognition particle(SRP): The signal recognition particle is a ribonucleoprotein complex that recognizes signal sequences during translation and guides the ribosome to the translocation channel. SRPS from different organisms may have different compositions, but all contain related proteins and RNAs.

信号识别颗粒：信号识别颗粒是一种核蛋白复合物，翻译时识别信号序列，指导核糖体结合翻译链。不同组织的信号识别颗粒有不同的组成，但所有的信号识别颗粒都包含相关的蛋白和RNA。

4 Golgi apparatus: The Golgi apparatus is an organelle that receives newly synthesized proteins from the endoplasmic reticulum and processes them for subsequent delivery to other destination. It is composed of several flattened membrane disks arranged in a stack. Plants contain multiple Golgi bodies, which are called dictyosomes.

高尔基复合体：高尔基复合体作为一种细胞器，接收内质网新合成的蛋白质并经过分选后把它们运送到不同的目的地。高尔基复合体由多个扁平的膜囊组成，植物细胞含有多个高尔基体，也叫网体。

第七节　细胞的能量转换——线粒体和叶绿体

Section 7　Energy Convertion of the Cell-Mitochondria and Chloroplast

叶绿体　chloroplast
叶绿体膜　chloroplast membrane
叶绿素　chlorophyll
叶绿素 a/b 结合蛋白　chlorophyll a/b binding protein
腺苷酸环化酶　A-cyclase
自磷酸化　autophosphorylation
ATP 合成酶　ATP synthetase
肌动蛋白　actin
顶体精子　acrosome
布朗棘伦模型　Brownian rachet model
卡尔文循环　Calvin cycle
詹纳斯绿 B　Janus green B
二氧化碳同化　CO_2 assimilation
碰撞假说　collision hypothesis
构象偶联假说　conformational coupling hypothesis
景天酸代谢　crassulaceae acid metabolism (CAM)
循环式光合磷酸　cyclic photophosphorylation
细胞色素　cytochrome
暗反应　dark reaction
电子传递链　electron-transport chain
铁氧还蛋白　ferredoxin(Fd)
基粒片层　grana lamella
基粒类囊体　granum thylakoid
前片层体　prolamellar body
膜间隙　inter membrane space
嵴内间隙　intracristal space
导肽，前导肽　leader peptide
基质　matrix
分子伴侣　molecular chaperone
非循环式光合磷酸化　noncyclic photophosphorylation
(细)胞核　nucleus
光合磷酸化　photophosphorylation

1 Chloroplast: Chloroplasts are the plant organelles which house photosynthesis.

叶绿体：植物细胞中发生光合作用的主要细胞器。

2 Nucleus: The nucleus is a cellular organelle that contains the genomic DNA. It is surrounded by a nuclear envelope consisting of a double layer of membranes and is the site of DNA replication, gene transcription, and most RNA processing.
细胞核：细胞核是含有基因组 DNA 的细胞器，由双层膜组成的核膜所包被，是 DNA 复制、基因转录和大部分 RNA 加工的场所。

3 Thylakoid: The thylakoids of plants are organelles in which the light reactions of photosynthesis, along with the proteins responsible for electron transfer, NADPH and ATP formation, are localized.
类囊体：叶绿体内的扁平膜囊，在光合作用中，是光反应的功能位点，其上具有和电子传递、NADPH 与 ATP 合成相关的蛋白质。

4 Photosystem: Photosystems are membrane-bound complexes within the thylakoid membrane of chloroplasts which absorb light and transport electrons.
光系统：光系统是叶绿体的内囊体膜上的膜结合复合物，具有吸收光能、传递电子的功能。

5 Electron transport chain(ETC): The electron transport chain is a series of proteins that transfer electrons in cellular respiration to generate ATP.
电子传递链：在细胞呼吸作用中由一系列蛋白组成的传递电子、产生 ATP 的电子传递途径。

第八节　细胞核与染色体
Section 8　The Nucleus and Chromosomes

染色质　chromatin
染色体　chromosome
染色单体　chromatid
染色单体连接蛋白　chromatid linking protein
染色中心　chromocenter
染色粒　chromomere
中央颗粒　central granule(CG)
中心成分　central domain
中央栓　central plug
中心粒　centriole
中心体　centrosome
中心体(或中心粒)周期　centrosome cycle/centriole cycle
有色体　chromoplast
染色体重排　chromosomal rearrangement
染色体列队　chromosome alignment
染色体支架　chromosomal scaffold
通道蛋白　channel protein
活性染色质　active chromatin
中央结构域　central domain
收缩环　contractile
锚定连接　anchoring junction
锚蛋白　ankyrin
环带亚单位　annular subunit
顺行转运　anterograde transport
对象运输　antiport
反义寡聚核苷酸　antisense oligonucleotide
构件因子　architectural factor
自主复制 DNA 序列　autonomously relicating DNA sequence
AR 着丝粒 DNA　centromere DNA sequence (CEN)
人造微小染色体　artificial minichromosome
顺式作用元件　Cis-acting element
亲核蛋白　karyophilic protein
核型　karyotype
关键序列　key sequence
动粒结构域　kinetochore domain
螺旋体　coiled body(CB)
柱状亚单位　column subunit
共有序列　consensus sequence
组成型异染色质　constitutive heterochromatin
细胞滞留因子　cytoplasmic retention factor
胞质环　cytoplasmic ring
致密纤维组分　dense fibrillar component(DFC)
双标记免疫荧光显微术　double label microscopy
脱皮激素　ecdysone
核内有丝分裂　endomitosis
二重对称轴　dyad axis
二分对称性　dyad symmetry
必需基因　essential gene
核仁外体　extranucleolar body
异染色质　heterochromatin
兼性异染色质　facultative heterochromatin
纤维中心　fibrillar center(FC)
纤维冠　fibrous corona
功能元件　functional element
遗传图谱　genetic map

基因组 genome
巨大染色体 giant chromosome
颗粒组分 granular component(GC)
螺旋-环-螺旋结构模式 helix-loop-helix motif
螺旋-转角-a 螺旋模式 helix-turn-a-helix motif
组蛋白乙酰转移酶 histone acetyltransferase
组蛋白去乙酰基化酶 histone deacetylase
激素答应元件 hormone response element
非活性染色质 inactive chromatin
抑制蛋白 inhibitor protein
内部着丝粒蛋白 inner centromere protein
内板 inner plate
间带 interband
核型模式图 ideogram
蛋白质印迹法,免疫印迹法 immunoblotting
亮氨酸拉链模式 leucine zipper motif
奢侈基因 luxury gene
连接 DNA linker DNA
腔内亚单位 luminal subunit
中着丝粒染色体 metacentric chromosome
非组蛋白 nonhistone
微球菌核酸酶 micrococcal nuclease
微带 miniband
微小染色体 minichromosome
分子货仓 molecular warehouse
小卫星染色体 minisatellite DNA
多级螺旋模型 multiple coiling model
负生长调节子 negative growth regulator
非必须基因 nonessential gene
核篮 nuclear basket
核被膜 nuclear envelope
核纤层 nuclear lamina
核基质 nuclear matrix
核孔复合体 nuclear pore complex(NPC)
核输出信号 nuclear export signal(ENS)
核定位信号 nuclear localization signal
NLS 核质环 nuclear ring
核骨架 nuclear skeleton
核仁 nucleolus
核仁周期 nucleolar cycle
核仁组织区 nucleolar organizing region (NOR)
核仁蛋白,核仁素 nuclein
核质蛋白 nucleoplasmin
核孔蛋白 nucleoporin(Nup)
核小体 nucleosome
核小体组蛋白 nucleosomal histone
核小体犁 nucleosome plow
核外膜 outer nuclear membrane
外板 outer plate
配对结构域 pairing domain
核周池 perinuclear cisternae
核周间隙 perinuclear space
核仁周边染色质 perinuclear chromatin
核仁相随染色质 nucleolar associated chromatin
多聚核糖体 polyribosome / polysome
多线染色体 polytene chromosome
聚乙烯吡咯烷酮 polyvinylpyrrolidone PVP
核膜区 pore membrane domain
优先敏感性 preferential sensitivity
前核仁体 prenuclear body(PNB)
胀泡 puff
反带 reverse band
核糖体 ribosome
核酶 ribozyme
随体 satellite
卫星 DNA satellite DNA
骨架-放射环结构模型 scaffold-radial loop structure model
主缢痕 primary constriction
次缢痕 secondary constriction
序列特异性 DNA 结合蛋白 sequence specific DNA-binding protein
小分子核仁核糖核蛋白 small nucleolar ribonucleo protein(snoRNP)
剪接体 spliceosome
螺线管 solenoid
辐 spoke
超螺线管 supersolenoid
近中着丝粒染色体 submetacentric chromosome
亚核结构域 subnuclear domain
近端着丝粒染色体 acrocentric chromosome
对称的同型二聚体 symmetric homodimer
应急因子 stringent factor
端着丝粒染色体 telocentric chromosome
端粒 DNA 序列 telomere DNA sequence
端粒酶 telomerase
四聚体 tetramer
转录辅阻抑制物 transcriptional co-repressor
上升调节 upregulation
锌脂模式 zinc finger motif

1 Chromatid: Chromatids are the copies of a chromosome produced by replication. The name is usually used to describe each of the copies in the period before they separate at the subsequent cell division.
染色单体:染色单体是复制产生的染色体的拷贝。通常用于描述细胞分裂间期染色单体分离前的染色体拷贝。

2 Centriole: A centriole is a small organelle found in animal cells. Each centriole is composed of nine specialized microtubules arranged to form a cylinder. Centrioles are found in the center of centrosomes and usually exist in pairs.
中心粒:中心粒是动物细胞中的小的细胞器。由 9 组特殊的微管组成的圆柱状结构,中心粒位于中心体的中央,通常成对存在。

3 Centrosome: A centrosome is a small organelle that acts as a microtubule-organizing center. Centrosomes form the poles of the mitotic spindle in animal cells. Each centrosome consists of a pair of centrioles surrounded by a large number of proteins.
中心体:中心体是作为微管组织中心的小的细胞器。在动物细胞中,中心体形成了有丝分裂纺锤丝的两极。每一个中心体都由一对中心粒及大量的蛋白质围绕而成。

4 Ribosome: The ribosome is a large complex of proteins and RNAs that directs the synthesis of proteins from messenger RNA(mRNA) in the process known as translation.
核糖体:核糖体是蛋白质和 RNA 组成的大分子复合物,在翻译过程中指导从 mRNA 到蛋白质的合成。

5 Nucleosome: The nucleosome is the basic structural unit of chromatin in the interphase chromosome. The nucleosome core particle consists of ～140bp of DNA and two copies each of the histones (H_2A, H_2B, H_3 and H_4).
核小体:核小体是细胞间期染色质的基本结构单位。核心颗粒由约 140bp 的 DNA 和组蛋白八聚体(H_2A, H_2B, H_3 and H_4 各两拷贝)组成。

6 Telomerase: Telomerase is a reverse transcriptase that contains its own RNA template, which it uses to create repeating units of one strand at the telomere by adding nucleotides to the 3' overhang.
端粒酶:一种反转录酶,能以自身 RNA 为模板,通过向 3' 增加核苷酸对 DNA 端粒序列进行延长而解决染色体末端复制问题。

7 Nuclear lamina: The nuclear lamina is an organized proteinaceous layer on the nucleoplasmic side of the nuclear envelope. It consists of (up to) three related intermediate filament proteins called lamins.
核纤层:核纤层是核膜核质面的纤维状蛋白层,由被称为核纤层蛋白的三种相关的中间丝组成。

8 Nuclear pore complex(NPC): A nuclear pore complex is a very large, proteinaceous structure that extends through the nuclear envelope, providing a channel for bidirectional transport of molecules and macromolecules between the nucleus and the cytosol.
核孔复合体:一个贯穿核膜的大的蛋白复合体,为细胞核与细胞质之间的分子和大分子的运输提供了一条双向通道。

第九节 细胞骨架
Section 9 Cytoskeleton

细胞骨架 cytoskeleton
基粒 basal body
基膜 basal lamina
肌动蛋白纤维 actin filament
微丝 microfilament
微管 microtubule
中间丝,中间纤维 intermediate filament
微管蛋白 tubulin
α-辅肌动蛋白 α-actinin
邻核帽 juxta-nuclear cap
角蛋白纤维 keratin filament
秋水仙素 colchicine
胞质环流 cyclosis
DNA 复制复合体 DNA replication complex
动力蛋白 dynein
动力蛋白臂 dynein arm
表染色质 epichromatin
纤维形肌动蛋白 F-actin

球型肌动蛋白　G-actin
基因缺失　gene deletion
基因剔除　gene knockout
GTP 酶活化蛋白　GTPase activating protein (GAP)
GTP 结合调节蛋白，G 蛋白　GTP binding regulatory protein
促旋酶，旋转酶　gyrase
重酶解肌球蛋白　heavy meromyosin(HMM)
异二聚体　heterodimer
同源岛　homology island
中间纤维结合蛋白　intermediate filament associated protein(IFAP)
轻酶解肌球蛋白　light meromyosin(LMM)
非组蛋白骨架　nonhistone protein scaffold
肌纤维　myofiber
肌原纤维　myofibrils
肌球蛋白　myosin
神经元纤维　neurofilament
神经胶质纤维　neuroglial filament
鬼笔环肽　philloidin
原丝　protofibril
原纤维　protofilament
应力纤维　stress fiber
波形纤维　vimentin filament
放射环模型　radial loop model
放射辐条　radial spoke
复制位点　replication site
复制灶　replication foci
复制叉　replication fork
紫杉酚　taxol
端网结构　terminal web
转基因　transgene
踏车现象　treadmilling
原肌球蛋白　tropomyosin(Tm)
肌钙蛋白　troponin
纽蛋白　vinculin
生物发生　biogenesis

1 Cytoskeleton：The cytoskeleton is the collection of structural filaments in the cytoplasm of eukaryotic cells. It confers the properties of cell shape，mechanical rigidity，and cell motility and is mainly composed of microtubules，actin filaments，and intermediate filaments.

细胞骨架：指真核细胞的细胞质中纤维网络结构的集合体。它赋予细胞在形状、机械压力和运动性等方面的特质，主要由微管、微丝和中间丝组成。

2 Microfilament：Microfilaments are dynamic polymers of actin that function in the cytoskeleton to confer cell shape and plasticity，as well to transport proteins，RNA and organelles to specific locations.

微丝：由肌动蛋白单体组装而成的细胞骨架纤维，赋予细胞一定的形状和可塑性，并且负责将蛋白质、RNA 和一些细胞器运送至特定位点。

3 Microtubule：Microtubules are thin，hollow filaments composed of the proteins α- and β-tubulin. Microtubules help organize the interior of the cell and determine its shape，in part by acting as tracks for the movement of the kinesin or dynein motor proteins.

微管：一种纤细的、中空的由 α 与 β 微管蛋白形成的纤维，在某种程度上，通过驱动蛋白和动力蛋白与其结合产生各种运动协助细胞内的有序组织，决定细胞的形状。

4 Intermediate filament：Intermediate filaments are one of three types of cytoskeletal filaments that run throughout the cytoplasm of an eukaryotic cell. Intermediate filaments are the strongest of the three types and appear devoted to helping give tissues mechanical strength.

中间丝：中间丝是三种细胞骨架纤维之一，贯穿真核细胞的细胞质。它是最强韧的骨架纤维，为组织提供机械稳定性。

5 Microtubule-organizing center(MTOC)：A microtubule-organizing center is a region from which microtubules emanate. In animal cells，the centrosome is the major microtubule-organizing center.

微管组织中心：在细胞中微管起始组装的地方，在动物细胞中，中心体是主要的微管组织中心。

6 Stress fiber：Stress fibers are contractile bundles of actin filaments，myosin filaments，and other proteins. They are anchored to protein complexes(focal adhesions) at the plasma membrane and function in cell adhesion.

应力纤维：由肌动蛋白丝、肌球蛋白丝和其他蛋白组成的具有收缩性的纤维束，在质膜上与蛋白复合物(黏合斑) 锚定，在细胞黏着中发挥作用。

第十节　细胞增殖及其调控
Section 10　Cell Proliferation and the Control

细胞分裂周期　cell division cycle
细胞分裂周期基因　cell division cycle gene
检验点　checkpoint
细胞周期蛋白　cyclin
后期　anaphase
后期促进因子复合物　anaphase-promoting complex(APC)
化学偶联假说　chemical coupling hypothesis
化学突触　chemical synapase
化学渗透偶联假说　chemiosmotic coupling hypothesis
交叉　chiasma/crossover
星体　aster
动粒　kinetochore
着丝粒　centromere
着丝粒-动粒复合体　centromere-kinetochore complex
驱动蛋白　kinesin
凝聚体　condensation stage
中板聚合　congression
周期蛋白框　cyclin box
周期中细胞　cycling cell
破坏框　destruction box
终变期　diakinesis
DNA 复制执照因子学说　DNA replication licence factor theory
DNA 合成期,S 期　DNA synthesis phase
双线期　diplotene/diplonema
赤道面　equatorial plane
赤道板　equatorial plate
鞭毛　flagella
分裂沟　furrow
发生泡　germinal vesicle
同源染色体　homologous chromosome
分裂间期　interkinesis
(联合复合体)侧成分　lateral element
细线期　leptotene
成熟促进因子　maturation-promoting factor (MPF)
灯刷染色体　lampbrush chromosome
减数分裂　meiosis
中期　metaphase
微丝结合蛋白　microfilament associated protein
微管相关蛋白　microtubule associated protein (MAP)
微管组织中心　microtubule organizing center (MTOC)
微绒毛　microvilli
中间体　midbody
细胞促分裂因子　mitosis promoting factor
有丝分裂　mitosis
复制起始点识别复合体　original recognition complex (ORC)
粗线期　pachytene/pachynema
中心粒外周基质　pericentriolar matrix
极性微管　polar microtubule
染色体超前凝集　premature chromosome condensation(PCC)
减数分裂前间期　premeiotic interphase
增殖细胞核抗原　proliferating cell nuclear antigen
前期　prophase
静止期细胞　quiescent cell
限制点　restriction point
纺锤体　spindle
纺锤体极体　spindle polar body
纺锤体装配检验点　spindle assembly checkpoint
标准细胞周期　standard cell cycle
起始点　start
监控机制　surveillance mechanism
联会　synapsis
联会复合体　synaptonemal complex
末期　telophase
端粒　telomere
端化　terminalization
温度敏感突变体　temperature-sensitive mutant
四分体　tetrad
原生质团,变形体　plasmodia
泛素(肽)　ubiquitin
泛素化途径　ubiquitination pathway
偶线期　zygoytene/zygonema

1 Centromere: The centromere is a constricted region of a chromosome that includes the site of attachment(the kinetochore) to the mitotic or meiotic spindle. The centromere consists of unique

DNA sequences and proteins not found anywhere else in the chromosome.
着丝粒:染色体上的主缢痕部位,包括与有丝分裂或减数分裂纺锤丝相连的区域(动粒),由特定的DNA序列和染色体其他部位不具有的蛋白组成。

2 Kinesin: Kinesins are a family of molecular motor proteins that bind and translocate along microtubules. Most kinesins move in a minus-end to plus-end direction along microtubules.
驱动蛋白:分子马达蛋白中的一个家族,与微管连接并沿其运动。大多数的驱动蛋白都沿微管从负极到正极的方向运动。

3 Lampbrush chromosome: Lampbrush chromosomes are the extremely extended meiotic bivalents of certain amphibian oocytes.
灯刷染色体:特定时期的两栖类动物的卵母细胞存在的极端伸展的减数分裂的二价体结构。

4 Telomere: A telomere is the natural end of a chromosome, consisting of a DNA sequence of a simple repeating unit with a protruding single stranded end.
端粒:位于正常染色体的末端,由携带突出的单链末端的含有简单重复单位的DNA序列组成。

5 Synapsis: Synapsis is the association of two pairs of sister chromatids that occurs at the start of meiosis.
联会:在减数分裂起始阶段,由姊妹染色单体配对的过程。

6 Synaptonemal complex: The synaptonemal complex refers to the morphological structure of synapsed chromosomes.
联会复合体:指发生联会的同源染色体之间形成的一种形态学结构。

第十一节 细胞分化与基因表达调控

Section 11 Cell Differentiation and the Control of Gene Expression

核仁内染色质 intranucleolar chromatin
组成型剪接 constitutive splicing
去分化,脱分化 dedifferentiation
缺失图法 deletion mapping
决定 determination
决定子 determinant
差别基因转录 differential gene transcription
DNA足迹技术 DNA foot printing
干细胞 stem cell
定向干细胞 directional stem cell
单能干细胞 monopotential cell
胚胎诱导 embryonic induction
胚胎干细胞 embryo stem cell
成红细胞 erythroblast
结蛋白纤维 desmin filament
桥板蛋白 desmoplakin
构建性甲基化酶 establishment methylase
成纤维细胞生长因子 flbroblast growth factor (FGF)
通用转录因子 general transcription factor
基因组印记 genomic imprinting
生殖质 germ plasm
生长因子 growth factor
生长因子受体 growth factor receptor
成虫盘 imaginal disc
维持性甲基化酶 maintenance methylase
恶性肿瘤 malignancy
恶性转化 malignant transformation
隐蔽mRNA masked mRNA
DNA甲基化 DNA methylation
甲基转移酶 methyl transferase
甲基化不足 under methylation
多级调控系统 multistage regulation system
多潜能性 pluripotency
位置效应 position effect
前起始复合体 preinitiation complex
加工水平的调控 processing-level control
蛋白质异形体 protein isoform
癌 cancer
癌基因 oncogene
原癌基因 proto-oncogene
假原基因 pseudoplasmodium
转移灶 metastases
再生 regeneration
调节基因 regulated gene
成视网膜细胞膜 retinoblastoma
劳氏肉瘤病毒 Rou sarcoma virus
对称序列 symmetrical sequence

终末分化 terminal differentiation
组织特异性基因 tissue-specific gene
全能性 totipotency
反式作用因子 transacting factor
转录因子 transcriptional factor
转录激活子 transcriptional activator
剪接增强子 splicing enchancer
转录水平的调控 transcriptional level control
转录阻抑 transcriptional repression
同源异型基因 homeotic selector gene
转分化 transdifferentiation
翻译水平的调控 translational level control
肿瘤细胞 tumor cell
抑癌基因 suppressor gene
管家基因 house-keeping gene
非翻译区 untranslated region(URT)

1 Differentiation: Differentiation is the process that occurs when a cell becomes more specialized.
分化:分化是细胞变得更加特定化的过程。

2 Stem cell: Stem cells are relatively undifferentiated cells that can divide continuously to generate more stem cells as well as more differentiated (specialized) cells. In many organisms, different populations of stem cells give rise to different cell lineages.
干细胞:干细胞是相对来说未分化的细胞,可以持续分裂成更多的干细胞和更多的分化细胞。在许多生物体中,不同的干细胞群可以产生不同的细胞系。

3 Proto-oncogene: A proto-oncogene is a normal gene that can be mutated to become an oncogene.
原癌基因:指由正常基因突变所产生的癌基因。

第十二节 细胞衰老与凋亡
Section 12 Cell Senescence and Apoptosis

衰老小体 apoptic body
细胞凋亡 apoptosis
凋亡小体 apoptosis body
程序性细胞死亡 programmed cell death
坏死 necrosis
癌细胞 cancer cell
凋亡蛋白酶活化因子 apoptosis proteinase-activating factor(Apaf)
铁调蛋白 iron regulatory protein(IRP)
铁应答元件 iron-response element(IRE)
彗星电泳法 comet assay
黄色素 yellow pigment
异性活化 hetero activation
致密体 dense body
同性活化 homoactivation
死亡效应结构域 death effector domain(DED)
死亡诱导信号复合体 death-inducing signaling complex(DISC)
选择性剪接 alternative splicing
DNA 梯状条带 DNA ladders
染色体外 rDNA 环 extrachromosomal rDNA circle(ERC)
铁蛋白 ferritin
Hayflick 界限 Hayflick limitation
活性氧基因或分子 reactive oxygen species (ROS)

1 Apoptosis: Apoptosis is the capacity of a cell to respond to a stimulus by initiating a signaling pathway that leads to its death through the activation of a characteristic set of reactions.
细胞凋亡:一种有序的细胞死亡方式,细胞应答于某种刺激,通过启动某种信号传递途径,激活一系列的特征性的反应最终导致死亡的过程。

2 Necrosis: Necrosis is a type of cell death after injury in which cells swell and break open, causing inflammation in the surrounding area.
细胞坏死:细胞受到损伤后的一种死亡方式,在这一过程中细胞膨胀、破裂,引起周围区域的炎症反应。

3 Programmed cell death: Programmed cell death is cell death occurring at a defined point in development.
程序性细胞死亡:细胞发育过程中在某一调控位点发生的死亡方式。

第十六章　遗传学
Chapter 16　Genetics

第一节　什么是遗传学
Section 1　What Is Genetics

遗传　heredity
遗传学　genetics
基因　gene

1 Genetics: The branch of biology that deals with heredity and the expression of inherited traits.
遗传学：遗传学是生物学的一个分支，研究遗传和获得性遗传特征的表达。

2 Gene: The fundamental physical unit of heredity whose existence can be confirmed by allelic variants and which occupies a specific chromosomal locus. A DNA sequence coding for a single polypeptide.
基因：基因是遗传的最小物质单位，在染色体上占有特定位点，其存在可由等位基因变异体来确定。基因是编码一个多肽的 DNA 序列。

第二节　孟德尔定律
Section 2　Mendel's Law

孟德尔定律　Mendel's law
真实遗传　breeding true
自由组合定律　law of independent assortment
分离定律　law of segregation
亲代　parent
子一代　first filial generation (F1)
正反交　reciprocal cross
显性性状　dominant character
隐性性状　recessive character
分离　segregation
颗粒式遗传　particulate inheritance
混合式遗传　blending inheritance
遗传因子　hereditary determinant
配子　gamete
合子　zygote
等位基因　allele
基因型　genotype
表型　phenotype
纯合体　homozygote
杂合体　heterozygote
测交　test cross
亲组合　parental combination
重组合　recombination
概率　probability
棋盘法　punnet square
分枝法　branching process
适合度　goodness of fit
孟德尔性状　Mendelian character

1 Testcross: A cross between an individual whose genotype at one or more loci may be unknown and an individual who is homozygous recessive for the genes in question.
测交：在一个或多个位点的基因型未知的个体与这些基因的隐性纯合体之间的杂交。

2 Phenotype: The observable properties of an organism that is genetically controlled.
表型：表型是指遗传上可控的生物体表现出的可观察到的特征。

3 Genotype: The specific allelic or genetic constitution of an organism; often, the allelic composition of one or a limited number of genes under investigation.
基因型：某一生物个体特定的等位基因或遗传组成。在研究中经常指一个或一定数量的基因的等位基因组成。

4 Homozygote: An individual with identical alleles at one or more loci. These individuals will produce identical gametes and will therefore breed true.
纯合子(纯合体)：是指一个或多个位点上的两个等位基因相同的基因型个体，这些个体产生相同的配子而且真实遗传。

5 Gamete: A specialized reproductive cell with a haploid number of chromosomes.
配子:具有单倍体数目的染色体的特定的生殖细胞。

6 Zygote: The diploid cell produced by the fusion of haploid gametic nuclei.
合子:通过单倍体的配子核融合而产生的二倍体细胞。

7 Allele: One of the possible mutational states of a gene, distinguished from other alleles by phenotypic effects.
等位基因:一个基因的可能突变状态,在表型效应上区别于其他的等位基因。

第三节 遗传的染色体学说
Section 3 Chromosome Theory of Inheritance

细胞核 nucleus
细胞质 cytoplasm
染色体 chromosome
细胞膜 cell membrane
细胞壁 cell wall
中心粒 centriole
微管 microtubule
线粒体 mitochondrium
吸能过程 endergonic process
核糖体 ribosome
内质网 endoplasmic reticulum
溶解体 lysosome
水解酶 hydrolase
原酶 proenzyme
高尔基体 Golgi body
质体 plastid
基粒 grana
原核类 prokayote
核膜 nuclear membrane
真核类 eukaryote
核液 nuclear sap
染色质 chromatin
核仁 nucleolus
二倍体 diploid
单倍体 haploid
染色单体 chromatid
着丝粒 centromere
初级缢痕 primary constriction
次级缢痕 secondary constriction
随体 satellite
核仁形成区 nucleolar organizer
组蛋白 histone
非组蛋白蛋白质 nonhistone protein
绳珠模型 beads on-a-string model
核体 nucleosome
连丝 linker
有丝分裂 mitosis
核区 nuclear area
纺锤体 spindle body
间期 interphase
前期 prophase
中期 metaphase
后期 anaphase
末期 telophase
胞质分割 cytokineesis
减数分裂 meiosis
细线期 leptotene
偶线期 zygotene
粗线期 pachytene
双线期 diplotene
浓缩期 diakinesis
同源染色体 homologous chromosome
双价体 bivalent
姐妹染色单体 sister chromosome
交叉端化 terminalization
性原细胞 gonium
精原细胞 spermatogonium
卵原细胞 oogonium
初级精母细胞 primary spermatocyte
初级卵母细胞 primary oocyte
次极精母细胞 secondary spermatocyte
次极卵母细胞 secondary oocyte
第一极体 first polar body
第二极体 second polar body
种系细胞 germ-line cell
精子形成 spermatogonium
小孢子母细胞 microsporocyte
大孢子母细胞 macrosporocyte
反足细胞 antipodal
极核 endosperm nuclei

双重受精 double fertilization
交配型 mating type
染色体学说 chromosome theory of inheritance
卵核 female gametic nuclei

1 Mitosis: A form of cell division resulting in the production of two cells, each with the same chromosome and genetic complement as the parent cell.
有丝分裂：一个细胞变成两个细胞的分裂方式，每一个细胞都含有和母细胞相同的染色体和遗传物质。

2 Meiosis: The process in gametogenesis or sporogenesis during which one replication of the chromosomes is followed by two nuclear divisions to produce four haploid cells.
减数分裂：配子或孢子形成的过程，在这一过程中染色体复制一次，随后核分裂两次，最终产生四个单倍体细胞。

3 Chromosome: In prokaryotes, one or more DNA molecules containing the genome; in eukaryotes, a DNA molecule complexed with RNA and proteins to form a threadlike structure containing genetic information arranged in a linear sequence and visible during mitosis and meiosis.
染色体：在原核生物中，一个或多个 DNA 分子构成染色体组；在真核生物中，一个 DNA 分子与 RNA、蛋白质形成细丝状的复杂结构，遗传信息以线性序列排列的方式包含其中，这一结构在有丝分裂和减数分裂中清晰可见。

4 Homologous chromosomes: Chromosomes that synapse or pair during meiosis. Chromosomes that are identical with respect to their genetic loci and centromere placement.
同源染色体：减数分裂时发生联会或配对的染色体。在等位基因位点和着丝点分布上一致的染色体。

5 Chromosome theory of inheritance: The idea put forward independently by Walter Sutton and Theodore Boveri that chromosomes are the carriers of genes and the basis for the Mendelian mechanisms of segregation and independent assortment.
遗传的染色体学说：在孟德尔的分离与自由组合定律的基础上，由 Walter Sutton 和 Theodore Boveri 提出的染色体是基因的携带者的遗传学说。

第四节 基因的作用及其与环境的关系
Section 4 Effect of Gene and Its Relations with Environment

表型效应 phenotypic effect
反应规范 reaction norm
修饰基因 modifier gene
多效现象 pleiotropism
表现度 expressivity
成骨不全 ostogenesis imperfect
外显率 penetrance
抗原 antigen
抗体 antibody
抗血清 antiserum
凝集反应 agglutination
并显性 codominance
镰形细胞贫血症 sickle cell anemia
表型模写 phenocopy
短肢畸形 phocomelia
反应停 thalidomide
显性致死 dominant lethal
配子致死 gametic lethal
合子致死 zygotic lethal
亚致死现象 partial lethality
遗传背景 genetic background
等位形式 allelic form
相 phase
复等位现象 multiple allelism
复等位基因 multiple alleles
凝集素 lectin
孟买型 Bombay phenotype
新生儿溶血症 erythroblastosis fetalis
互补基因 complementary genes
不相容现象 incompatibility
强化基因 intensifier
限制基因 restriction gene
抑制基因 inhibitor
上位效应 epistasis
血型 blood group

1 Penetrance: The frequency, expressed as a percentage, with which individuals of a given genotype manifest at least some degree of a specific mutant phenotype associated with a trait.

外显率:特定基因型的个体所占的比率,在某种程度上表明与某一性状相关的突变型的存在情况,用百分比表示。

2 Codominance:Condition in which the phenotypic effects of a gene are fully and simultaneously expressed in the heterozygote.

共显性:指在杂合体中等位基因的表型效应全部同时表达的状态。

3 Expressivity:The degree or range in which a phenotype for a given trait is expressed.

表现度:某一特定的表型所表现出的程度或范围。

4 Multiple alleles:Three or more alleles of the same gene.

复等位基因:同一基因的三个或更多的等位基因。

5 Epistasis:Nonreciprocal interaction between genes such that one gene interferes with or prevents the expression of another gene. In Drosophila for example, the recessive gene eyeless, when homozygous, prevents the expression of eye color genes present in the genome.

上位效应:基因间的不可逆的相互作用,例如一个基因干扰或阻止另一个基因的表达。以果蝇为例,隐性基因“无眼”纯合时阻止基因组中存在的眼的有色基因的表达。

6 Recessive: A term describing an allele that is not expressed in the heterozygous condition.

隐性:在杂合状态下不表达的等位基因称为隐性。

第五节 性别决定与伴性遗传

Section 5 Sex Determination and Sex-linked Inheritance

性染色体 sex-chromosome

常染色体 autosome

异配性别 heterogametic sex

同配性别 homogametic sex

伴性遗传 sex-linked inheritance

绞花遗传 criss-cross inheritance

X 连锁遗传 X-linked inheritance

A 型血友病 hemophilia A

X 连锁隐性遗传 X linked recessive inheritance

携带者 carrier

色盲 color blindness

限性遗传 holandric inheritance

毛耳 hairy ear

Y 连锁的组织相容性抗原 Y linked histocompatibility antigen

Z 连锁遗传 Z linked inheritance

偏母的 matroclinous

偏父的 patroclinous

初级例外子代 primary exceptional progeny

不分开 non-disjunction

发育性状 developmental character

单极纺锤体 monopolar spindle

1 Sex chromosome:A chromosome, such as the X or Y in humans, which is involved in sex determination.

性染色体:与性别决定有关的染色体,例如人类的 X 或 Y 染色体。

2 Autosomes:Chromosomes other than the sex chromosomes. In humans, there are 22 pairs of autosomes.

常染色体:区别于性染色体的染色体,在人类的染色体中有 22 对常染色体。

3 Sex-linked inheritance:The inheritance of a trait (phenotype) that is determined by a gene located on one of the sex chromosomes.

性连锁遗传:表型性状由位于某条性染色体上的基因所决定的遗传方式。

4 Carrier: An individual heterozygous for a recessive trait.

携带者:带有隐性基因的杂合个体。

5 Backcross: A cross involving an F1 heterozygote and one of the P1 parents (or organism with a genotype identical to one of the parents).

回交:子一代和两个亲本的任一个(或者和任一个亲本基因型相同的个体)进行杂交的方法叫作回交。

第六节 染色体和连锁群
Section 6 Chromosome and Linkage Group

交叉型假设 chiasmatype hypothesis
交叉 chiasma
对应片段 homologous segment
交换 crossing over
基因重组 recombination
连锁群 linkage group
三点侧交 three-point test cross
图距 map distance
重组子 recombinant
三杂合体 trihybrid
干涉 interference
并发率 coincidence
连锁图 linkage map
遗传学图 genetic map
重组率 recombination fraction
交换率 crossing over percentage
基因座位 gene loci
图距单位 map unit
作图函数 mapping function
四分子分析 tetrad analysis
顺序四分子 ordered tetrad
座位 locus
交互过程 reciprocal process
基因转变 gene conversion
原养型 prototroph
菌株 strain
营养缺陷型 auxotroph
赖氨酸依赖型 lysine dependent
赖氨酸缺陷型 lysine deficient
着丝粒作图 centromere mapping
互为镜影 mirror image
第一次分裂分离 first division segregation
第二次分裂分离 second division segregation
随机趋向 random orientation
单体干扰 chromatid interference
双交换四分子的分析 analysis of double exchange
连锁相 linkage phase
耦合 coupling
相斥 repulsion
细胞学图 cytological map
标记基因 gene marker

1 Linkage: Condition in which two or more nonallelic genes tend to be inherited together. Linked genes have their loci along the same chromosome; they do not assort independently but can be separated by crossing over.
连锁:两个或更多的非等位基因一起遗传的现象。连锁的基因处于同一染色体上,它们不能独立分配只能通过杂交分离。

2 Linkage group: A group of genes that have their loci on the same chromosome.
连锁群:位于同一染色体上的基因群称为一个连锁群。

3 Chiasma: The crossed strands of nonsister chromatids seen in diplonema of the first meiotic division. Regarded as the cytological evidence for exchange of chromosomal material, of crossing over.
交叉:在减数分裂 I 前期的双线期,在非姊妹染色单体间的某些部位可见交叉缠结的现象,可作为染色体基因交换重组的细胞学证据。

4 Locus: The site or place on a chromosome where a particular gene is located.
基因座:特定的基因在染色体上占有其特定的位置,称为基因座。

5 Interference: A measure of the degree to which one crossover affects the incidence of another crossover in an adjacent region of the same chromatid. Negative interference increases the chances of another crossover; positive interference reduces the probability of a second crossover event.
干涉:用来衡量同一染色体上一次单交换对邻近另一次单交换的发生率的影响程度。负干涉提高另一次交换的几率,正干涉减少另一次交换的可能性。

6 Recombination: The process that leads to the formation of new gene combinations on chromosomes.
重组:导致染色体上的基因重新组合的形成过程。

第七节 细菌和噬菌体的重组和连锁

Section 7 Recombination and Linkage of Bacteria and Phage

原核生物 prokaryote
基因带 genophore
菌落 clone
遗传组成 genetic constitution
突变 mutation
大肠杆菌 Escherichia coli
污染 contamination
噬菌体 bacteriophage
基本培养基 minimal medium
敏感 sensitive
抗性 resistance
遗传信息 genetic information
受体 recipient
供体 donor
性因 sex factor
致育因子 fertility factor
高频重组 high frequency of recombinant
影印培养 replica plating
中断杂交技术 interrupted mating technique
原点 origin
转移过程 transfer process
附加体 episome
基因群 gene family
插入序列 insertion sequence
外基因子 exogenote
内基因子 endogenote
部分二倍体 partial diploid
相反的重组子 reciprocal recombinant
性导 sexduction
雄性菌株 male strain
烈性噬菌体 virulent phage
裂解 lysis
噬菌斑 plaque
宿主范围 host range
溶菌阻碍现象 lysis inhibition
快速溶菌 rapid lysis
混合感染或复感染 mix or double infection
温和噬菌体 temperate phage
诱导 induction
溶源性 lysogeny
分离子 segregant
溶源菌 lysogenic bacteria
转导 transduction
过滤因子 filteranle agent
普遍性转导 generalized transduction
特异性转导 specialized transduction
局限性转导 restricted transduction
特异转导者 transducer
转导子 transductant
超数感染 superinfection

1 Mutation: The process that produces an alteration in DNA or chromosome structure; the source of most alleles.

突变:DNA 或染色体的结构发生变化的过程,是大多数等位基因产生的根源。

2 Hfr: A strain of bacteria exhibiting a high frequency of recombination. These strains have a chromosomally integrated F factor that is able to mobilize and transfer part of the chromosome to a recipient F-cell.

高频重组品系:表现出高频重组的菌株。这些菌株中带有一个整合的 F 因子,从而能够运动和转移部分染色体 DNA 到 F- 受体细胞中。

3 Transduction: The transfer of virally mediated genes from one bacterium to another or the transfer of eukaryotic genes mediated by retrovirus.

转导:以病毒为载体将基因从一个细菌细胞传递到另一个细菌细胞或者通过逆转录病毒转移真核基因的过程。

4 Sexduction: Transmission of chromosomal genes from a donor bacterium to a recipient cell by the F factor.

性导:利用 F 因子将供体细胞的染色体基因导入受体细胞的过程。

5 Lysogenic bacterium: a bacterial cell carrying the DNA of a temperate bacteriophage integrated into its chromosome.

溶源性细菌:携带温和噬菌体的 DNA 并整合到自身染色体上的细菌细胞。

6 Temperate phage:A bacteriophage that can become a prophage and confer lysogeny upon the host bacterial cell.

温和噬菌体:能够成为前噬菌体并且在宿主细胞中溶源性表达的噬菌体。

7 Plaque: A clear area on an otherwise opaque bacterial lawn, caused by the growth and reproduction of phages.

噬菌斑:由于噬菌体的生长和繁殖导致形成的不同于无光泽的菌苔的空斑。

第八节 数量性状遗传

Section 8 Quantitative Character Inheritance

质量性状 qualitative character
数量性状 quantitative character
平均数 mean
平均平方和 mean square
多基因假说 multiple factor hypothesis
全或无性状 all or none attribute
方差 variance
标准误差 standard error
广义遗传率 heritability in the broad sense
狭义遗传率 heritability in the narrow sense
株系 line
纯系 pure line
近亲繁殖或近交 inbreeding
纯合性 homozygosity
衰退现象 recession phenomenon
杂合性 heterozygosity
自交 self cross
近交系数 coefficient of inbreeding
系统的近交 systematic inbreeding
杂种优势 heterosis
显性说 dominance hypothesis
偏态分布 skewed distribution
正态分布 normal distribution
超显性假说 overdominance hypothesis
等位基因的相互作用 intra-allelic interaction
多基因 polygenes

1 Overdominance:The phenomenon where heterozygotes have a phenotype that is more extreme than either homozygous genotype.

超显性:杂合体的某种表型明显超越与之相应的纯合基因型的现象。

2 Selfing:In plant genetics, the fertilization of ovules of a plant by pollen produced by the same plant; reproduction by self-fertilization.

自交:在植物遗传学中,某一植株的胚珠与同一植株产生的花粉发生受精称为自交,也叫自交繁殖。

3 Inbreeding:Mating between closely related organisms.

近交:亲缘关系比较近的生物体间的交配称为近交。

4 Coefficient of inbreeding: The probability that two alleles present in a zygote are descended from a common ancestor.

近交系数:近交系数就是合子中的两个等位基因由同一祖先遗传的可能性。

第九节 遗传物质的改变(一)染色体畸变

Section 9 Change of Genetic Matter(1) Chromosomal Aberrations

突变 mutation
基因突变或点突变 gene or point mutation
染色体变异或畸形 chromosomal variation or aberration
染色单体的断裂 chromosomal breakage
染色体片段 chromosomal segment
重新愈合或重建 restitution
缺失 deletion or deficiency
重复 duplication or repeat
倒位 inversion
易位 translocation
球节 knob
染色粒 chromomere
多线染色体 polytene chromosome
缺失环 deletion loop
断裂-融合-桥的循环 breakage-fusion-bridge cycle
拟显性现象 pseudodominance

猫叫综合征　cri-du-chat syndrome
染色体组　chromosome complement
不等交换　unequal crossover
相互易位　reciprocal translocation
邻近分离　adjacent segregation
交换分离　alternate segregation
假连锁现象　pseudolinkage
平衡易位携带者　balanced translocation carrier
染色体病　chromosomal disease
部分不育　partial sterility
倒位杂合体　inversion heterozygote
倒位环　inversion loop
交换抑制因子　crossing over repressor
永久杂种　permanent hybrid
平衡致死品系　balanced lethal system
游离端　telomere
多倍体　polyploid
同源多倍体　autopolyploid
二显体　duplex
三显体　triplex
单显体　simplex
同源三倍体　autotriploid
异源多倍体　allopolyploid
染色体组分析　genome analysis
非整倍体　aneuploid
先天愚型　down syndrome
发生率　incidence
表型性别　phenotypic sex
整倍体　euploid

1 Chromosomal aberration: Any change resulting in the duplication, deletion, or rearrangement of chromosomal material.
染色体畸变：导致染色体结构发生重复、删除或者重排的任何变化都称为染色体畸变。

2 Deficiency: A chromosomal mutation involving the loss or deletion of chromosomal material.
缺失：导致染色体物质丢失或删除的染色体变异方式。

3 Duplication: A chromosomal aberration in which a segment of the chromosome is repeated.
重复：染色体上某一片段发生重复的染色体畸变。

4 Inversion: A chromosomal aberration in which the order of a chromosomal segment has been reversed.
倒位：染色体上某一个片段颠倒后重接产生的染色体畸变。

5 Translocation: A chromosomal mutation associated with the transfer of a chromosomal segment from one chromosome to another.
易位：染色体上的某一片段从一条染色体转移到另一条染色体上的现象。

6 Autopolyploidy: Polyploid condition resulting from the duplication of one diploid set of chromosomes.
同源多倍体：由同一物种的二倍体染色体组加倍所形成的多倍体。

7 Allopolyploid: Polyploid condition formed by the union of two or more distinct chromosome sets with a subsequent doubling of chromosome number.
异源多倍体：两个或两个以上的不相同物种杂交产生的染色体组加倍形成的多倍体称为异源多倍体。

8 Balanced lethal genes: Recessive, nonallelic lethal genes, each carried on different homologous chromosomes. When organisms carrying balanced lethal genes are interbred, only organisms with genotypes identical to the parents(heterozygotes) survive.
平衡致死基因：位于一对同源染色体的两个不同成员上的非等位致死基因。当带有平衡致死基因的个体杂交时，只有和父母本（杂合体）基因型相同的子代个体才能够存活。

9 Euploid: Polyploid with a chromosome number that is an exact multiple of a basic chromosome set.
整倍体：指具有精确的一套或多套基本染色体组的多倍体。

10 Polyploidy: A cell or individual having more than two sets of chromosomes.
多倍体：具有两套以上染色体组的细胞或个体。

第十节 遗传物质的改变(二)基因突变
Section 10 Change of Genetic Matter (2) Gene Mutation

点突变 point mutation
自然突变或自发突变 spontaneous mutation
诱发突变 induced mutation
形态突变 morphological mutation
生化突变 biochemical mutation
可见突变 visible mutation
半致死 semilethal
低活性 subvital
条件致死突变 condition lethal mutation
枝变或芽变 bud sport
软骨发育不全 achondroplasia
世代时间 generation time
正突变 forward mutation
回复突变 back mutation
抑制因子突变 suppressor mutation
回复品系 revertant stock
筛选 screening
生物素 biotin
家系分析 pedigree analysis
性连锁平衡致死剂 balanced sex-linked lethal
致死突变 lethal mutation

1 Point mutation: A mutation that can be mapped to a single locus. At the molecular level, a mutation that results in the substitution of one nucleotide for another.
点突变:单个位点上发生的突变。在分子水平上指一个碱基对另一个碱基的替代所造成的突变。

2 Spontaneous mutation: A mutation that is not induced by a mutagenic agent.
自发突变:在无突变诱导物存在的情况下自然发生的基因突变。

3 Pedigree: In human genetics, a diagram showing the ancestral relationships and transmission of genetic traits over several generations in a family.
系谱:在人类遗传学中,系谱指用图解表明一个家族中的亲缘关系和遗传性状在代与代之间的传递情况。

4 Lethal gene: A gene whose expression results in death.
致死基因:表达后导致死亡的基因。

第十一节 遗传的分子基础
Section 11 Molecular Basis of Genetics

烟草花叶病毒 tobacco mosaic virus, TMV
败血症 septicemia
核酸 nucleic acids
核苷酸 nucleotide
核苷 nucleoside
腺嘌呤 adenine, A
鸟嘌呤 guanine, G
胞嘧啶 cytosine, C
胸腺嘧啶 thymine, T
尿嘧啶 uracil, U
脱氧核糖 deoxyribose
核糖 ribose
多核苷酸 polynucleotide
极性 polarity
多核苷酸链 polynucleotide chain
原子团 group of atoms
双螺栓 double helix
反向平行 antiparallel
弱氢键 weak hydrogen bond
碱基对 pairs of bases
核苷酸对 nucleotide pairs
异源双链 heteroduplexes
变性 denaturation
复性 renaturation
退火 annealing
解链温度 melting temperature
同源性 homology
半保留复制 semi-conservative replication
聚合酶 polymerase
引物 primer
模板 template
重带 heavy band
轻带 light band
热变性 thermal denaturation

色差染色体 harlequin chromosome
复制叉 replication fork
复制子 replicon
起始点 origin of replication
多复制子 multi-replication
不连续复制 discontinuous replication
冈崎片断 okazaki fragment
引物酶 primase
核酸外切酶活性 exonuclease activity
连接酶 ligase
拓扑异构酶 topoisomerase
促旋酶 helicase
单链结合蛋白 single-stranded binding
运载蛋白质 transport protein
促性腺激素 gonadotropic hormone
睾酮 testosterone
球蛋白血症 agammaglobulinemia
胞液受体 cytosol receptor
着色性干皮症 xeroderma pigmentosum
自体催化 autocatalysis
转录 transcription
依赖 DNA 的 RNA 聚合酶 DNA-dependent RNA polymerase
全酶 holoenzyme
核心酶 core enzyme
启动子 promoter
终止区 terminator
游离的三磷酸核苷 ribonucleoside triphosphate
初级 RNA 的转录物 primary RNA transcript
甲基化 methylation
内含子 intron
编码顺序 coding sequence
外显子 extron
密码子 codon
三联体 triplet
简并 degeneracy
无义密码子 nonsense codon
起始信号 start signal
核仁的组织中心 nucleolar organizer
多核糖体或多体 polyribosome or polysome
核糖核酸酶 ribonuclease
三叶草结构 clover-leaf structure
外环 extra loop
摆动假说 wobble hypothesis
起始蛋白质 initiation protein
延伸因子 elongation factor
释放因子 release factor
前导序列 leader sequence
拖尾序列 trailer sequence
转译装置 translation apparatus
生化先成论 biochemical predestination
分泌型蛋白质 secretory protein
疏水性序列 hydrophobic sequence
受体糖蛋白 receptor glycoprotein
中心法则 central dogma
原生质体 protoplast
反转录病毒 retrovirus
反转录酶 reverse transcriptase
出芽 budding
转化细胞 transformed cell
互补 DNA complementary DNA (cDNA)
质粒 plasmid
分子克隆 molecular cloning
基因间区域 intergenic region
通读框 open reading frame
隔裂基因 split gene
千碱基对 kilobase pair
重叠基因 overlapping gene
高度重复序列 highly repetitive sequence
结构基因 structural gene
异染色体区域 heterochromatic region
串联重复 tandem repeats
一基因一酶说 one-gene-one-enzyme theory
白化病 albinism
苯酮尿症 phenylketonuria
半乳糖血症 galactosemia
念珠学说 bead theory
异核体 heterokaryon
体细胞交换 somatic crossing
有丝分裂交换 mitotic crossing over
顺反子 cistron
位点 site
顺反试验 cis-trans test
快速溶菌 rapid lysis
突变位点 mutational site
复感染 double infection
超感染 superinfection
两点杂交 two-point crosses
互补 complementation
重组作图法 recombinational mapping

广谱质粒 promiscuous plasmid
遗传工程 genetic engineering
重组 DNA 技术 recombinant DNA technique
嵌镶 DNA 分子 chimeric DNA molecule
核酸内切酶 restriction endonuclease
基因操作 gene manipulation
修饰酶 modification enzyme
甲基化 methylation
同裂酶 isoschizomers
黏粒 cosmid
转化或转染 transfection
噬菌斑杂交法 plaque hybridization
寡聚脱氧核糖核酸 oligodeoxynucleotide
限制性酶切片段 restriction fragment
菌落杂交 colony hybridization
珠蛋白 globin
蛋白基因簇 globin gene cluster
原位杂交 in site hybridization
限制性片段长度多态 restriction fragment length polymorphism
地中海贫血 thalassemia

1 Replication fork: The Y-shaped region of a chromosome associated with the site of replication.
复制叉:DNA 复制时在 DNA 链上形成的 Y 字形结构称为复制叉。

2 Extron: The DNA segments of a gene that are transcribed and translated into proteins.
外显子:基因中被转录和翻译成蛋白质的序列称为外显子。

3 Intron: A portion of DNA between coding regions in a gene that is transcribed but does not appear in the mRNA product.
内含子:基因中编码序列间的 DNA 序列,能够转录但不会出现在 mRNA 产物中。

4 Degenerate code: The genetic code, where a given amino acid may be represented by more than one codon. for example, some amino acids (leucine) have six codons, whereas others (isoleucine) have three.
简并密码:为同一种氨基酸由超过一种以上的三联体密码子编码,例如,某种氨基酸(亮氨酸)由 6 个密码子编码,而其他的(异亮氨酸)由 3 个密码子编码。

5 Central dogma: The concept that information flow progresses from DNA to RNA to proteins. Although exceptions are known, this idea is central to an understanding of gene function.
中心法则:是指遗传信息从 DNA 传递给 RNA,再从 RNA 传递给蛋白质的观点,尽管有例外,这一观点仍是理解基因功能的核心。

6 Cis-trans test: A genetic test to determine whether two mutations are located within the same cistron.
顺反测验:一种遗传学测验,可确定两个突变位点是否位于同一个顺反子内。

7 Cistron: That portion of a DNA molecule coding for a single polypeptide chain; defined by a genetic test as a region within which two mutations cannot complement each other.
顺反子:DNA 分子中编码一条多肽链的部分,可通过遗传测验确定,基于这一区域内的两个突变位点不能互补。

8 Structural gene: A gene that encodes the amino acid sequence of a polypeptide chain.
结构基因:编码组成多肽链的氨基酸序列的基因。

9 Cosmid: A vector designed to allow cloning of large segments of foreign DNA. Cosmids are composed of the cos sites of phage λ inserted into a plasmid. In cloning, the recombinant DNA molecules are packaged into phage protein coats, and after infection of bacterial cells, the recombinant molecule replicates and can be maintained as a plasmid.
黏粒:为克隆大量外源 DNA 而设计的载体。将 λ 噬菌体黏性末端的 cos 位点插入质粒而构建。克隆时,重组 DNA 分子被包进噬菌体的衣壳蛋白中,侵染细菌细胞后,重组分子进行复制并可作为质粒被保留。

第十二节　突变和重组机理

Section 12　Mechanism of Mutation and Recombination

碱基替换　base substitution
移码突变　frameshift mutation
碱基类似物　base analogue
转换　transition
颠换　transversion
烷化剂　alkylating agents
形成交键　cross-linkage
同义突变　same sense mutation
错义突变　missense mutation
无义突变　nonsense mutation
血红蛋白病　hemoglobinopathy
一基因一多肽说　one gene-one polypeptide theory
断裂愈合模型　breaking joining model
模写选择模型　copy-choice model
相关螺旋　relational coiling
异常分离　abnormal segregation
吡哆醇　pyridoxine
染色单体转变　chromatid conversion
半染色单体转变　half-chromatid conversion
减数后分裂　post-meiotic segregation
断裂点　breakage point
交联桥　cross bridge
中间体　Holliday intermediate
不相称碱基对　mismatch base pair
核酸外切酶　exonuclease
转座遗传因子　transposable genetic element
控制因子　controlling element
解离因子　dissociation
颠倒重复序列　inverted repeat sequences
插入序列　insertion sequences, IS
转座子　transposon
抗性转移功能基因　resistance-transfer functional gene
靶部位　target site
前病毒　provirus
末端重复序列　terminal repeats
致癌基因　retroviral oncogene
光复活　photoreactivation
暗修复或切除修复　excision repair
重组修复　recombination repair
光复活酶　photoreacting enzyme

1 Base substitution: A single-base change in a DNA molecule that produces a mutation. There are two types of substitution: transitions, in which a purine is substituted for a purine or a pyrimidine for a pyrimidine; and transversions, in which a purine is substituted for a pyrimidine, or vice versa.
碱基替换:DNA 分子上单个碱基的变化所产生的变异。有两种替代方式:转换,嘌呤和嘌呤、嘧啶和嘧啶间的替换;颠换,嘌呤与嘧啶间的替换,或是相反。

2 Frameshift mutation: A mutational event leading to the insertion of one or more base pairs in a gene, shifting the codon reading frame in all codons that follow the mutational site.
移码突变:基因中一个或多个碱基对插入所造成的突变事件,所有密码子中开放阅读框发生漂移造成突变位点的产生。

3 Transition: A mutational event in which one purine is replaced by another or one pyrimidine is replaced by another.
转换:一种嘌呤替换另一种嘌呤,一种嘧啶替换另一种嘧啶的突变方式。

4 Transversion: A mutational event in which a purine is replaced by a pyrimidine or a pyrimidine is replaced by a purine.
颠换:一种突变方式,即嘌呤替换成嘧啶,或是相反。

5 Photoreactivation repair: Light-induced repair of damage caused by exposure to ultraviolet light. Associated with an intracellular enzyme system.
光复活修复:光诱导的对紫外线曝光引起的损伤进行修复,与细胞内的酶系统偶联。

6 Excision repair: Removal of damaged DNA segments followed by repair. Excision can include the removal of individual bases (base repair) or a stretch of damaged nucleotides (nucleotide repair). The gap created by excision is filled by DNA polymerase, and the ends are ligated to form an intact molecule.

切除修复：将DNA分子受损伤的部分切除并修复的方式。切除部分包括某个碱基或一段受损的核苷酸序列。由DNA聚合酶合成切除的部分，末端通过连接使DNA恢复正常结构。

7 Missense mutation: A mutation that alters a codon to that of another amino acid, causing an altered translation product to be made.

错义突变：一个密码子发生突变产生另外一种氨基酸，从而导致翻译产物改变的突变方式。

8 Nonsense codon: The nucleotide triplet in an mRNA molecule that signals the termination of translation. Three such termination codons are known: UGA, UAG and UAA.

无义密码子：mRNA分子中作为翻译终止信号的三联体密码子，已知有三个这样的密码子：UGA、UAG和UAA。

9 Nonsense mutation: A mutation that changes an amino acid codon into a termination codon: UAG, UAA, or UGA. Leads to premature termination during translation of mRNA.

无义突变：某一氨基酸的密码子突变为终止密码子(UAG、UAA或UGA)，导致mRNA的翻译提前终止的突变方式。

10 Transposon: A mobile piece of DNA that is flanked by terminal repeat sequences and typically bears genes encoding transposition functions.

转座子：一段可以移动的DNA序列，侧翼由末端重复序列组成，具有编码换位功能的基因。

第十三节 细胞质和遗传
Section 13 Cytoplasm and Inheritance

前质体 proplastid

毒素蛋白质 killer toxin protein

卡巴粒 kappa particle

雄性不育 male sterility

1 *k*-particles: DNA-containing cytoplasmic particles found in certain strains of *Paramecium aurelia*. When these self-reproducing particles are transferred into the growth medium, they release a toxin, paramecin, that kills other sensitive strains. A nuclear gene, *K*, is responsible for maintaining kappa particles in the cytoplasm.

卡巴粒：存在于某些品系的草履虫细胞质中的放毒性颗粒，当这些自我复制的颗粒转入生长培养基中，其释放出毒性物质——草履虫素，可杀伤细胞质中不含此种颗粒的敏感型草履虫。一个核基因K，控制细胞质中保持这种卡巴粒。

2 Maternal effect: Phenotypic effects on the offspring produced by the maternal genome. Factors transmitted through the egg cytoplasm that produces a phenotypic effect in the progeny.

母体影响：由母体中基因组决定的子代表型效应，通过卵细胞的细胞质传递一些遗传因子控制子代的某种表型效应。

3 Maternal inheritance: The transmission of traits via cytoplasmic genetic factors such as mitochondria or chloroplasts.

母体遗传：通过细胞质中的遗传因子控制某些性状表达的现象，例如线粒体和叶绿体。

第十四节 遗传与个体发育
Section 14 Inheritance and Individual Development

分裂球 blastomere

分化 differentiation

胚盘 blastoderm

极细胞质 pole plasm

极细胞 pole cell

极粒 polar granule

异固缩现象 heteropycnosis

异染色质化 heterochromatinization

全能性 totipotency

血红蛋白 hemoglobin

四聚体 tetramer

基因簇 gene families

早期胚胎　early embryo
同工酶　isoenzymes or isozymes
操纵子学说　operon theory
结构基因　structural gene
调整基因　regulator gene
操纵基因　operator
操纵子　operon
启动基因　promoter
乳糖操纵子　lactose operon
诱导物　inducer
降解物激活蛋白　catabolite activator protein
增强子　enhancer
降钙素　calcitonin
半衰期　half life
信息储存所　informational store house
信使　messenger
放线菌素　actinomycin
干细胞　stem cell
红细胞母细胞　erythroblast
表达部位　expression site
抗体产生细胞　antibody-producing cell
上位性的雄性决定因子　epistatic male-determing factors
自发装配过程　self-assembly process
浓缩蛋白质　condensing protein
溶菌酶　lysozyme
孢子形成　sporogenesis
前孢子　prespore or forespore
内生孢子　endospore

1 Enhancer: Originally identified as a 72-bp sequence in the genome of a virus, SV40, that increases the transcriptional activity of nearby structural genes. Similar sequences that enhance transcription have been identified in the genomes of eukaryotic cells. Enhancers can act over a distance of thousands of base pairs and can be located upstream, downstream, 5′, 3′ or internal to the gene they affect, and thus are different from promoters.

增强子：增强子是最早在SV40病毒中发现的72bp的序列，在结构基因附近增强转录活性。后来发现真核细胞基因组中也有类似的提高转录活性的序列。与启动子不同，增强子是增强转录的一种远端性控制元件，可位于基因上游、下游、5′端、3′端或者基因内部。

2 Operon: A genetic unit consisting of one or more structural genes encoding polypeptides, and an adjacent operator gene that controls the transcriptional activity of the structural gene or genes.

操纵子：操纵子是由编码多肽的一个或多个结构基因以及相邻的操纵基因组成的遗传单位，操纵基因控制结构基因的转录活性。

3 Promoter: Region having a regulatory function and to which RNA polymerase binds prior to the initiation of transcription.

启动子：启动子具有调节转录的功能，是RNA聚合酶结合并起始转录的位点。

4 Expressivity: The degree or range in which a phenotype for a given trait is expressed.

表现度：具有相同基因型的个体之间基因表达的变化程度称为表现度。

5 Isozyme: Any of two or more distinct forms of an enzyme that have identical or nearly identical chemical properties but differ in some property such as net electrical charge, pH optimal, number and type of subunits, or substrate concentration.

同工酶：构成上明显不同但具有相同的功能或相似的化学特性的一组酶称为同工酶。同工酶在某些理化性质如电荷、最适pH、亚基类型与数量、底物浓度等方面有所不同。

第十五节　遗传和进化
Section 15　Inheritance and Evolution

群体遗传学　population genetics
分子遗传学　molecular genetics
细胞色素　cytochrome
相互分歧　divergence
纤维蛋白肽　fibrinopeptide
单被基因组　haploid genome
免疫球蛋白　immunoglobin, IG
分子杂交　molecular hybridization
分子变性　molecular denaturation
熔解温度　melting temperature

温度稳定性　thermal stability
重复DNA　repetitive DNA
灵长类　primate groups
编码制度　coding system
复合染色体　compound chromosome
用进废退学说　theory of the inheritance of acquired character
痕迹奇观　rudimentary organs
自然选择学说　theory of natural selection
催产素　oxytocin
黑色素细胞激素　melanocyte stimulating hormone
促肾上腺皮质激素　adrenocorticotropic hormone
遗传平衡定律　law of genetic equilibrium
融合遗传概念　concept of blending inheritance
群体密度　population density
适合度　fitness
适应值　adaptive value
选择系数　selective coefficient
中性基因　neutral gene
随机遗传漂变　random genetic drift
发病率　incidence
遗传多肽　genetic polymorphism
隔离屏障　isolation barrier
隔离机制　isolation mechanism
异源多倍体或双二倍体　amphidiploid

1 Population: a local group of individuals belonging to the same species, which are actually or potentially interbreeding.
群体:指种群,即同种生物在特定空间内自由交配的所有个体的集群。

2 Fitness: A measure of the relative survival and reproductive success of a given individual or genotype.
适合度:适合度是个体产生能存活并具有繁衍能力的后代的衡量指标。

3 Adaptation: A heritable component of the phenotype that confers an advantage in survival and reproductive success. The process by which organisms adapt to the current environmental conditions.
适应性:表型的可遗传成分,是生存和繁衍成功的有力保证,即生物体对所处生态环境的适应性。

4 Isolating mechanism: Any barrier to the exchange of genes between different populations of a group of organisms. In general, isolation can be classified as spatial, environmental, or reproductive.
隔离机制:由多个个体组成的不同群体间基因交换的障碍,一般来说,隔离产生的原因包括空间、环境和生殖等方面。

5 Natural selection: Differential reproduction of some members of a species resulting from variable fitness conferred by genotypic differences.
自然选择:基因型的不同造成适应性的变化从而使得某一物种的一些成员产生差异的现象。

6 Repetitive DNA sequences: DNA sequences present in many copies in the haploid genome.
重复DNA序列:单倍体基因组中存在的有多个拷贝的DNA序列。

第十七章　微生物学
Chapter 17　Microbiology

第一节　绪论
Section 1　Introduction

微生物学　microbiology
微生物　microorganism
扫描电子显微镜　scanning electron microscope
原核生物　prokaryote
细菌　bacterium/bacteria
放线菌　actinomycete
蓝细菌　blue algae
支原体　mycoplasma
立克次氏体　rickettsia
衣原体　chlamydia
古生菌　archaea
真核微生物　eukaryotic microorganism
真菌　fungus/fungi
酵母菌　yeast
霉菌　mold
蕈菌　mushroom
藻类　alga/algae
非细胞微生物　acellular microorganism
病毒　virus
亚病毒　subvirus
类病毒　viroid
拟病毒　virusoid/viroid-like
朊病毒　prion
人类免疫缺陷病毒　human immunodeficiency virus(HIV)
艾滋病　acquired immunodeficiency syndrome (AIDS)
毒素　toxin
癌症流感　flu/influenza
生物工程学　biotechnology
模式生物　model organism
生物转化　bioconversion /biotransformation
嗜极菌　extremophile
普通微生物学　general microbiology
应用微生物学　applied microbiology
微生物生理学　microbial physiology
微生物遗传学　microbial genetics
微生物生态学　microbial ecology
微生物分类学　microbial taxonomy
食品微生物学　food microbiology
真菌学　mycology
细菌学　bacteriology
病毒学　virology

1 Microbiology: The study of organisms that are usually too small to be seen with the naked eye. Special techniques are required to isolate and grow them.
微生物学：研究那些因为太小而肉眼看不见的生物的学科。分离培养这些生物需要特殊技术。

2 Microbial ecology: The study of microorganisms in their natural environments, with a major emphasis of physical conditions, processes, and interactions that occur on the scale of individual microbial cells.
微生物生态学：研究自然环境中微生物的学科，着重强调物理条件、过程和发生在个体微生物细胞水平上的相互作用。

3 Actinomycete: An aerobic, grampositive bacterium that forms branching filaments and asexual spores.
放线菌：好氧的革兰氏阳性细菌，形成分枝菌丝和无性孢子。

4 Mycoplasma: Bacteria that are members of the class Mollicutes and order Mycoplasmatales; they lack cell walls and cannot synthesize peptidoglycan precursors; most require sterols for growth; they are the smallest organisms capable of independent reproduction.
支原体：柔膜体纲支原体目的细菌，缺乏细胞壁并不能合成肽聚糖前体，大多数需固醇生长。它们是能独立繁殖的最小生物。

5 Fungus: Achlorophyllous, heterotrophic, spore-bearing eucaryotes with absorptive nutrition; usually, they have a walled thallus.
真菌:无叶绿素、异养、产孢子、具吸收性营养的真核生物,菌丝具有细胞壁。

6 Yeast: A unicellular fungus that has a single nucleus and reproduces either asexually by budding or fission, or sexually through spore formation.
酵母菌:具有单核的单细胞真菌,通过芽殖或裂殖进行无性生殖或通过形成孢子进行有性生殖。

7 Mold: Any of a large group of fungi that cause mold or moldiness and that exist as multicellular filamentous colonies; also the deposit of growth caused by such fungi. Molds typically do not produce macroscopic fruiting bodies.
霉菌:致霉真菌的一大类群。以多细胞丝状群体形式生存,能引起沉积物或生长物。霉菌不产生肉眼可见的子实体。

8 Alga: A common term for a series of unrelated groups of photosynthetic eukaryotic microorganisms lacking multicellular sex organs (except for the charophytes) and conducting vessels.
藻类:指一系列不相关类群的光合真核微生物,这些微生物缺乏多细胞的性器官(轮藻植物例外)和导管。

9 Virus: An infectious agent having a simple acellular organization with a protein coat and a single type of nucleic acid, lacking independent metabolism, and reproducing only within living host cells.
病毒:具有简单的非细胞结构的感染因子,具有蛋白外壳和一种核酸,缺乏独立的代谢机制,只在活体宿主细胞内繁殖。

10 Viroid: An infectious agent of plants that is a single-stranded RNA not associated with any protein; the RNA does not code for any proteins and is not translated.
类病毒:一种植物感染因子。为一条与蛋白质无关的单链 RNA 分子,不编码任何蛋白质,不被转译。

11 Toxin: A microbial product or component that can injure another cell or organism at low concentrations. Often the term refers to a poisonous protein, but toxins may be lipids and other substances.
毒素:微生物产物或组分,在低浓度情况下就能伤害另一细胞或生物。这一术语往往是指有毒蛋白质,但毒素也可能是脂类或其他物质。

12 Biotransformation: The use of living organisms to modify substances that are not normally used for growth.
生物转化:用活体生物改进(修饰)通常不用于生长的物质。

13 Tuberculosis: An infectious disease of humans and other animals resulting from an infection by a species of *Mycobacterium* and characterized by the formation of tubercles and tissue necrosis, primarily as a result of host hypersensitivity and inflammation. Infection is usually by inhalation, and the disease commonly affects the lungs(pulmonary tuberculosis), although it may occur in any part of the body.
结核:由一种分枝杆菌属细菌感染引起的人和其他动物的传染病,特征为形成结核结节和组织坏死,主要原因是由于宿主超敏反应和炎症。通常由吸入病原而感染。这种病一般感染肺部,也可以在身体的任何部位发生。

第二节 原核生物的形态、构造和功能

Section 2 Morphology, Structure and Function of Prokaryotes

核区 nuclear region
杆菌 bacillus
螺菌 spirillum
弧菌 vibrio
微米 micron
纳米 nanometer
纳米细菌 nano-bacteria
革兰氏染色法 Gram stain
革兰氏阳性细菌 Gram positive bacteria(G+)
革兰氏阴性细菌 Gram negative bacteria(G—)

肽聚糖　peptidoglycan
N-乙酰葡糖胺　N-acetylglucosamine
N-乙酰胞壁酸　N- acetylmuramic acid
β-1,4-糖苷键　β-1,4－glucosidic bond
双糖单位　disaccharide unit
溶菌酶　lysozyme
四肽尾　tetrapeptide side chain
肽桥　peptide interbridge
磷壁酸　teichoic acids
脂多糖　lipopolysaccharide(LPS)
脂质 A　lipid A
核心多糖　core polysaccharide
O-特异侧链　O-specific side chain
周质空间　periplasmic space
假肽聚糖　pseudomurein
N-乙酰塔罗糖胺糖醛酸　N-acetyltalosaminouronic acid
缺壁细菌　cell wall deficient bacteria
L 型细菌　L-form of bacteria
磷脂　phosphatide, phospholipid
周边蛋白　peripheral protein
细胞内含物　inclusion body
储藏物　reserve material
聚-β-羟丁酸　poly-β-hydroxy-butyrate(PHB)
聚羟链烷酸　polyhydroxy-alkanoate(PHA)
异染粒　metachromatic granule
磁小体　magnetosome
羧酶体　carboxysome
气泡　gas vacuole
糖萼　glycocalyx
荚膜　capsule
微荚膜　microcapsule
黏液层　slime layer
菌胶团　zoogloea
鞭毛　flagellum
鞭毛钩　hook
鞭毛丝　filament
鞭毛蛋白　flagellin
趋氧性　oxygentaxis
趋磁性　magnetotaxis
旋转论　rotation theory
挥鞭论　bending theory
“拴菌”试验　tethered-cell experiment
单端鞭毛菌　monotricha
端生丛毛菌　lophotricha
两端鞭毛菌　amphitricha
周毛菌　peritricha
菌毛　fimbria, fimbriae(pl)
性菌毛　sex pilus
芽孢　endospore
孢囊　cyst
伴孢晶体　parasporal crystal
二分裂　binary fission
三分裂　trinary fission
复分裂　multiple fission
芽殖　budding
孢子丝　spore-bearing mycelium
分生孢子　conidiospore
藻胆素　phycobilin
蓝细菌肽　cyanophycin
水华　water bloom
类支原体　mycoplasma-like organism(MLO)
植原体　phytoplasma
异形胞　heterocyst
静息孢子　akinete
链丝段　hormogonium
赤潮　red tide
克隆　clone
菌膜　scum
土腥味素　geosmin
基内菌丝　substrate mycelium
气生菌丝　aerial mycelium
原体　elementary body
始体　initial body
甾醇　sterol
类立克次氏体　Rickettsia-like bacteria(RLB)

1 Gram stain: A differential staining procedure that divides bacteria into gram-positive and gram-negative groups based on their ability to retain crystal violet when decolorized with an organic solvent such as ethanol.
革兰氏染色:一种鉴别染色法,根据细菌用有机溶剂(如乙醇)脱色后保留结晶紫的能力,把细菌分为革兰氏阳性和革兰氏阴性两个类群。

2 Peptidoglycan: A large polymer composed of long chains of alternating N-acetylglucosamine and N-acetylmuramic acid residues. The polysaccharide chains are linked to each other through connections between tetrapeptide chains attached to the N-acetylmuramic acids. It provides much of the strength and rigidity possessed by bacterial cell walls.

肽聚糖：由交替的N-乙酰葡糖胺和N-乙酰胞壁酸残基长链组成的大聚合体。附着在N-乙酰胞壁酸上四肽链互相连接成的多糖链。它使细菌细胞壁增加刚性和强度。

3 Lipopolysaccharide: A molecule containing both lipid and polysaccharide, which is important in the outer membrane of the gram-negative cell wall.

脂多糖：既含脂类也含多糖的分子，在革兰氏阴性菌细胞壁外膜中具重要性。

4 Teichoic acids: Polymers of glycerol or ribitol joined by phosphates, they are found in the cell walls of gram-positive bacteria.

磷壁酸：由磷酸二酯键连接甘油或核糖醇而成的多聚体，见于革兰氏阳性细菌细胞壁。

5 Pseudomurein: A modified peptidoglycan lacking D-amino acids and containing N-acetyltalosaminuronic acid instead of N-acetylmuramic acid; found in methanogenic archaea.

假肽聚糖：被修饰的肽聚糖，缺乏D-氨基酸，以N-乙酰塔罗糖胺糖醛酸取代N-乙酰胞壁酸。见于产甲烷古生菌。

6 Glycocalyx: A network of polysaccharides extending from the surface of bacteria and other cells.

糖萼：从细菌和其他细胞表面扩展的多糖网络。

Capsule: A layer of well-organized material, not easily washed off, lying outside the bacterial cell wall.

荚膜：一层结构精细的物质，不宜洗掉，位于细菌细胞壁外部。

7 Flagellum: A thin, threadlike appendage on many prokaryotic and eukaryotic cells that is responsible for their mobility.

鞭毛：许多原核和真核细胞上细线状附器，能使这些细胞游动。

9 Fimbria: A fine, hairlike protein appendage on some gram-negative bacteria that helps attach them to surfaces.

菌毛：某些革兰氏阴性细菌的纤细丝状蛋白附器，有助于附着。

9 Sex pilus: A thin protein appendage required for bacterial mating or conjugation. The cell with sex pili donates DNA to recipient cells.

性菌毛：细菌配对或接合形成的细微蛋白附器。具有性菌毛的细胞提供DNA给受体细胞。

10 Endospore: An extremely heat- and chemical-resistant, dormant, thick-walled spore that develops within bacteria.

芽孢：细菌内产生的极耐热耐化学制品的厚壁休眠孢子。

11 Budding: A vegetative outgrowth of yeast and some bacteria as a means of asexual reproduction; the daughter cell is smaller than the parent.

芽殖：酵母和某些细菌营养期的突起，作为一种无性生殖方式。子细胞比亲本小一些。

12 Conidiospore: An asexual, thin-walled spore borne on hyphae and not contained within a sporangium; it may be produced singly or in chains.

分生孢子：产在菌丝上而不含在孢子囊里的无性薄壁孢子，可为单个的或呈链状。

第三节 真核微生物的形态、构造和功能

Section 3 Morphology, Structure and Function of Eukaryotes

菌物界 mycetalia/fungi

真菌 eumycetes

黏菌 myxomycetes fungi-like protozoa

假菌 pseudofungi

原生动物 protozoa

甘露聚糖 mannan

葡聚糖 glucan

几丁质 chitin

纤毛 cilium/cilia

鞭杆 shaft

微管连丝蛋白 nexin

细胞基质 cytomatrix

膜边体 lomasome

几丁质酶体 chitosome

氢化酶体　hydrogenosome
解旋酶　helicase
麦角甾醇　ergosterol
2μm 质粒　2μm plasmid
工程菌　engineering bacteria
假酵母(拟酵母)　pseudo-yeast
真酵母　euyeast
有性繁殖　sexual multiplication
假菌丝　pseudohyphae
真菌丝　euhyphae
芽痕　bud scar
蒂痕　birth scar
裂殖　fission
掷孢子　ballistospore
厚垣孢子　chlamydospore
子囊　ascus
子囊孢子　ascospore
质配　plasmogamy
核配　caryogamy
生命周期　life cycle
单倍体　haploid
丝状真菌　filamentous fungi
菌丝　hypha/hyphae(复数)
菌丝体　mycelium/mycelia(复数)
营养菌丝　vegetative mycelium
气生菌丝　aerial mycelium
假根　rhizoid
匍匐菌丝　stolon
吸器　haustorium
附着胞　adhesive cell
附着枝　adhesive branch
菌核　sclerotium
菌索　rhizmorph/funiculus
菌网　net
子实体　fruiting body/sporocarp/fructification
分生孢子头　conidial head
孢子囊　sporangium
有性孢子　sexual spore
无性孢子　asexual spore
分生孢子器　pycnidium
分生孢子座　sporodochium
分生孢子盘　acervulus
分生孢子梗　conidiophore
子囊果　ascocarp
闭囊壳　cleistothecium
子囊壳　perithecium
子囊盘　apothecium
菌丝球　mycelial bead
游动孢子　zoospore
孢囊孢子　sporangiospore
分生孢子　conidiospore
节孢子　arthrospore
卵孢子　oospore
接合孢子　zygospore
担孢子　basidiospore
食用菌　edible mushroom
锁状联合　clamp connection
一级菌丝　primary mycelium/primary hyphae
二级菌丝　second mycelium
三级菌丝　third mycelium
菌盖　cap/pileus
菌褶　gill/lamella
菌环　collar/ring
菌蕾　button
菌托　volva

1 Glucans: Polysaccharides composed of glucose units held together by glycosidic linkages. Some types of glucans have α(1-3) and α(1-6) linkages and bind bacterial cells together on teeth forming a plaque ecosystem.

葡聚糖：由糖苷键连接葡萄糖组成的多糖。某些葡聚糖具 a(1-3) 和 a(1-6) 键合，把牙齿上细菌细胞结合到一起，形成牙斑生态系统。

2 Chitin: A tough, resistant, nitrogen-containing polysaccharide forming the walls of certain fungi, the exoskeleton of arthropods, and the epidermal cuticle of other surface structures of certain protists and animals.

几丁质：硬且有抵抗力的含氮多糖，形成某些真菌细胞壁、节肢动物的外骨骼和某些原生生物和动物的外表结构表皮角质层。

3 Helicases: Enzymes that use ATP energy to unwind DNA ahead of the replication fork.

解旋酶：用 ATP 能量使复制叉前的 DNA 解链的酶。

4 Chlamydospore: An asexually produced, thick-walled resting spore formed by some fungi.

厚垣孢子：某些真菌无性生殖的厚壁休眠孢子。

5 Ascus：A specialized cell，characteristic of the ascomycetes，in which two haploid nuclei fuse to produce a zygote，which immediately divides by meiosis；at maturity an ascus will contain ascospores.

子囊：具有子囊菌特征的特化细胞。子囊中两个单倍体核融合产生一个合子，合子随即进行减数分裂。成熟时子囊会含有子囊孢子。

6 Hypha：The unit of structure of most fungi and some bacteria，a tubular filament.

菌丝：大多数真菌和某些细菌的结构单位，管形丝状体。

7 Fruiting body：A specialized structure that holds sexually or asexually produced spores，found in fungi and in some bacteria.

子实体：携带由有性或无性生殖产生的孢子的特化结构，见于真菌和某些细菌。

8 Ascocarp：A multicellular structure in ascomycetes lined with specialized cells called asci in which nuclear fusion and meiosis produce ascospores. An ascocarp can be open or closed and may be referred to as a fruiting body.

子囊果：子囊菌的多细胞结构，排列着称为子囊的特化细胞，子囊中核融合和减数分裂产生子囊孢子。子囊果可开放或封闭，也可以成为子实体。

第四节　病毒和亚病毒
Section 4　Virus and Subvirus

寄生物　parasite
专性寄生物　obligate parasite
潜伏性侵染　latent infection
核心　core
衣壳　capsid
衣壳粒　capsomer
亚单位　subunit
核衣壳　nucleocapsid
包膜　envelope
包涵体　inclusion body
噬菌斑　plaque
枯斑　lesion
空斑　plaque
五邻体　penton
六邻体　hexon
T偶数噬菌体　T even phage
单链　single strand
双链　double strand
碱基(对)　base
噬菌体　phage
噬放线菌体　actinophage
噬蓝细菌体　cyanophage
烈性噬菌体　virulent phage
裂解性周期　lytic cycle
增殖性周期　productive cycle
吸附　adsorption/attachment
侵入　penetration/injection
释放　release
效价　titre/titer
敏感宿主　sensitive host
噬菌斑形成单位　plaque-forming unit (pfu)
感染中心数　infective centre
双层平板法　two layer plating method
成斑率　efficiency of plating (EOP)
一步生长曲线　one-step growth curve
潜伏期　latent phase
隐晦期　eclipse phase
胞内累计期　intracellular accumulation phase
裂解期　rise phase
裂解量　burst size
平稳期　plateau
溶源性　lysogeny
溶原菌　lysogenic bacteria
温和噬菌体　temperate phage
前噬菌体　prophage
裂解性循环　lytic cycle
溶源性循环　lysogenic cycle
脱壳　uncoating
人畜共患病　zoonosis
多角体　polyhedron
核型多角体病毒　nuclear polyhedrosis virus (NPV)
质型多角体病毒　cytoplasmic polyhedrosis virus (CPV)
颗粒体病毒　granulosis virus (GV)

病毒卫星 satellite
蛋白侵染子 protein infection
杀菌剂 germicide
免疫原性 immunogenicity
发酵 fermentation
柯斯质粒 cosmid
克隆载体 charon
杆状病毒 baculovirus

1 Capsid: The protein coat or shell that surrounds a virion's nucleic acid.
衣壳:包围毒粒核酸的蛋白外壳。

2 Capsomer: The ring-shaped morphological unit of which icosahedral capsids arc constructed.
衣壳粒:构成二十面体壳体的环状形态学单位。

3 Envelope: 1. All the structures outside the plasma membrane in bacterial cells .
2. In virology it is an outer membranous layer that surrounds the nucleocapsid in some viruses.
包膜:1. 细菌细胞质膜外结构。
2. 病毒学中,包围某些病毒核壳的膜质外层。

4 Lysogeny: The state in which a phage genome remains within the bacterial host cell after infection and reproduces along with it rather than taking control of the host and destroying it.
溶源性:噬菌体感染后其基因组存留在细菌宿主细胞中的一种状态,噬菌体基因组随宿主细胞分裂进行复制,而不是接管并裂解细胞。

第五节 微生物的营养和培养基
Section 5 Nutrition and Medium of Microbe

大肠菌群 coliform
营养 nutrition
营养物 nutrient
培养基 medium
营养要素 nutrient key element
碳源 carbon source
大量营养物 macronutrient
碳源谱 spectrum of carbon sources
异养微生物 heterotrophic microorganism
自养微生物 autotrophic microorganism
有机碳源 organic carbon source
无机碳源 inorganic carbon source
有机酸 organic acid
明胶 gelatin
蔗糖 sucrose/saccharose
甘露糖 mannitose/mannose/seminose
半乳糖 galactose
糖蜜 molasses
琼脂 agar
双功能营养物 difunctional nutrient
氮源 nitrogen source
氮源谱 spectrum of nitrogen sources
牛肉膏 beef extract
蛋白胨 peptone
酵母膏 yeast extract
玉米浆 corn steep liquor
肝浸液 liver infusion
麦芽汁 malt extract
尿素 urea/carbamide
硝酸盐 nitrate
铵盐 ammonium cuprate
氨基酸自养型生物 amino acid autotrophs
氨基酸异养型生物 amino acid heterotrophs
菌体蛋白 single cell protein (SCP)
能源 energy source
生长因子自养型微生物 auxoautotrophs
生长因子异养型微生物 auxoheterotrophs
维生素 vitamin
无机盐 mineral salts
微量元素 microelements
溶剂 solvate/ dissolvent
天然气 natural gas (NG)
石蜡油 liquid paraffin mineral syrup/ paraffin oil
光能有机营养型 organotrophic
无机化能营养型 chemolithotrophy
有机化能营养型 organolithotrophy
单纯扩散 simple diffusion
被动运送 passive transport
促进扩散 facilitated diffusion
渗透酶 permease
移位酶 translocase
移位蛋白 translocator protein
主动运送 active transport

离子泵 ion pump
贫养菌(寡养菌) oligophyte
基团移位 group translocation
磷酸转移酶系统 phosphotransferase system
热稳载体蛋白 heat-stable carrier protein (HPr)
难养菌 fastidious microorganism
水活度 water activity
相对湿度值 equilibrium relative humidity(ERH)
氧化还原势 redox potential
巯基乙酸 mercaptoacetic acid thioglycollic acid
刃天青 resazurin
琼脂平板 agar plate
摇瓶培养 shake culture
发酵罐 fermenter
天然培养基 complex media
牛肉膏蛋白胨培养基 beef extract peptone medium
麦芽汁培养基 malt extract medium
组合培养基 chemical defined media
半组合培养基 semi-defined media
半合成培养基 semi-synthetic media
马铃薯蔗糖培养基 potato sucrose medium
液体培养基 liquid medium
固体培养基 solid medium
固化培养基 solidified medium
凝固剂 gelling agent
海藻酸胶 alginate
多聚醇 pluronic polyol
硅胶 silica gel
滤膜 membrane filter
半固体培养基 semi-solid medium
稀琼脂 sloppy agar
脱水培养基 dehydrated culture medium
脱水商品培养基 dehydrated commercial medium
预制干燥培养基 pre-fabricated dried culture medium
选择性培养基 selected medium
加富性选择培养基 enriched selected medium
富集 enrichment
抑制性选择培养基 inhibited selected media
鉴别性培养基 differential medium
伊红美蓝乳糖培养基 Eosin Methylene Blue

1 Autotrophic microorganism: An organism that uses CO_2 as its sole or principal source of carbon.
自养微生物:利用二氧化碳作为唯一或主要碳源的生物。

2 Heterotrophic microorganism: An organism that uses reduced, preformed organic molecules as its principal carbon source.
异养微生物:利用还原的、预先形成的有机分子作为主要碳源的生物。

3 Differential medium: Culture media that distinguish between groups of microorganisms based on differences in their growth and metabolic products.
鉴别培养基:根据生长的差异和代谢产物的不同,区分不同类型微生物的培养基。

第六节 微生物的新陈代谢
Section 6 Metabolism of Microbes

腺苷三磷酸 (ATP)adenosine/triphosphate
还原力 reducing power
同化 assimilation
最初能源 primary energy sources
通用能源 universal energy resource
生物氧化 biological oxidation
EMP 途径 Embden-Meyerhof-Parnas pathway
糖酵解途径 glycolysis pathway
己糖二磷酸途径 hexose diphosphate pathway (HDP)
HMP 途径 hexose monophosphate pathway
戊糖磷酸途径 pentose phosphate pathway
氧化磷酸化 oxidative phosphorylation
硝酸盐呼吸 nitrate respiration
反硝化作用 denitrification
同化性硝酸盐还原作用 assimilative nitrate reduction
异化性硝酸盐还原作用 dissimilative nitrate reduction
硫酸盐呼吸 sulfate respiration
硫呼吸 sulphur respiration
铁呼吸 iron respiration
碳酸盐呼吸 carbonate respiration
延胡索酸呼吸 fumarate respiration
同型酒精发酵 homoalcoholic fermentation
同型乳酸发酵 homolactic fermentation

V. P. 试验 Vogos-Prouskauer test
异型乳酸发酵 heterolactic fermentation
化能自养微生物 chemoautotrophs
生长速率 growth rate
硝化细菌 nitrifying bacteria
亚硝酸氧化酶 nitrite oxidase
光能营养细菌 phototrophs
光合细菌 photosynthetic bacteria
菌绿素 bacteriochlorophyll (Bchl)
不产氧光合作用 anoxygenic photosynthesis
嗜盐菌 halophile
紫膜蛋白 purple membrane protein
光介导 ATP 合成 light-mediated ATP synthesis
细菌视紫红质 bacteriorhodopsin
紫膜光合磷酸化 photophosphorylation by purple membrane
两用代谢途径 amphibolic pathway
葡糖异生作用 gluconeogenesis
代谢物回补顺序 anaplerotic sequence
代谢物补偿途径 replenishment pathway
草酰乙酸 oxaloacetate (OA)
乙醛酸 循环 glyoxylate cycle
厌氧乙酰-CoA 途径 anaerobic acytyl-CoA pathway
活性乙酸途径 activated acetic acid pathway
逆向 TCA 循环 reverse TCA cycle
还原性 TCA 循环 reductive TCA cycle
羟基丙酸途径 hydroxyl propionate pathway
生物固氮 biological nitrogen fixation
固氮微生物 nitrogen-fixing organism
自生固氮菌 free-living nitrogen-fixer
共生固氮菌 symbiotic nitrogen-fixer
联合固氮菌 associative nitrogen-fixer
固氮酶 nitrogenase
放氧性光合生物 oxygenic phototroph
超氧化物歧化酶 superoxide dismutase (SOD)
类菌体 bacteroid
选择毒力 selective toxicity
自溶素 autolysin
次生代谢物 secondary metabolite
生物药物素 biopharmaceutin
代谢调节 regulation of metabolism
诱导酶 induced enzyme
组成酶 constitutive enzyme

1 ATP: The triphosphate of the nucleoside adenosine, which is a high energy molecule or has high phosphate group transfer potential and serves as the cell's major form of energy currency.
腺苷三磷酸:腺苷的三磷酸为高能分子,或具有高磷酸基团转移的潜能,作为细胞的主要能流形式。

2 Gluconeogenesis: The synthesis of glucose from noncarbohydrate precursors such as lactate and amino acids.
葡糖异生:从非糖类前体(如乳酸和氨基酸)合成葡糖糖。

3 Glyoxylate cycle: A modified tricarboxylic acid cycle in which the decarboxylation reactions are bypassed by the enzymes isocitrate lyase and malate synthase, it is used to convert acetyl-CoA to succinate and other metabolites.
乙醛酸循环:一种改变的三羧酸循环,脱羧反应被异柠檬酸和苹果酸合成酶引入旁路。用于将乙酰辅酶 A 转化成琥珀酸和其他代谢产物。

第七节 微生物的生长及其控制
Section 7 Growth and Control of Microbes

繁殖 reproduction
培养物 culture
高密度培养 high cell-density culture(HCDC)
菌落计数法 colony-counting method
平板浇注 pour plate
涂布平板 spread plate
菌落形成单位 colony forming unit(cfu)
同步培养 synchronous culture
同步生长 synchronous growth
生长曲线 growth curve
生长速率常数 growth rate constant
接种物 inoculum, inocula(pl)
指数期 exponential phase
对数期 logarithmic phase
代时 generation time
倍增时间 doubling time
平衡生长 balanced growth
生长限制因子 growth-limited factor
生长得率 growth yield
初生代谢物 primary metabolite

代谢产物合成期 idiophase
生物测定 bioassay
衰亡期 decline phase/ death phase
自溶 autolysis
连续培养 continuous culture
开放培养 open culture
单批培养 batch culture
密闭培养 closed culture
连续生长 continuous growth
恒浊器 turbidostat
恒化器 chemostat
单级连续培养器 one-step continuous fermentor
多级连续培养器 multi-step continuous fermentor
菌体生长期 trophophase
产物合成期 idiophase
稀释率 dilution rate
连续发酵 continuous fermentation
下游工程 down-stream processing
生长温度三基点 three cardinal points
最适生长温度 optimum growth temperature
好氧微生物 aerobe
厌氧微生物 anaerobe
专性好氧菌 obligate aerobe
兼性厌氧菌 facultative anaerobe
兼性好氧菌 facultative aerobe
耐氧菌 aerotolerant anaerobe
专性厌氧菌 strict anaerobe
产甲烷菌 methanogen
过氧化物酶 peroxidase
试管斜面 test-tube slant
琼脂平板 agar plate
克氏扁瓶 Kolle flask
高层琼脂柱 high agar column
韦荣氏管 Veillon tube
厌氧培养皿 anaerobic Petri dish
恒盖特滚管技术 Hungate roll-tube technique
预还原无氧灭菌培养基 pre-reduced anaerobically sterilized medium
厌氧罐 anaerobic jar
厌氧手套箱 anaerobic glove box
台式发酵罐 bench top fermentor
传感器 sensor
固体曲 solid mouldy bran
浅盘培养 shallow pan cultivation
生物反应器 bioreactor
固定化细胞 immobilized cell
污染 contamination
灭菌 sterilization
杀菌 bacteriocidation
溶菌 bacteriolysis
消毒 disinfection
防腐 antisepsis
制菌作用 bacteriostasis
除氧剂 deoxidizer
防腐剂 antiseptic
化疗 chemotherapy
选择毒力 selective toxicity
干热灭菌法 dry heat sterilization
灼烧 incineration/ combustion
湿热灭菌法 moist heat sterilization
巴氏消毒法 Pasteurization
低温维持法 low temperature holding method (LTH)
高温瞬时法 high temperature holding method (HTS)
间歇灭菌法 fractional sterilization/ tyndallization
加压蒸汽灭菌法 normal autoclaving
连续加压蒸汽灭菌法 continuous autoclaving
热死时间 thermal death time
热死温度 thermal death point
滤器 filter
最低抑制浓度 minimum inhibitory concentration (MIC)
半致死剂量 lethal dose 50 (LD50)
最低致死剂量 minimum lethal dose (MLD)
表面消毒剂 surface disinfectant
石炭酸系数 phenol coefficient (P. C.)
供试菌 test organism
抗代谢药物 antimetabolite
磺胺药 sulphonamide/sulfa drug
对氨基苯甲酸 para-amino benzoic acid (PABA)
竞争性拮抗作用 competitive antagonism
牛津单位 oxford unit
广谱型抗生素 broad-spectrum antibiotic
半合成抗生素 semi-synthetic antibiotic
抗药性 antibiotic resistance

1 Colony forming units: The number of microorganisms that can form colonies when cultured using spread plates or pour plates, an indication of the number of viable microorganisms in a sample.

菌落形成单位:用涂平板或倒平板的方法培养时,微生物形成的菌落数,指示样品中活的微生物数。

2 Exponential phase: The phase of the growth curve during which the microbial population is growing at a constant and maximum rate, dividing and doubling at regular intervals.

指数期:生长曲线的一个阶段,其间微生物群体以恒定的最大速率生长,以固定的间隔时间分裂和倍增。

3 Lethal dose 50: Refer to the dose or number of organisms that will kill 50% of an experimental group of hosts within a specified time period.

半致死剂量:在指定时间内,使实验组宿主机体的 50% 死亡的生物剂量或数量。

第八节　微生物的遗传变异和育种

Section 8　Heredity Variation and Breeding of Microbes

变异　variation
遗传型　genotype
表型　phenotype
饰变　modification
致病菌　pathogenic bacteria
卡巴颗粒　Kappa particle
千碱基对　kilo base pair
兆碱基对　mega base pair
操纵子　operon
结构基因　structure gene
操纵基因　operator
启动基因　promoter
阻遏物　repressor
密码子　codon
松弛性复制控制　relaxed replication control
严紧型复制控制　stringent replication control
质粒消除　curing elimination
整合　integration
克隆载体　cloning vector
聚丙烯凝胶电泳　polyacrylamide gel electrophoresis (PAGE)
密度梯度离心法　density gradient centrifugation
致育因子　fertility factor
性因子　sex factor
转移区　transfer region
转座因子　transposable element
抗性转移因子　resistance transfer factor (RTF)
抗性决定因子　r-determinant
Col 质粒　Colicin plasmid
Col plasmid
细菌素　bacteriocin
Ti 质粒　tumor inducing plasmid
冠瘿质粒　crown gall plasmid
冠瘿碱类　opines
Ri 质粒　root inducing plasmid
Mega 质粒　Mega plasmid
基因突变　gene mutation
野生型菌株　wild type strain
突变株　mutant
选择性突变株　selectable mutant
非选择性突变株　non-selectable mutant
营养缺陷型　auxotroph
基本培养基　minimum medium (MM)
抗性突变型　resistant mutant
条件致死突变型　conditional lethal mutant
温度敏感突变株　temperature sensitive mutant/Ts mutant
形态突变型　morphological mutant
抗原突变型　antigenic mutant
产量突变型　metabolite quantitative mutant
突变率　mutation rate
回复突变株　back mutant reverse mutant
抗药性突变株　resistance mutant
正突变　plus-mutant
负突变　minus-mutant
影印平板培养法　replica plating
点突变　point mutation
诱发突变　induced mutation
置换　substitution
转换　transition
颠换　transversion
移码突变　fram-shift mutation, phase-shift mutation
移码突变株　frame-shift mutant
烷化剂　alkylating agent
染色体畸变　chromosomal aberration
缺失　deletion
重复　duplication
插入　insertion
易位　translocation

倒位 inversion
转座 transposition
转座子 transposon (Tn)
转座噬菌体 transposable phage
可移动基因 moveable gene
可移动遗传因子 mobile genetic element
插入序列 insertion sequence (IS)
跳跃基因 jumping gene
正向重复 direct repeat
反向重复 inverted repeat
转座酶基因 transposase gene
靶序列 target sequence
自发突变 spontaneous mutation
紫外线 ultraviolet ray (UV)
嘧啶二聚体 pyrimidine dimmer
光复活作用 photoreactivation
单体 monomer
光解酶 photolyase
切除修复 excision repair
暗修复 dark repair
诱变育种 breeding by induced mutation
离子束 ion beam
拟辐射物质 radiomemetic chemical
艾姆氏试验 Ames test
黄曲霉毒素 aflatoxin
出发菌株 original strain
表型延迟 phenotype lag
协同效应 synergism
春日霉素 kasugamycin
选择性遗传标记 selective genetic marker
梯度平板法 gradient plate
代谢拮抗物 antimetabolite
琼脂块培养法 agar block method
完全培养基 complete medium (CM)
补充培养基 supplemental medium (SM)
野生型 wild type, wild strain
原养型 prototroph
夹层培养法 layer plating method
生长谱法 auxanography
基因重组 gene recombination
遗传重组 genetic recombination
杂交 hybridization
受体菌 recipient cell, receptor
供体菌 donor cell
球状体 spheroplast
感受态 competence
膜相关DNA结构蛋白 membrane-associated DNA binding protein
转化因子 transforming principle
转染 transfection
缺陷噬菌体 defective phage
转导子 transductant
完全转导(普遍转导) complete transduction
转导颗粒 transducing particle
流产转导 abortive transduction
局限转导 specialized transduction restricted transduction
误切 abnormal excesion
低频转导 low frequency transduction (LFT)
低频转导裂解物 LFT lysate
高频转导 high frequency transduction (HFT)
高频转导裂解物 HFT lysate
双重溶原菌 double lysogen
溶源转变 lysogenic conversion
接合 conjugation mating
接合子 conjugant
附加体 episome
滚环模型 rolling circle model
高频重组菌株 high frequency recombination strain
环状染色体图 chromosome map
接合中断法 interrupted mating experiment
F因子转导 F-duction
性导 sexduction
F质粒媒介转导 F-mediated transduction
原生质体融合 protoplast fusion
亲本菌株 parent strain
聚乙二醇 polyethylene glycol
有性杂交 sexual hybridization
准性杂交 parasexual hybridization
半知菌类 fungi imperfecti
菌丝联结 anastomosis
核融合 nuclear fusion
体细胞交换 somatic crossing-over
有丝分裂交换 mitotic crossing-over
准性杂交育种 breeding by parasexuality
遗传工程 genetic engineering
限制性核酸内切酶 restriction endonuclease
黏性末端 cohesive end
连接酶 ligase
嵌合体 chimaera
干扰素 interferon
白细胞介素 interleukin

转基因动物 transgenic animal
转基因植物 transgenic plant
衰退 degeneration
纯种分离法 pure culture isolation
离异 dissociation
菌种保藏 preservation, conservation maintenance
冷冻干燥保藏法 lyophilization freeze-drying method
原种 stock culture
典型培养物 type culture
二甲亚砜 dimethy sulfoxide (DMSO)

1 Codon: A sequence of three nucleotides in mRNA that directs the incorporation of an amino acid during protein synthesis or signals the start or stop of translation.
密码子:mRNA 中 3 个核苷酸的顺序指导蛋白质合成中氨基酸的并入,或发出转译起始或终止的信号。

2 Auxotroph: A mutated prototroph that lacks the ability to synthesize an essential nutrient and therefore must obtain it or a precursor from its surroundings.
营养缺陷型:突变的原养型。缺乏合成必需营养物的能力,因此必须从环境中获取这种营养物或前体。

3 Replica plating: A technique for isolating mutants from a population by plating cells from each colony growing on a nonselective agar medium onto plates with selective media or environmental conditions, such as the lack of a nutrient or the presence of an antibiotic or a phage; the location of mutants on the original plate can be determined from growth patterns on the replica plates.
影印平板培养法:从微生物群体中选择突变型的一种方法。用影印板接种非选择琼脂培养基上的菌落到具有选择性的培养基或环境条件中(如缺乏一种营养物或存在有一种抗生素或噬菌体),从影印平板上的生长图形可以确定被影印平板上突变型的位置。

4 Point mutation: A mutation that affects only a single base pair in a specific location.
点突变:只影响特意位置中单一碱基对的突变。

5 Photoreactivation: The process in which blue light is used by a photoreactivating enzyme to repair thymine dimmers in DNA by splitting them apart.
光复合作用:光复合酶用蓝光拆开 DNA 中的胸腺嘧啶二聚体使 DNA 得以修复的过程。

6 Protoplast fusion: The joining of cells that have had their walls weakened or completely removed.
原生质体融合:细胞壁已经弱化的或完全去掉的细胞的结合。

第九节 微生物的生态
Section 9 Ecology of Microbes

生态系统 ecosystem
群落 community
种群 population
个体 individual
器官 organ
组织 tissue
细胞器 organelle
贫营养细菌 oligotrophic bacteria
自净作用 self cleaning
滤膜培养法 filter membrane culture method
材料劣化 material deterioration
霉变 mildew/mould
腐朽 decay
腐烂 putrefaction, rot
腐蚀 corrosion
真菌毒素 mycotoxin
嗜热微生物 thermophile
耐热菌 thermotolerant bacteria
兼性嗜热菌 facultative thermophile
专性嗜热菌 obligate thermophile
极端嗜热菌 extremothermophile
超嗜热菌 hyperthermophile
嗜极酶 extreme enzyme
嗜冷微生物 psychrophile
耐冷微生物 psychrotolerant
嗜酸微生物 acidophile
耐酸的 acidtolerant
嗜碱生物 alkalinophile
洗涤剂 detergent
嗜盐微生物 halophile

耐盐的　halotolerant
嗜压生物　barophile
正常菌群　normal flora
微生态学　microecology
条件致病菌　opportunist pathogen
内源感染　endogenous infection
益生菌剂　probiotics
无菌动物　germ-animal
悉生生物　gnotobiotic
悉生生物学　gnotobiology
根际微生物　rhizosphere microorganism
附生微生物　epibiotic microorganism epibiont
共生　symbiosis
互利共栖　mutualism
互养共栖　syntrophism
协同共栖　synergism
寄生　parasitism
捕食　predation
拮抗　antagonism
偏利共栖　commensalism
卫星状共栖　satellitism
互生　metabiosis
偏害共栖　amensalism
混合培养　mixed cultivation
混合发酵　mixed fermentation
地衣　lichen
菌根　mycorhiza
菌套　mantle
泡囊一丛枝状菌根　vesicular-arbuscular mycorrhiza (VAM)
内共生生物　endosymbiont
外共生生物　ectosymbiont
瘤胃微生物　rumen microflora
寄生物　parasitism
宿主　host
兼性寄生物　facultative parasite
微生物杀虫剂　microbial pesticide
生物农药　bio-pesticide
猎食　predatism/ predation
食物链　food chain
分解者　decomposer
碳素循环　carbon cycle
氮素循环　nitrogen cycle
亚硝化细菌　nitrosobacteria
硝酸化细菌　nitrobacteria
氨化作用　ammonification
铵盐同化作用　assimilation of ammonium
反硝化作用　denitrification
硫素循环　sulfur cycle
脱硫作用　desulfuration
硫化作用　sulfur oxidation
细菌沥滤　bacterial leaching
磷素循环　phosphorus cycle
植酸盐　phytate
富营养化　eutrophication
赤潮　red tide
生化需氧量　biochemical oxygen demand
化学需氧量　chemical oxygen demand
总需氧量　total oxygen demand
溶解氧量　dissolved oxygen
悬浮物含量　suspend solid
总有机碳量　total organic carbon
异生物质(外生物质)　xenobiotic
共代谢　co-metabolism
难降解化合物　recalcitrant compound
活性污泥　activated sludge
有机垃圾好氧生物反应器　aero-bioreactor of organic garbage
可再生资源　renewable resource
沼气　marsh gas，swamp gas
生物气　biogas
甲烷形成(沼气发酵)　methanogenesis
发光细菌　luminescent bacteria
自诱导　autoinduction
自诱导物　autoinducer

1 Ecosystem：A self-regulating biological community and its associated physical and chemical environment.

生态系统：自我调节的生物群落及其相关的物理、化学环境。

2 Community：An assemblage of different types of organisms or a mixture of different microbial populations.

群落：不同类型的生物集群，或不同微生物种群的混合物。

3 Population：An assemblage of organisms of the same type.

种群：同类生物的集聚。

4 Endogenous infection: An infection by a member of an individual's own normal body microbiota.
内源感染:一个体自身一些正常微生物区系成员引起的感染。

5 Gnotobiotic: Animals that are germfree (microorganism free) or live in association with one or more known microorganisms.
悉生生物:生长在无菌(无微生物)环境中的动物或与一种或多种已知微生物生活在一起的动物。

6 Denitrification: The reduction of nitrate to gas products, primarily nitrogen gas, during anaerobic respiration.
反硝化作用:厌氧呼吸中硝酸盐还原成气态产物,主要为氮气。

7 Biochemical oxygen demand (BOD): The amount of oxygen used by organisms in water under certain standard conditions; it provides an index of the amount of microbially oxidizable organic matter present.
生化需氧量:标准条件下,生物在水中的用氧量,指示微生物可氧化的有机物量。

8 Chemical oxygen demand (COD): The amount of chemical oxidation required to convert organic matter in water and wastewater to CO_2.
化学需氧量:把水和废水中的有机物转变成二氧化碳所需要的化学氧化量。

第十节 传染与免疫
Section 10 Infection and Immunity

病原微生物 pathogenic microorganism
传染 infection
传染病 infectious disease/communicable disease/contagious disease
毒力 virulence
致病力 pathogenicity
侵袭力 invasiveness
定居因子抗原 colonization factor antigen (CFA)
胶原酶 collagenase
血浆凝固酶 coagulase
链激酶 streptokinase
卵磷脂酶(α 毒素) lecithinase
外毒素 exotoxin
内毒素 endotoxin
鲎试剂法 limulus assay
鲎变形细胞溶解物试验法 limulus amoebocyte lysate test (LAL test)
免疫力 immunity
免疫防御 immunologic defence
免疫稳定 immunologic homeostasis
免疫监视 immune surveillance
隐性传染 inapparent infection
带菌者 carrier
显性传染 apparent infection
急性传染 acute infection
慢性传染 chronic infection
局部感染 local infection
全身感染 systemic infection
毒血症 toxemia
菌血症 bacteremia
败血症 septicemia
脓毒血症 pyemia
非特异性免疫 nonspecific immunity
先天免疫 innate immunity
自然免疫 natural immunity
屏障结构 barrier structure
血脑屏障 blood-brain barrier
白细胞 leukocyte
红细胞 erythrocyte
血小板 blood platelet
多形核白细胞 polymorphonuclear leukocyte (PMN)
粒细胞 granulocyte
巨噬细胞 macrophage
骨髓干细胞 stem cell
白细胞介素 inter leukin-1 (IL-1)
炎症 inflammation
干扰素诱生剂 interferon inducer
植物血凝素 phytohaemoagglutinin (PHA)
抗病毒蛋白 antiviral protein
转移抑制蛋白 translation inhibitory protein
特异性免疫 specific immunity
获得性免疫 acquired immunity
体液免疫系统 humoral immunity
细胞免疫系统 cellular immunity

细胞介导免疫系统 cell-mediated immunity
中枢免疫器官 central immune organ
胸腺素 thymosin
胸腺生成素 thymopoietin
免疫细胞 immunocyte
免疫活性细胞 immunologically competent cell
T淋巴细胞 T lymphocyte
调节性T细胞 T regulator (TR)
效应性T细胞 T effector (TE)
大颗粒淋巴细胞 large granular lymphocyte (LGL)
免疫原 immunogen
抗原性 antigenicity
完全抗原 complete antigen
不完全抗原 incomplete antigen
抗原决定簇 antigenic determinant
异物性 foreignness
抗原表位 epitope
表面抗原 surface antigen
菌体抗原 somatic antigen
特异性抗原 specific antigen
共同抗原 common antigen
类属抗原 group antigen
交叉反应 cross-reaction antigen
凝集吸收反应 agglutination absorption
抗抗体 anti-antibody
多克隆抗体 polyclonal antibody
多发性骨髓瘤 multiple myeloma
重链 heavy chain
轻链 light chain
铰链区 hinge region
初次免疫应答 primary immune response
再次免疫应答 secondary immune response
滴度 titre
淋巴细胞杂交瘤技术 lymphocyte hybridoma
杂交瘤 hybridoma
骨髓瘤细胞 myeloma cell
血清学反应 serological reaction
免疫诊断学 immunodiagnostic
凝集反应 agglutination
间接凝集试验 indirect agglutination test
被动凝集反应 passive agglutination
沉淀反应 precipitation
环状沉淀反应 ring precipitation
絮状沉淀反应 flocculation precipitation
单向琼脂扩散法 simple agar diffusion
双向琼脂扩散法 double agar diffusion
对流免疫电泳法 counter immuno-eletrophoresis (CIE)
火箭电泳法 rocket electrophoresis
双向免疫电泳法 two dimentional immuno- electrophoresis
补体结合试验 complement fixation test (CFT)
中和试验 neutralization test
免疫标记技术 immunolabelling technique
免疫定位分析 immunolocalization
免疫亲和层析 immunoaffinity chromatography
免疫生物传感器技术 immunobiosensor technique
免疫酶技术 immunoenzymatic technique
酶免疫测定法 enzyme immunoassay
酶联免疫吸附法 enzyme linked immunosorbant assay (ELISA)
间接免疫吸附测定法 indirect immunosorbant assay/ indirect ELISA
斑点酶免疫吸附法 dot-ELISA
酶标记免疫定位 enzyme-labelling immunolocalization
免疫组织化学 immuno-histochemistry
免疫印迹 immunoblot
放射免疫测定法 radio immuno-assay (RIA)
免疫电镜技术 immuno-electron microscopy (IEM)
胶体金 colloidal gold
发光免疫测定法 luminescent immuno-assay (LIA)
免疫调节剂 immunoregulative preparation
活疫苗 live vaccine
减毒活疫苗 attenuated vaccine
死疫苗 dead vaccine
亚单位疫苗 subunit vaccine
化学疫苗 chemical vaccine
多肽疫苗 polypeptide vaccine
抗毒素 antitoxin
抗病毒血清 antiviral serum
抗菌血清 antibacterial serum

1 Virulence: The degree or intensity of pathogenicity of an organism as indicated by case fatality rates and/or ability to invade host tissues and cause disease.
毒力:一种机体致病性的程度或强度,以病例致死率和侵染宿主组织并致病的能力表示。

2 Invasiveness: The ability of a microorganism to enter a host, grow and reproduce within the host, and spread throughout its body.
侵袭力：微生物进入宿主机体生长繁殖并扩散到全身的能力。

3 Endotoxin: The heat-stable lipopolysaccharide in the outer membrane of the cell wall of gram-negative bacteria that is released when the bacterium lyses, or sometimes during growth, and is toxic to the host.
内毒素：革兰氏阴性细菌细胞壁外膜中耐热的脂多糖，细菌溶解时释放出来，有时在生长中也释放，对宿主具毒性。

4 Exotoxin: A heat-labile, toxic protein produced by a bacterium as a result of its normal metabolism or because of the acquisition of a plasmid of prophage that redirects its metabolism. It is usually released into the bacterium's surroundings.
外毒素：细菌产生的一种热不稳定的毒性蛋白。这是细菌正常代谢产生的，或是因为获得的质粒或原噬菌体改变了细菌的代谢而产生的。通常分泌到细菌周围的环境中。

5 Immune surveillance: The still somewhat hypothetical process in which lymphocytes such as natural killer cells recognize and destroy tumor cells; other cells with abnormal surface antigens (e. g., virus-infected cells) also may be destroyed.
免疫监视：这还是在假想中的过程：淋巴细胞（如自然杀伤(NK) 细胞）识别并破坏肿瘤细胞，带有异常表面抗原的其他细胞（如被病毒感染的细胞）也可被破坏。

6 Macrophage: The name for a large mononuclear phagocytic cell, present in blood, lymph, and other tissues. Macrophages are derived from monocytes. They phagocytose and destroy pathogens; some macrophages also activate B cells and T cells.
巨噬细胞：大的单核吞噬细胞，存在于血液、淋巴和其他组织。巨噬细胞来源于单核细胞，吞噬并破坏病原体，有些还激活 B 细胞和 T 细胞。

7 T lymphocyte: A type of lymphocyte derived from bone marrow stem cells that matures into an immunologically competent cell under the influence of the thymus. T cells are involved in a variety of cell-mediated immune reactions.
T 淋巴细胞：一类淋巴细胞。来源于骨髓干细胞，在胸腺影响下发育成熟为免疫感受态细胞。T 细胞参与多种细胞介导的免疫反应。

第十一节　微生物的分类和鉴定

Section 11　Classification and Identification of Microbes

微生物系统学　microbial systematics
分类　classification
鉴定　identification
命名　nomenclature
分类单元　taxon, taxa(pl), category
界　kingdom
门　phylum
纲　class
目　order
科　family
属　genus
种　species
典型菌株　type strain
模式种　type species
微生物基因组学　microbial genomics
新种　species nova/ sp. nov/ nov sp.
域　domain
双名法　binominal nomenclature
种名加词　specific epithet
属名　generic name
变种　variety (var.)
型　form
纯培养物　pure culture
纯分离物　pure isolate
菌物界　mycetalia
吸收式营养　absorption
摄食性营养　ingestion
光合式营养　photosynthesis
动物界　animalia
植物界　plantae

原生生物界　protista
真菌界　fungi
原核生物界　monera
病毒界　archetista
三域学说　three domain theory
细菌域　bacteria
真核生物域　eukarya
化学分类法　chemotaxonomy
数值分类法　numerical taxonomy
解链温度　melting temperature (Tm)
核酸分子杂交法　hybridization of nucleic acid
rRNA 寡核苷酸编目　oligonucleotide catalog
指纹图谱　finger print
缔合系数　associated coefficient
统计分类法　taxonometrics
操作分类单元　operational taxonomic units
简单匹配相关系数　simple matching coefficient
相似度矩阵　similarity matrices
树状谱　dendrogram
表元　phenon
表观群　phenotype group

1 Genus: A well-defined group of one or more species that is clearly separate from other genera.
属：一个定义明确的由一个或多个种组成的类群，明显不同于其他属。

2 Species: Species of higher organisms are groups of interbreeding of potentially interbreeding natural populations that are reproductively isolated. Bacterial species are collections of strains that have many stable properties in common and differ significantly from other groups of strains.
种：高等生物的种是能进行品种间杂交或有潜在品种间杂交能力的生殖隔离的自然种群。细菌的种是菌株的集合，这些菌株有很多稳定的共同特性明显不同于其他种的菌株群。

3 Binominal nomenclature: The system in which an organism is given two names; the first is the capitalized generic name, and the second is the uncapitalized specific epithet. The names of genus and species are all in Latin language.
双名法：一种生物取两个名的命名法，第一个名是属名，首字母要大写，第二个名是小写的种名形容词，属名种名均为拉丁文。

4 Pure culture: A population of cells that are identical because they arise from a single cell.
纯培养物：相同的细胞集聚成的群体，这些细胞都产生于一个单细胞。

5 Eukarya: The domain that contains organisms composed of eukaryotic cells with primarily glycerol fatty acyl diesters in their membranes and eukaryotic rRNA.
真核生物域：由真核细胞组成的生物，主要带有膜中的甘油脂肪酰基二酯和真核 rRNA。

第十二节　若干微生物的学名
Section 12　Scientific Name of Microbes

真核生物　Eukaryotes

八孢裂殖酵母　*Schizosacchromyces octosporus*
白地霉　*Geotrichum candidum*
鞭毛纲　Flagellata
捕虫菌目　Zoopagales
草履虫属　Paramecium
产黄青霉　*Penicillium chrysogenum*
粗糙脉孢菌　*Neurospora crassa*
发酵性酵母　*Saccharomyces fermentati*
根霉属　*Rhizopus*
构巢曲霉　*Aspergillus nidulans*
黑曲霉　*Aspergillus niger*
黄曲霉　*Aspergillus flavus*
假蜜环菌　*Armillaria mellea*
梨头霉　*Absidia*
路德类酵母　*Saccharomycodes ludwigii*
毛霉属　*Mucor*
米曲霉　*Aspergillus oryzae*
酿酒酵母　*Saccharomyces cerevisiae*
双孢蘑菇　*Agaricus bisporus*
纤毛纲　Ciliata

原核生物　Prokaryotes

巴氏梭菌　*Clostridium pasteurianum*
白喉棒杆菌　*Corynebacterium diphtheriae*
柄杆菌属　*Caulobacter*
肠杆菌属　*Enterobacter*

大肠埃希氏菌 *Escherichia coli*
德氏乳杆菌 *Lactobacillus delbruchii*
地衣芽孢杆菌 *Bacillus licheniformis*
放线菌属 *Actinomyces*
肺炎支原体 *Mycoplasma pneumoniae*
粪肠球菌 *Enterococcus faecalis*
弗兰克氏菌属 *Frankia*
根癌土壤杆菌 *Agrobacterium tumefaciens*
根瘤菌属 *Rhizobium*
固氮菌属 *Azotobacter*
龟裂链霉菌 *Streptomyces rimosus*
红螺菌目 *Rhodospirillales*
红霉素链霉菌 *Streptomyces erythreus*
红皮盐杆菌 *Halobacterium cutirubrum*
胡萝卜软腐欧文氏菌 *Eriwinia carotovora*
灰色链霉菌 *Streptomyces griseus*
活跃硝化细菌 *Nitrobacter agilis*
霍乱弧菌 *Vibrio cholerae*
甲基球菌科 *Methylococcaceae*
甲烷杆菌属 *Methanobacterium*
甲烷球菌属 *Methanococcus*
结核分枝杆菌 *Mycobacterium tuberculosis*
金黄色葡萄球菌 *Staphylococcus aureus*
巨大芽孢杆菌 *Bacillus magaterium*
克雷伯氏菌属 *Klebsiella*
枯草芽孢杆菌 *Bacillus subtilis*
枯草芽孢杆菌 *Bacillus subtilis*
蜡状芽孢杆菌 *Bacillus cereus*
痢疾志贺氏菌 *Shigella dysenteriae*
淋病奈瑟氏球菌 *Neisseria gonorrhoeae*
流感嗜血杆菌 *Haemophilus influenzae*
拟杆菌属 *Bacteroides*
破伤风梭菌 *Clostridium tetani*
球形芽孢杆菌 *Bacillus sphaericus*
肉毒梭菌 *Clostridium botulinum*
乳酸链球菌 *Stretococcus lactis*
乳酸乳杆菌 *Lactobacillus lactis*
沙雷氏菌属 *Sarratia*
沙门氏菌属 *Salmonella*
沙眼衣原体 *Chlamydia psittaci*
伤寒沙门氏菌 Salmonella typhi
生孢梭菌 *Clostridium sporogenes*
嗜酸乳杆菌 *Lactobacillus* acidophilus
双歧杆菌属 *Bifidobacterium*
苏云金芽孢杆菌 *Bacillus thuringiensis*
梭杆菌属 *Fusobacterium*
炭疽芽孢杆菌 *Bacillus anthracis*
铜绿假单胞 *Pseudomonas aeruginosa*
香菇 *Lentinus edodes*
硝化杆菌属 *Nitrobacter*
小单孢菌属 *Micromonospora*
盐生盐杆菌 *Halobacterium halobium*
洋葱假单胞菌 *Psedomonas cepacia*
幽门螺杆菌 *Helicobacter pylori*
詹氏甲烷球菌 *Methanococcus jannaschii*
植物乳杆菌 *Lactobacillus plantarum*

病毒 Virus

斑驳病毒 Velvet Tobacco Mottle Virus, VT-MoV
丁型肝炎病毒 Hepatitis D Virus
脊髓灰质炎病毒 Polio Virus
克一雅氏病(早老性痴呆病) Creutafeldt-Jakob Disease
库鲁病(震颤病) Kuru Disease
马铃薯纺锤形块茎病毒 Potato Spindle Tuber Disease, PSTD
牛痘苗病毒 Smallpox
牛海绵状脑病(疯牛病) Bovine Spongiform Encephalitis, Mad Cow Disease, BSE
腺病毒 Adeno Virus
烟草花叶病毒 Tobacco Mosaic Virus
羊瘙痒病 Scrapie in Sheep
乙瘙肝炎病毒 Hepatitis B Virus, HBV

第十八章 微生物工程
Chapter 18 Microorganism Engineering

第一节 绪论
Section 1 Introduction

需氧发酵 aerobic fermentation
厌氧发酵 anaerobic fermentation
生物反应器 bioreactor
浅盘发酵 shallow pan fermentation
微生物转化发酵 microbiological conversion fermentation

1. Pasteurization: The process of heating milk and other liquids to destroy microorganisms that can cause spoilage or disease.
巴氏消毒法:对奶和其他液体的加热过程,破坏引起酸败或疾病的微生物。
2. Fermentation: An energy yielding process in which an energy substrate is oxidized without an exogenous electron acceptor. Usually organic molecules serve as both electron donors and acceptors.
发酵:在没有外来电子受体的情况下,能量底物被氧化的产能过程,通常由有机分子起电子供体和受体作用。

第二节 工业微生物的生长与产物的生物合成
Section 2 Growth and Metabolite Synthesis of Industrial Microorganisms

初级代谢产物 primary metabolite
次级代谢产物 secondary metabolite
合成期 synthetic phase
代谢树 metabolic tree
聚酮 polyketone
甲羟戊酸 mevalonic acid
环多醇 cyclitol
氨基环多醇 amino cyclitol
聚合 polymerization
分叉中间体 bifurcation intermediate
代谢调节 metabolic control
合成调节 synthetic control
协同反馈抑制 concerted cooperative feedback inhibition
累计反馈抑制 cumulative feedback inhibition
增效反馈抑制 progression feedback inhibition
顺序反馈抑制 sequential feedback inhibition
顺序诱导 sequential induction
反馈阻遏 feedback repression
反馈调节 feedback control
葡萄糖效应 glucose effect
诱导调节 induction control
自身反馈抑制 autofeedback inhibition
前体 precursor

1. Chemostat: A continuous culture apparatus that feeds medium into the culture vessel at the same rate as medium containing microorganisms is removed; the medium in a chemostat contains one essential nutrient in a limiting quantity.
恒化器:连续培养装置。培养液加入培养容器的速率与含微生物的培养液流出速率相同,恒化器中的培养液含有一种限定量的必需营养物。
2. Turbidostat: A continuous culture system equipped with a photocell that adjusts the flow of medium through the culture vessel so as to maintain a constant cell density or turbidity.
恒浊器:一种连续培养系统,装有光电管,能调整通过培养瓶的培养液流量,使细胞密度和浊度保持不变。
3. Primary metabolites: Microbial metabolites produced during the growth phase of an organism.
初级代谢产物:微生物在菌体生长期产生的代谢产物。
4. Secondary metabolites: Products of metabolism that are synthesized after growth has been completed.
次生代谢产物:微生物在菌体生长期结束后合成的代谢产物。

5 Feedback inhibition：A negative feedback mechanism in which a pathway end product inhibits the activity of an enzyme in the sequence leading to its formation. When the end product accumulates in excess，it inhibits its own synthesis.
反馈抑制：一种负向的反馈机制。一个途径的终产物抑制这一系列中的一种酶。终产物积累过量时抑制终产物自身的合成。

第三节 工业微生物的菌种选育
Section 3 Screening and Breeding of Industrial Microorganisms

自然选育 natural breeding
出发菌株 original strain
杂交育种 hybrid breeding
原生质体融合 protoplast fusion
分子育种 molecular breeding

1 Protoplast fusion：The joining of cells that have had their walls weakened or completely removed.
原生质体融合：细胞壁已经弱化的或完全去掉的细胞的结合。

第四节 培养基
Section 4 Medium

孢子培养基 spore medium
种子培养基 seed medium
发酵培养基 fermentation medium

1 Selective media：Culture media that favor the growth of specific microorganisms，this may be accomplished by inhibiting the growth of undesired microorganisms.
选择性培养基：有利于特异性微生物生长的培养基，抑制非所要求的微生物生长。

2 Agar：A complex sulfated polysaccharide，usually extracted from red algae，that is used as a solidifying agent in the preparation of culture media.
琼脂：一种复合硫酸化多糖，通常从红藻中提取，在培养基的制备中用作固化剂。

第五节 灭菌
Section 5 Sterilization

分批灭菌 batch sterilization
热交换器 heat exchanger
连续灭菌系统 continuous sterilization system
补料系统 fed system
消沫剂系统 defoamer system
静电 electrostatic
纤维介质 fiber medium
肉汤培养基 broth medium
斜面培养基 slant medium

1 Sterilization：The process by which all living cells，viable spores，viruses，and viroids are either destroyed or removed from an object or habitat.
灭菌：从物体或栖息地消灭或除掉全部的活细胞、能萌发的孢子、病毒和类病毒的过程。

2 Autoclave：An apparatus for sterilizing objects by the use of steam under pressure. Its development tremendously stimulated the growth of microbiology.
高压灭菌器：使用加压蒸汽对物体进行灭菌的装置。它的开发极大地促进了微生物学的发展。

第六节 培养与保藏
Section 6 Culture and Preservation

培养时间 culture time
冷藏时间 cryopreserved time
接种 inoculate
接种环 inoculating hood
接种针 inoculating needle
种龄 seed age
菌种保藏 culture preservation

第七节 通气与搅拌
Section 7 Aeration and Agitation

溶解氧浓度 dissolving oxygen concentration
呼吸强度 respiratory intensity
呼吸临界氧浓度 respiratory critical oxygen concentration
接种体 inoculum
搅拌器 agitator/ mixer/stirrer
溶解氧的饱和度 saturation degree of dissolving oxygen
气膜阻力 gas film resistance
界面阻力 interfacial resistance
液膜阻力 liquid film resistance
传递阻力 transfer resistance
细胞膜阻力 cell membrane resistance
传递速率 transfer rate
比表面积 specific surface area
液相 liquid phase
氧传递系数 oxygen transfer coefficient
摄氧率 oxygen uptake rate
离子强度 ionic strength
流变类型 rheological type
剪切速度 shear rate
流体动力学 fluid dynamics hydrodynamics
牛顿液体 Newtonain liquid
屈服应力 yield stress
涨塑性流体 dilatant fluid
搅拌速率 stir speed
表观粘度 apparent viscosity
泡沫 foam/scum
泡沫发生器 foam generator
鼓沫器 sparger
单孔式鼓沫器 single orifice sparger
多孔环状鼓沫器 multiorifice ring sparger
溶氧 dissolved oxygen
氧传感器 oxygen sensor
氧化锆分析仪 zirconia analyzer

第八节 发酵工艺的控制
Section 8 Control of Fermentation Technique

搅拌功率 power of agitator
空气流量 air flow
浊度 cloudiness turbidity
基质浓度 substrate concentration
溶解氧浓度 dissolved oxygen concentration
氧化还原电位 oxidation-reduction potential
产物的浓度 product concentration
比消耗速率 specific consumption rate
分批培养 batch culture
补料分批培养(FBC) fed-batch culture
偶联 couple
生长期 growth phase
生产期 production phase
产物合成阶段 product synthesis phase
产物分泌期 product secretory phase
自溶阶段 autodigestion phase
菌体浓度 cell concentration
氧传递速率(OTR) oxygen transfer rate
温度系数(Q_{10}) temperature coefficient
生物热($Q_{生物}$) biothermal heat
搅拌热($Q_{搅拌}$) stirred heat
蒸发热($Q_{蒸发}$) evaporation heat
辐射热($Q_{辐射}$) radiant heat
发酵热($Q_{发酵}$) fermentation heat
比生产速率 specific production rate
咕林醇酰胺(B因子) cobinamide
半连续培养 semi-continuous culture
半连续发酵 semi-continuous fermentation
热致死 thermal death
反馈控制 feedback control
传感器 sensor transducer
控制器 controller
驱动器 actuator driver
放料和补料 withdraw and fill
流态泡沫 fluid foam
聚氧乙烯氧丙烯甘油 (GPE) polyoxyethylene oxypropylene glycerol
高碳醇 higher alcohol
天然油脂 natural oil and fat
氧化丙烯 oxypropylene, propene oxide/propylene epoxide
环氧乙烷 epoxyethane /ethylene epoxide/ ethylene oxide oxacyclopropane
惰性载体 inert carrier
吐温-80 tween-80/polysorbate 80
乳剂 emulsion

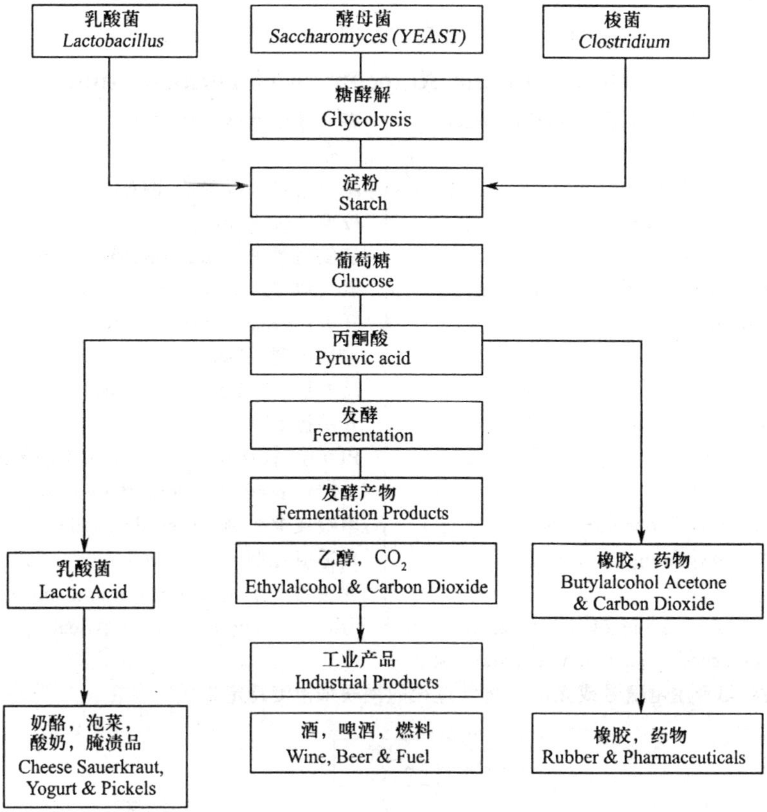

1 Oxidation-reduction reactions：Reactions involving electron transfers，the reductant donates electrons to an oxidant.

氧化还原反应:含有电子转移的反应。还原剂提供电子给氧化剂。

2 Thermal death time：The shortest period of time needed to kill all the organisms in a microbial population at a specified temperature and under defined conditions.

热致死时:在一指定的温度和规定的条件下,杀灭一个微生物群里中全部机体所需的最短时间。

第九节　发酵动力学

Section 9　Fermentation Dynamics

质量平衡法　mass balance method
内源代谢　endogenous metabolism
维持能　preservation energy
生长得率常数　growth yield coefficient
化学能　chemical energy
燃烧热　combustion heat
化学计量式　chemical equation
元素平衡　element conservation
物质平衡　matter conservation
饱和常数　saturation constant
生长限制基质浓度　substrate concentration for growth restriction
产物抑制　production inhibition
生长动力学　growth kinetics
细胞比死亡率　cell specific death rate
表观比生长率　apparent specific growth rate
生化模型　biochemical model
基质抑制模型　substrate inhibition model
氧限制模型　oxygen inhibition model
简单分批发酵　simple batch fermentation
基质比消耗率　substrate specific consumption rate
临界稀释率　critical dilution ratio

第十节　发酵过程检测与自控

Section 10　Detection and Autocontrol of Fermentation Process

可靠性可通过　(GEP)good engineering practice
离线传感器　off-line sensor
在线传感器　on-line sensor
原位传感器　in situ sensor
力敏元件　force sensor
热敏元件　thermal sensor thermal element
光敏元件　light sensor
磁敏元件　magnetic sensor
电化学传感器　electrochemical sensor
电极　electrode
生物传感器　biosensor
电信号　electric signal
干细胞浓度　stem cell concentration
耗氧率(OUR)　oxygen uptake rate
质谱　mass spectrum
高效液相色谱(HPLC)　high performance liquid chromatograph
流动注射分析系统(FIA)　fluid injection analysis
校准　calibrate
基质消耗率　substrate consumption rate
氧消耗率　oxygen consumption rate
呼吸商(RQ)　respiratory quotient
前馈控制　feedforward control
反馈控制　feedback control
开关控制　off-on control
PID控制(比例、积分、微分控制算法)　proportion integral differential control
串级反馈控制　cascade feedback control
自适应控制　adaptive control
变送器　transmitter/transmitting instrument

1 Biosensor: The coupling of a biological process with production of an electrical signal or light to detect the presence of particular substances.

生物传感器:籍产生电信号或光的生物学过程的耦联来检测特定物质的存在。

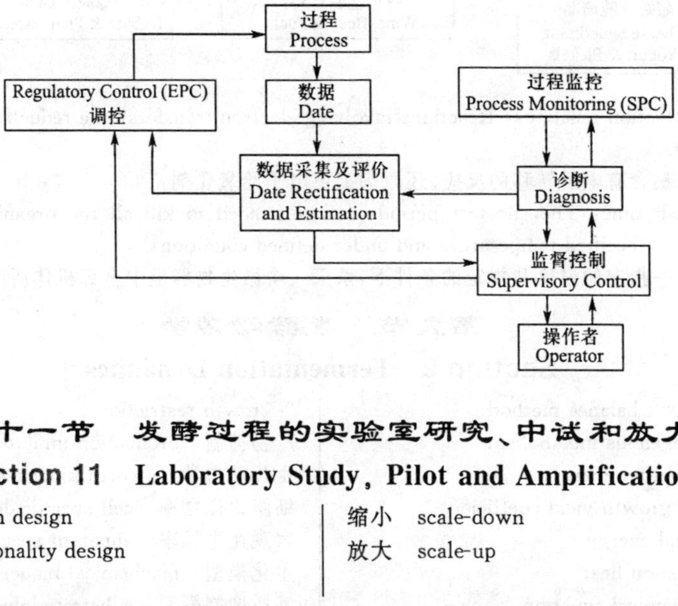

第十一节　发酵过程的实验室研究、中试和放大

Section 11　Laboratory Study, Pilot and Amplification

均匀设计　uniform design
正交设计　orthogonality design
中试　pilot
缩小　scale-down
放大　scale-up

1 Orthogonality: It is a systematic design which makes a complex designs feasible and compact.

正交:是使复杂的设计可行化和简洁化的一种系统设计。

第十二节 现代生物技术在发酵工业中的应用

Section 12 Applications of Modernistic Biotechnology in Fermentation Industry

基因工程 genetic engineering
细胞工程 cell engineering
发酵工程 fermentation engineering
生化工程 biochemical engineering
抑制素 somatostatin
悬浮培养 suspension culture
贴壁培养 anchorage culture
组织培养(愈伤组织培养) callus culture
吸附法 adsorption method
包埋法 embedding method
交联法 crosslinking method
生物反应器 bioreactor

1 Genetic engineering: The deliberate modification of an organism's genetic information by directly changing its nucleic acid genome.
基因工程:用直接改变生物的核酸基因组的方法来改变生物的遗传信息。

第十三节 发酵工业与环境保护

Section 13 Fermentation Industry and Environmental Protection

好氧生物 aerobe
厌氧生物 anaerobe
生物处理 biotreatment
生物膜 biofilmm/biological film/biomembrane
深井曝气 borehole aeration
氧化沟 oxidation channel
氧化塔 oxidation column, oxidation tower
上流式厌氧污泥床反应器(UASB) up-flow anaerobic sludge bed reactor
厌氧流化床(AFB) anaerobic fluidized bed
厌氧过滤器(AF) anaerobic filter
硬性填料 hard packing, hard filler
软性填料 soft packing
厌氧消化 anaerobic digestion

1 Aerobe: An organism that grows in the presence of atmospheric oxygen.
好氧生物:在大气氧存在的情况下生长的生物。

2 Anaerobe: An organism that grows in the absence of free oxygen.
厌氧生物:在缺乏游离态氧的环境中生长的生物。

3 Biofilms: Organized microbial systems consisting of layers of microbial cells associated with surfaces, often with complex structural and functional characteristics. Biofilms have physical/chemical gradients that influence microbial metabolic process. They can form on inanimate devices (catheters, medical prosthetic devices) and also cause fouling (e.g. of ships' hulls, water pipes, cooling towers).
生物膜:与表面相关的由微生物细胞层构成的有机微生物系统,往往具有复杂的结构功能特征。有物理或化学的梯度,影响微生物代谢过程。可在无生命的装置(导管、假器官装置)上形成,也可以引起淤积(如船体外壳、水管、冷却塔的淤积)。

4 Anaerobic digestion: The microbiological treatment of sewage wastes under anaerobic conditions to produce methane.
厌氧消化:在厌氧条件下用微生物学方法处理污水废物,产生甲烷。

第十四节 发酵过程经济学

Section 14 Fermentation Economics

固定成本 fixed cost
可变成本 variable cost
原料成本 input material cost
公用事业成本 utility cost
生产规模 production capacity
年产出计划 annual outgoing design
日常生产计划 daily production program
日产量 daily output, daily production, current yield
发酵周期 fermentation period

经济效益 economic effectiveness
输出功率 output power
额定功率 rated power
负荷率 load rate
产物浓度 product concentration
发酵批产量 fermentation batch output
发酵产率 fermentation production rate
基质转化率 substrate conversion
单位产量 specific duty
发酵成本 fermentation cost

第十五节 β-内酰胺类抗生素
Section 15 β-lactam Antibiotics

头霉素 cephamycin
克拉维酸 clavulanic
噻纳霉素 thienamycin
*β-内酰胺抗生素 β-lactam antibiotics
单菌胺 monobactam
6-氨基青霉烷酸 (6-APA)6-amino penicillanic acid

1 Antibiotic: A microbial product or its derivative that kills susceptible microorganisms or inhibits their growth.
抗生素:一种微生物产物或其衍生物,杀灭敏感微生物或抑制其生长。

第十六节 非β-内酰胺类抗生素
Section 16 Non-β-lactam Antibiotics

四环类抗生素(TCs) tetracycline
大环内酯抗生素 macrolide antibiotic
多烯大环内酯抗生素 polyene macrolide antibiotics
麦迪霉素 medemycin/midemycin
交沙霉素 josamycin
抑霉菌素 nystatin
两性霉素 amphotericin
曲古霉素 hachimycin, trichomycin
利福霉素 rifamycin
蒽环类抗生素 anthracene ring antibiotics
蒽环糖苷抗生素 anthracene ring glucoside antibiotics
链霉素 streptomycin
布鲁霉素 bluensomycin
2-脱氧链霉胺(2-DOS) 2-deoxy streptamine
新霉素 neomycin fradiomycin mycifradin
卡那霉素 kanamycin
庆大霉素 gentamycin/gentamicin
林可霉素 lincomycin
肽类抗生素 peptide antibiotics
短杆菌肽 gramicidins
短杆菌酪肽 tyrocidine
多粘菌素 polymyxin
杆菌肽 bacitracin
缬氨霉素 valinomycin
博来霉素 bleomycin

1 Tetracyclines: A family of antibiotics with a common four-ring structure, which are isolated from the genus *Streptomyces* or produced semisynthetically; all are related to chlortetracycline or oxytetracycline.
四环素:一类具有共同的四环结构的抗生素,从链霉菌分离得到或半合成产生,都与氯化四环素或氧化四环素有关。

2 Streptomycin: A bactericidal aminoglycoside antibiotic produced by *Streptomyces griseus*.
链霉素:灰色链霉菌产生的杀菌的氨基糖苷类抗生素。

3 Macrolide antibiotics: An antibiotic containing a macrolide ring, a large lactone ring with multiple keto and hydroxyl groups, linked to one or more sugars.
大环内酯抗生素:含有与一个或多个糖相连的大内酯环的抗生素,大内酯环带有多个酮基和羟基。

4 Nystatin: A polyene antibiotic from *Streptomyces noursei* that is used in the treatment of *Candida* infections of the skin, vagina, and alimentary tract.
抑霉菌素:由诺尔斯氏链霉菌产生的多烯抗生素,用于治疗皮肤、阴道和消化道假丝酵母属真菌感染。

第十七节 氨基酸和维生素
Section 17 Amino Acids and Vitamins

核黄素 lactochrome lactoflavine ovoflavin/riboflavin
维生素 B12 vitamin B12
钴维生素 cobalamin
β-胡萝卜素 β-carotene
腺苷钴啉醇酰胺 adenosyl cobinamide
腺苷钴啉酸 a,c-二酰胺 adenosyl cobyrinic acid a,c-diamide

Vitamin: An organic compound required by organisms in minute quantities for growth and reproduction because it cannot be synthesized by the organism, vitamins often serve as enzyme cofactors or parts of cofactors.

维生素:机体生长和繁殖所需的微量有机化合物,机体自身不能合成。维生素常作为酶的辅因子或辅因子组分。

第十八节 甾类化合物的微生物转化
Section 18 Microbial Conversion of Steroids

甾类化合物 steroid
胆甾醇 cholesterol
胆酸 cholic acid
肾上腺皮质激素 adrenal cortex hormone
性激素 sex hormone
可的松 cortisone
氢化可的松 hydrocortisone

第十九节 核酸 类物质及药用酶
Section 19 Nucleic Acid and Medical Enzyme

肌苷 carnine
鸟苷 guanosine
胞苷 cytidine
黄苷 xanthosine
尿苷 uridine
腺苷 adenosine
药用酶 medical enzyme
辅酶 coenzyme/ coferment
蛋白酶 pankrin
青霉素酰化酶 penicillin acylase

Coenzyme: A loosely bound cofactor that often dissociates from the enzyme active site after product has been formed.

辅酶:与酶结合松散的辅因子,产物形成后往往与酶活性位点分离。

第二十节 农牧用抗生素
Section 20 Agricultural Antibiotics

农用抗生素 agricultural antibiotics
春日霉素(春雷霉素) kasugamycin
有效霉素 validamycin
除草霉素(除莠霉素) herbimycin
赤霉素 gibberellin
阿维菌素 avermectin
莫能星 monensin
盐霉素 salinomycin

第二十一节 微生物产生的其他药物
Section 21 Other Pharmaceuticals Produced by Microbes

环孢菌素 cyclosporin
抑氨肽酶素 b bestatin
右旋糖苷 dextran
多抗甲素 polyactin
香菇多糖 lentinan
云芝多糖 krestin(PS-K)
茯苓多糖 pachyman
抑胃酶剂 pepstatin
胞内组织蛋白酶 D cathepsin D
亮肽素 leupeptin
萎蔫酸 fusaric acid

第十九章 微生物遗传学
Chapter 19 Microbial Genetics

第一节 微生物的遗传物质
Section 1 Genetic Material of Microbes

细菌 bacterium
原核微生物 prokaryotic microbe
遗传 heredity, inheritance
光滑型 smooth, S
粗糙型 rough, R
烟草花叶病毒 tobacco mosaic virus(TMV)
拟核类组蛋白 histone-like protein
骨架 scaffold
锯齿型 zigzag
夹环 clamp
负超螺旋 negative supercoil
拓扑异构酶 topoisomerase
切口 nick
复制因子 replication factor
复制蛋白 replication protein
滑动夹环 sliding clamp
夹环装卸器 clamp loader
ter 序列 termination sequence
染色质纤丝 chromatin fiber
复制周期 replicative cycle
端粒酶 telomerase
开放阅读框架 open reading frame(ORF)
内含子 intron
重叠基因 overlapping gene
移动基因 movable gene
基因组学 genomics
结构基因组学 structural genomics
功能基因组学 functional genomics
回文序列 repeated extragenic palinodrome(REP)
肠杆菌群 enterobacteria
肠杆菌基因间的重复的一致序列 enterobacteria repetitive intergenic consensus(ERIC)
缺陷性噬菌体 defective phage
隐蔽原噬菌体 cryptic prophage
内转录间隔区 internal transcribed spacer(ITS)
非转录间隔区 non-transcribed spacer(NTS)

1 Open reading frame: A reading frame sequence not interrupted by a stop codon; it is usually determined by nucleic acid sequencing studies.
开放阅读框架:不被终止密码子中断的读框顺序,通常由核酸序列研究决定。

2 Intron: A noncoding intervening sequence in a split or interrupted gene, which codes for RNA that is missing from the final RNA product.
内含子:断裂基因或间断基因内非编码的内插顺序,编码的 RNA 在最终 RNA 产物中被删除。

3 Genomics: The study of the molecular organization of genomes, their information content, and the gene products they encode.
基因组学:研究基因组的分子结构、信息含量及其编码的基因产物的学科。

第二节 基因突变和损伤 DNA 的修复
Section 2 Gene Mutation and Reparation of Damaged DNA

染色体畸变 chromosomal aberration
染色体突变 chromosomal mutation
生化突变型 biochemical mutant
致死突变型 lethal mutant
错义突变 missense mutation
同义突变 same/sense mutation
无义突变 nonsense mutation
琥珀突变 amber mutation
赭石突变 ocher mutation
乳白突变 opal mutation
碱基置换 base substitution
移码 frameshift
DNA 片段插入 DNA fragment insertion
DNA 缺失 DNA deletion
营养缺陷突变株 auxotrophic mutant
温度敏感突变株 temperature sensitive mutant

抗性突变株 resistant mutant
颠换 transversion
转换 transition
点突变 point mutation
缺失 deletion/deficiency
5-溴尿嘧啶 5-bromouracil (BU)
黄嘌呤 xanthine
2-氨基嘌呤 2-aminopurine(2-AP)
碱基类似物 base analogue
酮式-烯醇式互变异构 keto-enol tautomerization
辐射诱变 radiation mutagenesis
电离辐射 ionizing radiation
激光辐射 laser radiation
离子束诱变 ion beam mutagenesis
波动实验 fluctuation test
涂布实验 plate spread test
抗性基因 resistance gene
抗药性基因 drug-resistance gene
生理适应 physiological adaptation
互变异构移位 tautomeric shift
缺失突变 deficiency mutation
适应突变 adaptive mutation
光复活作用 photoreactivation
错配修复系统 mismatch repair system
切除修复 excision repair
核苷酸切除修复 nucleotide excision repair
短补丁修复 short-patch repair
长补丁修复 long-patch repair
N-糖基化酶 N-glycosylase
DNA 糖基化酶 DNA glycosylase
* 碱基切除修复 base excision repair
尿嘧啶-N-糖基化酶 uracil-N-glucosylase
重组修复 recombination repair
复制后修复 post replication repair
交联修复 repair of inter strand cross-links
SOS 修复系统 SOS-repair system
din 基因 damage inducible gene
自我切割 self-cleavage

1 Recombination repair: A DNA repair process that repairs damaged DNA when there is no remaining template, a piece of DNA from a sister molecule is used.
重组修复:没有存留的模板时损伤的 DNA 的修复过程,要用到一条姐妹链 DNA。

2 Frameshift mutations: Mutations arising from the loss or gain of a base or DNA segment, leading to a change in the codon reading frame and thus a change in the amino acids incorporated into protein.
移码突变:丢失或得到一个碱基或 DNA 片段而发生的突变,造成密码子读框的改变,并使蛋白质的氨基酸也改变。

3 Point mutation: The change of a purine displaced by another purine or a pyrimidine by another pyrimidine called transitions, whereas the change of a purine by a pyrimidine and vice versa are called transversion.
点突变:是指 DNA 中核苷酸的一个碱基被另一个碱基所取代,其中一个嘌呤被另一个嘌呤或是一个嘧啶被另一个嘧啶所取代,称转换,而一个嘌呤被另一个嘧啶或是一个嘧啶被另一个嘌呤所取代,叫做颠换。

4 Excision repair: The uvrABC endonuclease recognizes and removes damaged nucleotides from the DNA. DNA polymerase Ⅰ fills in the missing nucleotides and DNA ligase joins the gap. This mechanism of repair is used for other forms of DNA damage as well as UV damage.
切除修复:uvrABC核酸内切酶识别并切割DNA链中受损伤的核苷酸部位。DNA聚合酶Ⅰ填充合成缺损的核苷酸,DNA 连接酶封合新合成的 DNA 片段和原有 DNA 链之间的切沟。这一修复机理应用于像紫外损伤等形式的 DNA 损伤。

第三节 病毒遗传分析
Section 3 Genetic Analysis of Virus

病毒 virus
DNA 病毒 DNA virus
RNA 病毒 RNA virus
植物病毒 plant virus
动物病毒 animal virus
烈性噬菌体 virulent phage /lytic bacteriophage
温和噬菌体 temperate phage
一步生长曲线的实验 one-step-growth experiment
单菌释放实验 single burst experiment

重组测验 recombination test
突变子 muton
重组子 recon，recombinant
噬菌体突变体 phage mutant
快速溶菌 rapid lysis
互补测验 complementary test
突变位点 mutant site/ mutational site
突变率 mutation rate
顺反子 cistron
缺失作图 deletion mapping
缺失定位 deficiency mapping
突变点 mutant site
cos 位点 cohesive-end-site
原噬菌体 prophage
溶源细菌 lysogenic bacteria
溶源途径 lysogenic pathway
免疫性 immunity
裂解途径 lytic pathway
诱导 induction,derivation
基因簇 gene cluster
抗终止因子 anti-terminator
晚期基因 late gene
早期基因 early gene
同向重复序列 sequence of tandem repeats
病毒蛋白 viral protein
解旋酶 helicase
复性 renaturation/ annealing
滚环复制 rolling-circle replication
前早期 immediate early stage
后早期 delayed early stage
晚期 late stage
早期启动子 early promoter
正调节因子 positive regulator factor
终止因子 termination factor
复制基因 replicator
激活蛋白 activator protein
整合酶基因 integrase gene
切割酶基因 excision enzyme gene
重组基因 recombination gene
操纵基因 operator gene/ operator
晚期转录 late transcription
右向启动子 dextrad promoter
左向启动子 sinistrad promoter
阻遏物蛋白 repressor protein
转录整合酶 transcription integrase
保守序列 conserved sequence
整合酶 integrase
附着位点 attachment site
细菌位点 bacterial site
位点特异性重组 site-specific recombination
核心序列 core sequence
切割酶 excisase
反转录病毒 retrovirus
单链 RNA 病毒 single-strand RNA virus
原病毒 provirus
病毒核粒 nucleocapsid
基质蛋白 matrix protein/strome protein/ stromatin
衣壳蛋白 capsid protein
核酸核蛋白 nucleoprotein
引物结合位点 primer binding site
第一次跳跃 first jump
第二次跳跃 second jump
正链病毒 plus strand virus
亚基因组 subgenome

1 Virus: An infection agent having a simple acellular organization with a protein coat and a single type of nucleic acid, lacking independent metabolism, and reproducing only within living host cells.
病毒:具有简单的非细胞结构的感染因子,具有蛋白外壳和一种核酸,缺乏独立的代谢机制,只在活体宿主细胞内繁殖。

2 Prophage: The latent form of a temperate phage that remains within the lysogen, usually integrated into the host chromosome.
原噬菌体:温和噬菌体在溶原菌中的潜伏状态,通常将其 DNA 整合进宿主染色体。

第四节 细菌基因转移和基因重组
Section 4 Transfer and Recombination of Bacterial Gene

有性生殖 sexual reproduction
遗传重组 genetic recombination
基因定位 gene localization/ gene mapping
感受态细胞 competent cell
感受态因子 competence factor(CF)
感受态-特异蛋白质 competence-specific protein

转化小体 transformare
整合 integration
RecA 蛋白 Rec A protein
电穿孔 electroporation
基因枪 gene gun
接合作用 conjugation
性质粒 sex plasmid
F 因子 F factor
重组体 recombinant
回复突变 back mutation/ reversional mutation
复制区 replication region
严紧型质粒 stringent plasmid
插入区 insertion region
松弛性质粒 relaxed plasmid
解聚作用 disaggregation
再溶解作用 redissolving
性导 sexduction
接合型质粒 conjugative plasmid
性信息素-响应质粒 sex pheromone-response plasmid
凝集物质 aggregation substance
结合物质 binding substance
包装位点 packaging site
末端冗余 terminally redundant
环状排列的 circularly permuted
多联体分子 concatemer
共转导 cotransduction
共转导频率 cotransduction frequency
局限性转导 specialized transduction

1 Integration: The incorporation of one DNA segment into a second DNA molecule to form a new hybrid DNA. Integration occurs during such processes as genetic recombination, episome incorporation into host DNA, and prophage insertion into the bacterial chromosome.
整合:DNA的一个片段并入另一DNA分子中去,形成一个新的杂合的DNA。整合发生在诸如遗传重组、附加体并入宿主DNA和原噬菌体插入细菌染色体之类的过程中。

2 Conjugative plasmid: A plasmid that carries the genes for sex pili and can transfer copies of itself to other bacteria during conjugation.
接合质粒:携带性菌毛基因的质粒,在接合中能将自身拷贝转移到另一细菌里去。

3 Conjugation: 1. Gene transfer and reorganization of the bacteria need the direct contact between cells.
2. The protozoa usually adopts the complex way of sexual reproduction.
接合:1. 细菌基因转移和重组方式,需要细胞与细胞的直接接触。
2. 原生动物通常采用的复杂的有性生殖方式。

4 Composite transposon: They are made up of a central region containing the additional genes between two copies of an IS element. Examples of these are Tn10 which carries tetracycline resistance genes and Tn1681 which codes for a heat stable enterotoxin.
复合转座子:由两个IS插入序列拷贝中的两个附加基因的中心区域组成。例如运送四环素抗性基因的Tn10和编码一个热稳定的肠毒素的Tn1681。

第五节 质 粒
Section 5 Plasmid

质粒 plasmid
共价、闭合、环状双链 DNA covalent closed circular DNA(cccDNA)
抗性质粒 resistance plasmid
致育质粒 fertility plasmid
冠瘿瘤 crown gall
非接合性质粒 nonconjugative plasmid
隐蔽质粒 cryptic plasmid
消除 curing
遗传转移 genetic transfer
分子杂交 molecular hybridization
质粒拷贝数 plasmid copy number
广寄主范围 broad host range
抑制物-靶位调控 inhibitor-target regulation
重复子-竞争结合调控 iteron-binding regulation
重复子 iteron
拷贝数突变型 copy number mutant
重复子质粒 iteron plasmid
转录自体调控 transcriptional autoregulation
偶联模型 coupling/handcuffing model

不同的不相容群 different incompatibility group
质粒的不稳定性 plasmid instability
结构的不稳定性 structural instability
随机分配 random distribution
主动分配 active distribution
分配区 partition region
分配座位 partition locus
反式作用基因 trans-acting gene
顺式作用位点 cis-acting site
类着丝位点 centromere-like site
分配复合体 partition complex
寄主致死体 host-killing mechanism
蛋白水解酶 protease
多聚体 multimer/ polymer
自主转移质粒 self-transmissible plasmid
可移动质粒 mobilizable plasmid
转移起点 origin of transfer(ori T)
细菌人工染色体 bacterial artificial chromosome (BAC)
染色体转移能力 chromosomal mobilization ability
"4×4 杂交"方法 four-on-four cross
高频重组 high frequency of recombination(Hfr)
转染子 transfectant

1 Plasmid: A double-stranded DNA molecule that can exist and replicate independently of the chromosome or may be integrated with it. A plasmid is stably inherited, but is not required for the host cell's growth and reproduction.

质粒:环状双链DNA分子,能独立于染色体存在和复制或与其整合。质粒可稳定遗传,但宿主细胞的生长和繁殖不需要它。

第六节 放线菌遗传
Section 6 Actinomycetes Heredity

脉冲电泳 Pulsed-field gel electrophoresis
末端反向重复序列 terminal inverted repeat (TIR)
扩增 DNA 序列 amplification DNA sequence (ADS)
接头 junction
球拍框架 racket frame
异质系克隆 heteroclone
超致育型 ultra/fertility (UF)
原始致育型 initial fertility
正常致育型 normal fertility (NF)
进入劣势 entry disadvantage

第七节 酵母菌遗传
Section 7 Yeast Heredity

酵母菌 yeast
着丝粒 centromere/ point centromere
染色体步行法 chromosome walking
着丝粒序列 centromeric sequence (CEN)
区域着丝粒 regional centromere
起始识别复合物 origin recognition complex (ORC)
小菌落突变株 petite mutant
放线菌酮 cycloheximide
嗜杀株 killer
类病毒颗粒 virus-like particle
反式作用元件 trans-acting element
通用转录因子 general transcription factor
转录调控因子 transcriptional control factor
接合型转变 mating type conversion
整合型载体 integrative plasmid vector (Yip)
附加体载体 episomal plasmid vector (Yep)
复制型载体 replicative plasmid vector (YRp)
着丝粒载体 centromere plasmid vector (YCp)
酵母人工染色体 yeast artificial chromosome (YAC)
分泌载体 secretion vector/ excretion vector

1 Yeast: A unicellular fungus that has a single nucleus and reproduces either asexually by budding or fission, or sexually through spore formation.

酵母菌:具有单核的单细胞真菌,通过芽殖或裂殖进行无性生殖或通过形成孢子进行有性生殖。

第八节 丝状真菌遗传
Section 8 Heredity of Filamentous Fungi

第一次分裂分离 first division segregation
第二次分裂分离 second division segregation
着丝粒距离 centromere distance
亲代双亲型 parental ditype (PD)
非亲代双亲型 non-parental ditype (NPD)
四型 tetratype
准性生殖 parasexual cycle

第九节 原核生物基因表达的调控
Section 9 Regulation of Prokaryotic Gene Expression

激活物 activator
螺旋-转角-螺旋 helix-turn-helix(HTH)
共阻遏物 corepressor
效应物 effector
核心酶 core enzyme
上升突变 up mutation
转录泡 transcription bubble
热激 heat shock
热激蛋白 heat shock protein
发夹结构 hairpin
近基因 proximal gene
远基因 distal gene
抗终止作用 anti-termination
β-半乳糖苷透性酶 β-galactoside permease
β-半乳糖苷转乙酰酶 β-thiogalactoside transacetylase
异丙基-β-D-硫代半乳糖苷 isopropyl-β-D-thiogalactoside(IPTG)
组成型突变体 constitutive mutant
分解代谢物激活蛋白 catabolite activator protein(CAP)
分解代谢产物阻遏 catabolite repression
半乳糖转移酶 galactose transferase
调节子 regulon
自体调控 autoregulation
弱化作用 attenuation
前导肽 leader peptide
氮代谢调节系统 nitrogen regulation control system
严紧反应 stringent control
严紧因子 stringent factor
报警素 alarmone

1 Regulon: A collection of genes or operons that are controlled by a common regulatory protein.
调节子:由共同的调节蛋白控制的基因群或操纵子群的集合。

2 Attenuation: 1. A mechanism for the regulation of transcription of some bacterial operons by aminoacyl-tRNAs.
2. A procedure that reduces or abolishes the virulence of a pathogen without altering its immunogenicity.
弱化作用:1. 某些细菌操纵子通过氨酰 tRNA 的转录调节机制。
2. 减少或消除病原毒力而不改变其免疫原性的方法。

3 Heat shock protein: Proteins produced when cells are exposed to high temperatures or other stressful conditions. They protect the cells from damage and often aid in the proper folding of proteins.
热激蛋白:细胞接触高温或其他刺激产生的蛋白质,保护细胞免遭破坏并往往有助于蛋白质折叠。

第十节 微生物中的转座因子
Section 10 Transposable Element of Microbes

转座因子 transposable element(TE)/transposon
控制成分 controlling element
反转录病毒 retrovirus
DNA 转座子 DNA transposon
反转座子 retrotransposon
反向重复序列 inverted repeat sequence(IR)

转座酶 transposonase(Tnp)
正向重复序列 direct repeat sequence(DR)
细菌转座子 bacterial transposon(Tn)
肠毒素 enterotoxin
复合转座子 composite transposon
复杂转座子 complex transposon
剪-贴型转座 cut-and-paste transposition
复制型转座 replicative transposition
假复制叉 pseudoreplication fork
共整合体 cointegrate
解离酶 resolvase
保守型转座 conservative transposition
Mu 噬菌体 mutator phage
接合型转座子 conjugative transposon
靶序列 target sequence
反转座子 retrotransposon
长散布重复序列 long interspersed repeated sequence(LINE)
短散布重复序列 short interspersed repeated sequence(SINE)
酵母转座子 transposon in yeast(Ty)
转位 translocation

1 Transposon: A DNA segment that carries the genes required for transposition and moves about the chromosome. If it contains genes other than those required for transposition, it may be called a composite transposon. The name is often reserved only for transposable elements that also contain genes unrelated to transposition.

转座子:携带转座所需基因并在染色体上移动位置的DNA片段。如果这个片段除了带有转座所需基因以外的其他基因,叫作复合转座子。这个名称往往只专用于还带有与转座无关的基因的转座因子。

2 Retrovirus: A group of viruses with RNA genomes that carry the enzyme reverse transcriptase and form a DNA copy of their genome during their reproductive cycle.

反转录病毒:一组具有 RNA 基因组的病毒,携带有反转录酶,在病毒的繁殖周期中产生其基因组的 DNA 拷贝。

第十一节 遗传重组

Section 11 Genetic Recombination

同源重组 homologous recombination
一般重组 general recombination
位点特异性重组 site-specific recombination
分支迁移 branch migration
双链断裂 double-strand break
顺式作用重组位点 cis-acting recombination hotspot
重组酶 recombinase
预联会 pry/synapsis
联会 synapsis
链交换 strand exchange
补丁型重组 patch recombination
剪接型重组 splice recombination
减数分裂重组 meiotic recombination
有丝分裂重组 mitotic recombination
重组频率 recombination frequency
敲除 knock-out
附着点 attachment site
核心序列 core sequence
切除酶 excisionase
核心重组位点 core recombination site
异常重组(非同源性重组) illegitimate recombination
末端的连接 end-joining

1 General recombination: General recombination is mediated by RecA protein which causes the exchange of genetic information between DNA strands with very similar sequences.

一般重组:是以RecA蛋白或类似的RecA蛋白为媒介,在同源DNA分子序列相同或接近相同间引起遗传信息的交换。

2 RecA protein: The protein encoded by the RecA gene which is essential for homologous recombination. The RecA protein is also involved in the SOS response and the induction of lambda prophage in response to DNA damaging agents.

RecA 蛋白:由 RecA 基因编码的蛋白对同源重组很重要。RecA 蛋白也与 SOS 应答和 λ-原噬菌体对 DNA 损伤应答的诱导有关。

第十二节　基因工程
Section 12　Gene Engineering

基因工程　gene engineering
限制性核酸内切酶　restriction endonuclease
连接酶　ligase
甲基化酶　methylase
限制-修饰体系　restriction-modification system
回纹序列　palindrome
黏性末端　sticky end
同裂酶　isoschizomer
同尾酶　iso
切口　nick
缺口　gap
cos 位点　cohesive end site
插入型载体　insertion vector
替换型载体　replacement vector
侧翼序列　flanking sequence
柯斯质粒　cos-site-carrying plasmid/ cosmid
基因组文库　genomic library
感染　infection
互补　complementary
筛选　screening
原位杂交　hybridization in site/ in site hybridization/
杂交组蛋白化学　hybridization histochemistry
DNA 扩增　DNA amplification
多聚酶链式反应　polymerase chain reaction
退火　annealing
延伸　extension
定位诱变　localized mutagenesis
盒式诱变　cassette mutagenesis
重组 PCR　recombinant PCR
定位突变　localized mutation

第十三节　微生物育种
Section 13　Microbial Breeding

诱变育种　mutation breeding
杂交育种　cross breeding
遗传工程育种　genetic engineering breeding
激光诱变　laser mutagensis
离子束诱变　ion beam mutagensis
伴胞晶体　parasporal crystal
杀虫晶体蛋白　insecticidal crystal proteins(ICP)
乙型肝炎病毒　hepatitis virus(HBV)

第二十章　应用微生物学
Chapter 20　Applied Microbiology

第一节　绪论
Section 1　Introduction

微生物，微小动植物　microorganism
固氮作用　nitrogen fixation
重组　recombination
遗传性状　hereditary character
发酵食品　fermented food
酵母　ferment
抗生素　antibiotic
真菌　fungus
嗜热菌　thermophile
聚酯　polyester
疫苗　vaccine
木质素　lignose
细胞膜质，纤维质，纤维素　cellulose
半纤维素　hemicellulose
降解，递降分解(作用)，退化　degradation

1 Nitrogen fixation：The metabolic process in which atmospheric molecular nitrogen is reduced to ammonia；carried out by cyanobacteria，*Rhizobium*，and other nitrogen-fixing bacteria.
固氮作用：大气分子氮还原成氨的代谢过程，由蓝细菌、根瘤菌属细菌和其他固氮细菌进行。

2 Recombination：The process in which a new recombinant chromosome is formed by combining genetic material from two organisms.
重组：来自两种生物体的遗传物质结合成新的重组染色体的过程。

3 Antibiotic：A microbial product or its derivative that kills susceptible microorganisms or inhibits their growth.
抗生素：一种微生物产物或其衍生物，杀灭敏感微生物或抑制其他生长。

4 Fungus：Achlorophyllous，heterotrophic，spore-bearing eucaryotes with absorptive nutrition；usually，they have a walled thallus.
真菌：无叶绿素、异养、产孢子、具吸收性营养的真核生物，通常有一个带有壁的原植体。

5 Thermophile：A microorganism that can grow at temperatures of 55℃ or higher；the minimum is usually around 45℃.
嗜热菌：这种微生物能在55℃或更高的温度条件下生长，最低生长温度往往为45℃。

6 Vaccine：A preparation of either killed microorganisms；living，weakened (attenuated) microorganisms；or inactivated bacterial toxins(toxoids). It is administered to induce development of the immune response and protect the individual against a pathogen or a toxin.
疫苗：一种杀灭的、减毒的或毒性失活(类病毒)的微生物制剂。用疫苗能诱发免疫应答并保护接种疫苗个体不受病原体或毒素的危害。

第二节　微生物生物技术
Section 2　Microorganism Biological Technique

临床　clinic
面包酵母　baker yeast
甾族化合物，类固醇　steride
柠檬酸　citric acid
柠檬酸盐　citrate
添加剂　addition agent
甘味剂　sweetener
强化剂　reinforcer
防腐蚀添加剂　anticorrosion agent
饲料，饲料中的营养成分　feedstuff
洗涤剂，清洁剂　abluent
安定剂，稳定剂　stabilizer

治疗学，疗法　therapeutics
基因工程　gene engineering
胰岛素抗体　insulin antibody
糖尿病　diabetes
脑下垂体　hypophysis
白痴病，呆小症　crctinism
纤溶酶原　profibrinolysin
纤维蛋白溶酶　fibrinolysin
肺气肿　emphysema
尿激酶　urokinase
淋巴毒素　lymphotoxin
过氧化物　superoxide
凝血因子　coagulation factors
白血球，白细胞　leucocyte
溃疡　ulceration
肝炎　hepatitis
疱疹　herpes
脊髓灰质炎，小儿麻痹症　poliomyelitis
艾滋病　acquired immuno deficiency syndrome (AIDS)
口蹄疫　aftosa
苯基丙氨酸　phenylalanine
苯乙烯的　cinnamic
查耳酮　chalcone
黄烷酮　flavanone
类黄酮　flavonoid
花青甙素，花色素甙　anthocyanin
色素　pigment
(品)系，菌株，变种　strain
絮凝物　flocculate
(浮悬的)絮状物　floc
半胱氨酸，巯基丙氨酸　cysteine
多氯联苯　polychlorobiphenyl
催化作用　catalysis
水合物，水合(作用)　hydrate
木薯　cassava
再生，(核燃料的)回收　reprocessing
异丙醇　isopropanol
丙酮　acetone
甘油　glycerol
正丁醇 normal　butyl alcohol
山梨醇　sorbierite
丙酸　propionic acid
反丁烯二酸，富马酸，延胡索酸　fumaric acid
乙烯　ethene
二甲苯　xylene
丙烷　propane
乙烷　ethane
丁二烯　butadiene
正丁烷　normal butane
环己烷　cyclohexane
异丁烷，三甲基甲烷　isobutane
碳氢化合物　hydrocarbon
干燥，脱水(作用)　dehydration
分解代谢酶，降解酶　catabolic enzymes
结晶度　crystallinity
活化　activate
细胞密度　cell density
放射菌病　actinomycosis
链霉素　streptomycin
靶细胞　target cell
类异戊二烯　isoprenoid
牵引丝　drag line silk
多肽，缩多氨酸　polypeptide
侧链　lateral chain
活体，在体内　in vivo
突变，突变种　mutation
筛(选)，筛分(法)　sieving
驯化　domesticate
杆状病毒　baculovirus
亲缘　consanguinity
阿尔法淀粉酶　Alpha amylase
酪朊，酪蛋白，干酪素　casein

1 Hepatitis: Any infection that results in inflammation of the liver, also refers to liver inflammation as such.
肝炎：引起肝脏炎症的感染；也只肝脏本身发炎。

2 AIDS: An infectious disease syndrome caused by the human immunodeficiency virus and is characterized by the loss of a normal immune response, followed by increased susceptibility to opportunistic infections and an increased risk of some cancers.
艾滋病：由人类免疫缺陷病毒引起的传染病综合征，其特征为失去正常免疫应答，随后易感染机会性传染病和某些癌症。

第三节 微生物多样性
Section 3 Multiplicity of Microorganism

喜盐生物 halophile
原核类超界,原核生物 procaryote
真核生物 eukaryote
细胞器 cell organ
肽聚糖 peptidoglycan
胞壁酸 muramic acid
甘油酯 glyceride
醚键 ether linkage
革兰氏染色法 Gram stain
交联反应 cross linking reaction
结晶紫 crystal violet
磷脂 phospholipids
青霉素,盘尼西林 penicillin
光传导 optical conductance
厌氧呼吸 anaerobic respiration
代谢途径 metabolic pathway
质子梯度 proton gradient
后色胆素 bilin
视黄醛 retina
细胞色素 cytochrome
紫色细菌,蛋白细菌 purple bacteria
大肠杆菌 *Escherichia coli*
鞭节,鞭毛 flagella
内生孢子 endospore
乳酸杆菌素 lactobacillin
链球菌 *Streptococcus*
子囊果 ascocarp
子囊 ascus
无性繁殖 asexual reproduction
担子 basidium
分生孢子 conidium
配子囊 gametangium
真菌毒素,霉菌毒素 mycotoxin
吞噬生物 phagotroph
腐生物 saprophyte
隔片 septum
孢子果,子实体 sporocarp
叶状体,原植体 thallus
接合孢子 zygospore
芽接(法),芽殖 budding
曲霉菌 *Aspergillus*
限制性内切酶 restriction enzyme
水解酶 hydrolase
甲烷细菌 methane bacteria

1 Anaerobic respiration: An energy-yielding process in which the electron transport chain acceptor is an inorganic molecule other than oxygen.
厌氧呼吸:一个产能过程,其中电子传递链受体是无机分子而不是氧。

2 Cytochrome: Heme proteins that carry electrons, usually as members of electron transport chains.
细胞色素:携带电子的血红素蛋白,通常作为电子传递链分子。

3 Ascocarp: A multicellular structure in Ascomycetes lined with specialized cells called asci in which nuclear fusion and meiosis produce ascospores. An ascocarp can be open or closed and may be referred to as a fruiting boy.
子囊果:子囊菌的多细胞结构,排列着称为子囊的特化细胞,子囊中核融合和减数分裂产生子囊孢子。子囊果可开放或封闭,也可以成为子实体。

4 Basidium: A structure that bears on its surface a definite number of basidiospores(typically four) that are formed following karyogamy and meiosis. Basidia are found in the Basidiomycetes and are usually club-shaped.
担子:核配和减数分裂后在其表面形成一定数目的(典型是 4 个)担孢子结构。担子见于担子菌并通常呈棒状。

5 Thallus: A type of body that is devoid of root, stem, or leaf; characteristic of some algae, many fungi, and lichens.
原植体:原植体缺乏根、茎或叶的结构,是一些藻类、许多真菌和地衣的特征。

6 Saprophyte: An organism that takes up nonliving organic nutrients in dissolved form and usually grows on decomposing organic matter.
腐生生物:这种生物吸收非活体的溶解状态有机营养物,通常生长在腐败有机物上。

第四节 利用细菌生产蛋白质
Section 4 Production of Protein by Bacteria

淋巴因子 lymphokine
垂体生长激素 pituitary growth hormone
电转化 electroporation
转移 mobilization
普遍性转导 generalized transduction
黏粒 cosmid
复制起始区 replication origin
四环素 tetracycline
分裂反应 cleavage reaction
低(聚)核苷酸,寡核苷酸 oligonucleotide
噬粒 phagemid
互补分析 complementation assay
转位子,转座子 transposon
表达载体 expression vector
基因剂量效应 gene dosage effect
内含体 inclusion bodies
层析 chromatography
伴侣蛋白 chaperonin
缀合朊 conjugated protein
异构酶 isomerase
polyU-葡聚糖 polyU-sepharose

1 Cosmid: A plasmid vector with lamda phase cos sites that can be packaged in a phage capsid; it is useful for cloning large DNA fragments.
粘粒:带有λ噬菌体 cos 位点的质粒载体,可包装到噬菌体壳体中。对克隆大片段的 DNA 有用。

2 Transposon: A DNA segment that carries the genes required for transposition and moves about the chromosome. If it contains genes other than those required for transposition, it may be called a composite transposon. The name is often reserved only for transposable elements that also contain genes unrelated to transposition.
转座子:携带转座所需基因并在染色体上移动位置的 DNA 片段。如果这个片段除了带有转座所需基因,还带有其他基因,叫做复合转座子。这个名称往往只专用于还带有与转座无关的基因的转座因子。

3 Inclusion bodies: Granules of organic or inorganic material lying in the cytoplasmic matrix of bacteria.
内含体:存在于细菌细胞基质中的有机物或无机物颗粒体。

第五节 利用酵母生产蛋白质
Section 5 Production of Protein by Yeast

糖基化 glycosylation
高尔基体 Golgi apparatus
乙型肝炎 hepatitis
甘露糖 mannitose
血清蛋白 serum protein
唾液酸 sialic acid

1 Golgi apparatus: A membranous eukaryotic organelle composed of stacks of flattened sacs (cisternae), which is involved in packaging and modifying materials for secretion and many other process.
高尔基体:膜质真核细胞器,由许多扁平囊组成。参与分泌物质的包装和修改以及许多其他过程。

第六节 重组疫苗与合成疫苗
Section 6 Vaccine Recombination and Synthesis

传染病,传染性疾病 communicable disease
恐水病,狂犬病 hydrophobia
伤寒症 typhoid
白喉 diphtheria
结核 subcutaneous nodes
霍乱 cholera
百日咳 bronchocephalitis
黄热病 yellow fever
脑膜炎 cephalomeningitis
活疫苗 live vaccine

灭活疫苗 killed vaccine
血吸虫 bilharzia
保护性抗原 protective antigens
亚单位疫苗 subunit vaccine
免疫应答 immune response
致病原因 pathogeny
免疫监视 immune surveillance
肿瘤 tumo(u)r
化学诱变 chemomorphosis
肽疫苗 peptide vaccine
免疫原决定簇 immunogenic epitope
装配型拓扑位点 assembled topographic sites
不连续抗原决定簇 discontinuous epitope
免疫原性 immunogenic
短序列 short sequence
亲水图谱 hydrophilicity plot
辅助细胞 accessory cell
麻风(病) leprosy
结核样型麻风 tuberculoid leprosy
瘤型麻风病 lepromatous leprosy
孢子期 sporozoite
配子(母)细胞 gametocyte
合子 ookinete
环子孢子 circumsporozoite
肝细胞 hepatocyte
裂殖子 merozoite

第七节 微生物杀虫剂
Section 7 Microbial Insecticide

有机氯化合物 organic chlorine compound
双翅目 diptera
结合位点浓度 binding site concentration
离心法,离心过滤 differential centrifugation
毒素,毒质 toxin
核多角体病毒 nuclear polyhedrosis virus
病毒菌素 virusin
颗粒体病毒 granulosis virus
独角仙病毒 orytestype virus
芽生病毒 budded virus
包含体病毒 occluded virus
血淋巴 haemolymph
病原体,病原菌 pathogen
上皮 epithelium
苏云金杆菌 thuricide
微生物杀虫剂 microbial insecticide
包埋 imbed

1 Toxin: A microbial product or component that can injure another cell or organism at low concentrations. Often the term refers to a poisonous protein, but toxins may be lipids and other substances.
毒素:微生物产物或组分,在低浓度情况下就能伤害另一细胞或生物。这一术语往往是指有毒蛋白质,但毒素也可能是脂类或其他物质。

2 Pathogen: Any virus, bacterium, or other agent that causes disease.
病原体:致病的病毒、细菌或其他因子。

第八节 微生物酶
Section 8 Microbial Enzymes

酿造 brewing
酶制剂 zymine
黄糖浆 golden syrup
转化酶 invertase
葡糖苷酸 glucosiduronate
配糖键 glycosidic linkage
致病性 pathogenicity
纤维素酶 cellulase
木糖 xylose
诱导物 inducer
胞外酶,外酶 exoenzyme
发酵桶,发酵槽 fermentor
固态发酵 solid state fermentation
氧化酶 oxidase
泡罩塔 bubble captower
异构化度 degree of isomerization
麦芽糖浆 malt syrup
糊精 dextrin

1 Inducer: A small molecule that stimulates the synthesis of an inducible enzyme.
诱导物:刺激诱导酶合成的小分子。

第九节 微生物多糖和聚酯

Section 9 Microbial Polysaccharides and Polyester

悬浮(液) suspension
叶鞘 sheath
海草酸 alginic acid
免疫系统 immune system
细菌萜醇 bactoprenol
化学稳定器,恒化器 chemostat
构象 conformation
葡萄糖 dextrose
共聚多酯 copolyester
共聚多醚 copolyether
羟基丁酸 hydroxy butanoic acid
细菌荚膜 bacterial capsule
黄原酸盐 xanthogenate

第十节 植物与微生物的相互作用

Section 10 Interaction of Plants and Microorganism

共生生物 commensal
重组细胞,重组体[子] recombinant
冰成核蛋白 ice nucleation protein
毒力 virulence
磷酸化(作用) phosphorylation
末端核酸酶 border nuclease
生物碱 alkaloid
细胞分裂素 cytomin
吲哚乙酸 heteroauxin
章鱼(肉)碱 octopine
新霉素 neomycin
草甘膦 glyphosate
缺臂 disarmed
磺酰脲 sulfonylurea
超敏反应 hypersensitivity

1 Commensal: Living on or within another organism without injuring or benefiting the other organism.
共生生物:生活在另一个生物体上或内部,互相依存,共同生活。

2 Virulence: The degree or intensity of pathogenicity of an organism as indicated by case fatality rates and/ or ability to invade host tissues and cause disease.
毒力:一种机体致病性的程度或强度,以病例致死率或侵染宿主组织并致病的能力表示。

3 Hypersensitivity: A condition of increased immune sensitivity in which the body reacts to an antigen with an exaggerated immune response that usually harms the individual. Also termed an allergy.
超敏反应:免疫敏感性增高的状态,处于这一状态的个体与抗原作用产生过度的免疫应答,通常对个体造成损害。也称变态反应。

第十一节 生物量

Section 11 Biomass

生物量 biomass
结晶性 crystallinity
木质化 lignification
半纤维素 hemicellulose
微纤维 microfibril
裸子植物 gymnosperm
被子植物 angiosperm
松柏醇 coniferyl alcohol
木质醇 lignol
二聚木素 dilignol
加双氧酶 dioxygenase
白腐菌 white rot fungus
菌丝 hypha
还原酶 reductase
纤维素酶 cellulase

1 Hypha: The unit of structure of most fungi and some bacteria; a tubular filament.
菌丝:大多数真菌和某些细菌的结构单位,管形丝状体。

第十二节 乙醇、氨基酸、抗生素
Section 12 Ethanol, Amino Acid and Antibiotics

釜馏物 stillage
醛缩酶 aldolase
果己糖激酶 fructohexokinase
磷酸(酯)酶 phosphatase
呋喃果糖苷 fructofuranoside
何帕烷 hopane
营养缺陷型 auxotroph
野生型 wild type

1 Auxotroph: A mutated prototroph that lacks the ability to synthesize an essential nutrient and therefore must obtain it or a precursor from its surroundings.
营养缺陷型:突变的原养型。缺乏合成必需营养物的能力,因此必须从环境中获取这种营养物或前体。

2 Phosphatase: An enzyme that catalyzes the hydrolytic removal of phosphate from molecules.
磷酸酶:催化从分子中解离磷酸的酶。

第十三节 有机合成
Section 13 Organic Synthesis

原(营养)型微生物 prototroph
磺胺醋酰 sulfacetamide
磺胺药类 sulfa drugs
抗细菌因子 antibacterial agent
红霉素 erythromycin
金霉素 aureomycin
土霉素 terramycin
氯霉素 chloromycetin
庆大霉素 gentamicin
抗真菌剂 antifungal agent
放线菌素 dactinomycin
肽酶 peptidase
免疫抑制剂 immunodepressant
卡那霉素 kanamycin
匀霉素、潮霉素 homomycin
妥布霉素 tobramycin
先锋霉素 II cephaloridine II
肠球菌 *Enterococcus*
头孢霉菌素 cephalosporin
连续发酵 continuous fermentation
分解代谢产物 katabolite
反密码子 anticodon
诱变 mutagenesis
对映(结构)体 enantiomer
手性中心 chiral center
氧化还原酶 oxidoreductase
转移酶 transferase
水解酶 hydrolase
裂解酶 lyase
异构酶 isomerase
合成酶 synthetase
酰胺酶 amidase
脂(肪)酶 lipase
氨解 aminolysis
辅酶 coenzyme
降解质粒 degradative plasmid
沥滤(法) leaching
金属硫化物 metal sulfides

1 Prototroph: A microorganism that requires the same nutrients as the majority of naturally occurring members of its species.
原养型微生物:与同种自然发生的大多数个体一样,具有相同的营养需求的微生物。

2 Erythromycin: An intermediate spectrum macrolide antibiotic produced by *Streptomyces erythreus*.
红霉素:由红霉素链霉菌产生的中谱大环内酯抗生素。

3 Tetracyclines: A family of antibiotics with a common four-ring structure, which are isolated from the genus *Streptomyces* or produced semisynthetically; all are related to chlortetracycline or oxytetracycline.
四环素:一类具有共同的四环结构的抗生素。从链霉菌分离得到或半合成产生,都与氯化四环素或氧化四环素相关。

4 Coenzyme: A loosely bound cofactor that often dissociates from the enzyme active site after product has been formed.
辅酶:与酶结合松散的辅因子,产物形成后往往与酶活性位点分离。

第二十一章　分子生物学
Chapter 21　Molecular Biology

第一节　DNA的结构
Section 1　Structure of DNA

分子生物学　molecular biology
一级结构　primary structure
超螺旋DNA　supercoiled DNA
核苷酸　nucleotide
遗传信息　genetic information
共线性　colinear
主链　main chain
双螺旋　double helix
碱基堆积力小沟　minor groove
螺距　pitch
疏水键　hydrophobic bond
空间构象　space conformation
操纵子　operon
复性　renaturation
拓扑异构酶　topoisomerase
遗传物质　genetic material
等位基因　allele
左手螺旋　left helix
右手螺旋　right helix
配子　gamete
组蛋白　histone
蛋白　protein
DNA酶　DNA polymerase
螺线管　solenoid
脱氧核糖核苷酸　deoxyribonucleic acid
连接数　linking number
盘绕数　twisting number
超螺旋数　writhing number
自我复制　self-duplication
半保留复制　semiconservative replication
滚环型　rolling circle
沉默　silencing
引发前体　trigger precursor
引发酶　primase
可移动因子　transposable element
转座子　transposon
复合式转座子　composite transposon
转座酶　transposase
解离酶　resolvase
控制因子　controllable factor
杂种不育　hybrid dysgenesis
细胞质因子　cytoplasmic factor

1 Histones are conserved DNA-binding proteins that form the basic subunit of chromatin in eukaryotes. Histones H2A, H2B, H3, and H4 form an octameric core around which DNA coils to form a nucleosome. Histone HI is external to the nucleosome.
组蛋白:真核生物中构成基本单位保守的DNA结合蛋白。H2A、H2B、H3和H4形成八聚体核心,然后DNA盘绕其上形成核小体。核小体不包括组蛋白H1。

2 Operon is a unit of bacterial gene expression and regulation, including structural genes and control elements in DNA recognized by regulator gene products.
操纵子:细菌基因表达的一个单位,包括结构基因和能被调控基因产物识别的DNA控制元件。

3 Renaturation describes the reassociation of denatured complementary single strands of a DNA double helix.
复性:指的是DNA双螺旋的两条互补链重新结合。

4 Transposon is a DNA sequence able to insert itself (or a copy of itself) at a new location in the genome.
转座子:能将自身(或它的拷贝)插入到基因组新位置的DNA序列。

5 Silencing describes the repression of gene expression in a localized region, usually as the result of a structural change in chromatin.
沉默:通常由于染色质的结构改变,DNA某一区域基因被抑制的现象。

6 Linking number is the number of times the two strands of a closed DNA duplex cross over each other.
连接数:是指闭合 DNA 双螺旋的两条链相互缠绕的次数。

7 Allele is one of several alternative forms of a gene occupying a given locus on a chromosome.
等位基因:在染色体上占据给定位点的不同基因形式的其中一种。

8 Writhing number is the number of times a duplex axis crosses over itself in space.
扭曲数:双链中轴在空间绕过自身的次数。

第二节 基因的组织与结构
Section 2 Organization and Structure of Gene

编码蛋白质 encoded protein
调控单位 regulated unit
重叠序列 overlap sequence
蛋白质复合物 protein complex
调控蛋白 regulatory protein
编码序列 coding sequence
重复频度 repetition frequency
同源性 homology
浮力密度 buoyant density
梯度离心 gradient centrifugation
基因组 genome
重复结构单元 constitutional repeating unit
不连续基因 split gene
内含子 intron
蛋白质结构域 protein domains
假基因 pseudogene
转位因子 translocator
原核生物基因组的结构特点 characteristics of prokaryote genome structure
真核生物基因组的结构特点 structural features of eukaryotic genome
简单的多基因家族 simple multiple gene family
复杂的多基因家族 complex multiple gene family

1 Genome is the complete set of sequences in the genetic material of an organism. It includes the sequence of each chromosome plus any DNA in organelles.
基因组:生物体遗传物质的全部序列,它包含着每一条染色体和任何亚细胞器的 DNA 序列。

2 Coding region is a part of the gene that represents a protein sequence.
编码区:是编码蛋白质序列的其中一段。

3 Intron is a segment of DNA that is transcribed, but later removed from within the transcript by splicing together the sequences (exons) on either side of it.
内含子:是一段 DNA 序列,它能被转录,但在转录中由于其将两端的序列(外显子)发生剪切而被去除。

4 Pseudogene is inactive but stable components of the genome derived by mutation of an ancestral active gene. Usually they are inactive because of mutations that block transcription or translation or both.
假基因:由原始活性基因突变产生的,存在于基因组中的不活泼但稳定的 DNA 序列。由于发生了能阻止转录或翻译或二者的变异,它们通常不活泼。

第三节 蛋白质与 DNA 的相互作用
Section 3 Interaction of Protein and DNA

非特异性结合 nonspecific binding
特异性结合 specific binding
纯化 purification
可译框架 open reading frame
核酸酶 nuclease
修饰碱基 modified base
基本结构 fundamental structure
结合蛋白 binding protein
拉链结构 zipper
普遍性结合 universal combination
放射自显影 autoradiography
探针 probe
残基 residue
反向平行 antiparallel
复合物 complex
转移 RNA tRNA

编码链 coding strand
反义链 antisense strand
琥珀 amber
同义密码子 synonymous codon
摆动假说 wobble hypothesis
受体臂 acceptor arm
二氢尿嘧啶 dihydrouracil
肽基 phthalidyl
单顺反子 monocistron
多顺反子 polycistron
外显子 exon
尾巴 tail
不均一 RNA heterogeneous RNA
启动子 promoter

1 Open reading frame is a sequence of DNA consisting of triplets that can be translated into amino acids starting with an initiation codon and ending with a termination codon.
可译框架：由编码氨基酸的三联体组成的连续的 DNA 序列，能翻译成蛋白质，有起始密码子开始，终止密码子结束。

2 Promoter is a region of DNA where RNA polymerase binds to initiate transcription.
启动子：RNA 聚合酶聚合并起始转录的 DNA 区域。

3 Exon is any segment of an interrupted gene that is represented in the mature RNA product.
外显子：在成熟 RNA 中存在的断裂基因的任何片段。

4 Transfer RNA (tRNA) is the intermediate in protein synthesis that interprets the genetic code. Each tRNA can be linked to an amino acid. The tRNA has an anticodon sequence that is complementary to a triplet codon representing the amino acid.
转移 RNA：在蛋白质合成中，解释遗传密码的中间体，每个 tRNA 都要连有氨基酸，它与 t RNA 中代表氨基酸的三联体密码子互补。

5 The antisense strand of DNA is complementary to the sense strand, and is the one that acts as the template for synthesis of mRNA.
反义链：在基因的 DNA 双链中，转录时作为合成模板 mRNA 的那条单链叫作反义链。

第四节 DNA 的复制

Section 4 The Replication of DNA

复制 replication
基因调节 gene regulation
半保留复制 semiconservative replication
引物酶 primase
RNA 聚合酶 RNA polymerase
解螺旋酶 helicase
旋转酶 gyrase
复制起始点 origin of replication
全保留复制 conservative replication
模板链 template strand
环状分子 ring molecule
正超螺旋 positive supercoil
冈歧片段 Okazaki of fragment
DNA 损伤 DNA damage
置换环 displacement loop
引物 primer
DNA 复制的调控 regulation of DNA replication
染色体复制的调控 regulation of chromosome replication
染色体复制 chromosome duplication
双向复制 bidirectional replication
模板 templet
半不连续复制 semidiscontinuous replication

1 Silencing describes the repression of gene expression in a localized region, usually as the result of a structural change in chromatin.
沉默：通常指由于染色质结构的改变，而导致某一区域基因表达受抑制的现象。

2 Semiconservative replication is accomplished by separation of the strands of a parental duplex, each strand then acting as a template for synthesis of a complementary strand.
半保留复制：将亲本双链 DNA 分开，将每一条链做模板合成新的互补链的复制模式。

3 RNA polymerase are enzymes that synthesize RNA using a DNA template。
RNA 聚合酶：它是一种以 DNA 为模板用于合成 RNA 的酶。

4 Primase is a type of RNA polymerase that synthesizes short segments of RNA that will be used as primers for DNA replication.

引物酶:一种 RNA 聚合酶,用来合成短的 RNA。引物酶被用作 DNA 合成的引物。

第五节 基因的转录和转录后加工

Section 5 Gene Transcription and Post-Transcriptional Processing

转录 transcription
中心法则 central dogma
基因转录 genetic transcription
编码区 coding region
双链超螺旋 double helix
基因活性 gene activity
管家基因 housekeeping gene
阻遏蛋白 repressor protein
起始复合物 initiation complex
调控区域 regulatory regions
增强子 enhancer
加工 process
识别 recognition
剪接 splicing
编辑 edit
选择性加工 selective processing
相互作用 interreaction
元件 element

1 Transcription describes synthesis of RNA on a DNA template.

转录:以 DNA 为模板合成 RNA 的过程。

2 Central dogma describes the basic nature of genetic information: sequences of nucleic acid can be perpetuated and interconverted by replication, transcription, and reverse transcription, but translation from nucleic acid to protein is unidirectional, because nucleic acid sequences cannot be retrieved from protein sequences.

中心法则:它描述了遗传信息的准则,核酸序列可以通过复制转录和反转录做到永存,但核酸翻译成蛋白质是单向的,因为核酸序列不能从蛋白质中得到。

3 Housekeeping genes are those (theoretically) expressed in all cells because they provide basic functions needed for sustenance of all cell types.

管家基因:是那些在所有细胞都表达的基因。因为它提供了任何细胞类型都必需的功能。

4 Enhancer is a cis-acting sequence that increases the utilization of (some) eukaryotic promoters, and can function in either orientation and in any location (upstream or downstream) relative to the promoter.

增强子:是一种顺式作用元件,可以提高启动子的效率,并且可以在另一个起始位点作用或者是启动子的上游或下游作用。

5 Repressor is a protein that inhibits expression of a gene. It may act to prevent transcription by binding to an operator site in DNA or to prevent translation by binding to RNA.

阻遏蛋白:是基于抑制基因表达的蛋白质,它可以表现为通过绑定 DNA 的转录起始位点,或者通过结合 RNA 组织翻译。

第六节 蛋白质的生物合成

Section 6 Biosynthesis of Protein

翻译 translate
共同序列 consensus sequence
帽子结构 cap structure
开放阅读框架 open reading frame
发夹结构 hairpin structure
氨基酸活化 activation of amino acids
起始密码子 initiation codon
终止密码子 termination codon
核糖体 ribosome
可溶性蛋白质因子 soluble protein factor
肽链 peptide chain
肽键 peptide bond
结构性调控 structural control
翻译区域 translational domain
非翻译区域 untranslated region
质膜 plasma membrane

跨膜运输 transmembrane transport
多顺反子 polycistron
反密码子 anticodon
自我组装 self-assembly

1 Initiation codon is a special codon (usually AUG) used to start synthesis of a protein.
起始密码子：是一种特殊的密码(通常为 AUG)，被用作蛋白质合成的开始。

2 Self-assembly refers to the ability of a protein (or of a complex of proteins) to form its final structure without the intervention of any additional components (such as chaperones). The term can also refer to the spontaneous formation of any biological structure that occurs when molecules collide and bind to each other.
自我组装：指的是蛋白质或蛋白质复合体在没有任何外来成分(如伴侣蛋白)的干预下，形成最终结构的能力。它也指当分子碰撞或分子结合时，自发形成任何生物学结构的能力。

3 Anticodon is a trinucleotide sequence in tRNA that is complementary to the codon in mRNA and enables the tRNA to place the appropriate amino acid in response to the codon.
反密码子：tRNA 中的核苷酸三联体，它与 mRNA 的密码子互补，使 tRNA 对应的氨基酸放在对应的密码子上。

4 Cap structure is the structure at the 5′ end of eukaryotic mRNA, and is introduced after transcription by linking the terminal phosphate of 5′ GTP to the terminal base of the mRNA. The added G (and sometimes some other bases) are methylated, giving a structure of the form 7MeG5′ppp5′Np.
帽子结构：是指在真核生物 mRNA 在 5′ 端的一个结构。在转录后通过末端 5′ GTP 的磷酸集团和 mRNA 末端碱基而引入，增加的鸟嘌呤是甲基化的，产生了 7MeG5′ppp5′Np 的结构。

第七节 原核生物基因表达调控
Section 7 Regulation of Gene Expression in Prokaryotes

诱导反应 induced reaction
阻遏作用 repression
正调节 positive accommodation
负调控 negative regulation
结构基因 structural gene
控制模式 control model
操纵基因 operator gene
乳糖操纵子 lac operon/lactose operon
转录调控 transcriptional control
乳糖代谢 lactose metabolism
调控模型 control mode
显性 dominance
顺式作用元件 cis acting element
活化蛋白 activating protein
衰减子 attenuater
前导序列 leader peptide
序列分析 sequential analysis
诱导物 inducer
特异性元件 specific element
正调控模型 positive regulation model
负调控模型 negative regulation model
基因表达 gene expression
基因表达调控 gene expression regulation
RNA 加工成熟水平上的调控 maturity level regulation of RNA processing
翻译水平上的调控 regulation on the translation level
环境因素 environmental factor
激素水平 hormone level
发育阶段 development phase
转录产物 transcript
转录单元 transcription unit
终止子 terminator
起点 origin
上游 upstream
下游 downstream
增强子 enhancer
启动区 promoter
操纵区 operator
阻遏物 repressor
色氨酸操纵子 trp operon
辅阻遏蛋白 corepressor proteins
半乳糖操纵子 gal operon
异构酶 isomerase
半乳糖 galactosyl
磷酸尿嘧啶核苷转移酶 uracil nucleoside phosphate transferase
半乳糖激酶 galactokinase

半乳糖差向异构酶 galactose epimerase
阿拉伯糖 arabinose
核酮糖激酶 ribulokinase
阿拉伯糖异构酶 arabinose isomerase
核酮糖 ribulose
警报素 alarm pheromone

1 Attenuator is a terminator sequence at which attenuation occurs.
衰减子是衰减发生处的一种内部终止子序列。

2 Operator is the site on DNA at which a repressor protein binds to prevent transcription from initiating at the adjacent promoter.
操纵子是细菌基因表达的一个单位,包括结构基因和能被调控基因产物识别的 DNA 控制元件。

3 Promoter is a region of DNA where RNA polymerase binds to initiate transcription.
启动子是 RNA 聚合酶聚合并起始转录的 DNA 区域。

4 Terminator (t) is a sequence of DNA that causes RNA polymerase to terminate transcription.
终止子是给予 RNA 聚合酶转录终止信号的 DNA 序列。

5 Structural gene: codes for any RNA or protein product.
结构基因:是一类编码蛋白质或 RNA 的基因。

6 Transcription unit is the sequence between sites of initiation and termination by RNA polymerase, it may include more than one gene.
转录单元是 RNA 聚合酶识别的起始和终止的序列,它可能包括多于一个基因。

7 Upstreamidentifies sequences in the opposite direction from expression, for example, the bacterial promoter is upstream of the transcription unit, and the initiation codon is upstream of the coding region.
上游:基因表达相反方向的序列,例如,细菌的启动子在转录起始位点上游,起始位点在下游。

8 Downstream: identifies sequences proceeding farther in the direction of expression, for example, the coding region is downstream of the initiation codon.
下游:基因表达更远方向的序列,例如,编码区在起始位点下游。

第八节 真核生物基因表达
Section 8 Expression of Eukaryotic Genes

结构蛋白 structural protein
常染色质 euchromatin
蛋白质修饰 protein modification
遗传操作 genetic operation
启动子结构 promoter architecture
甲基化 methylation
靶序列 target sequence
衰减作用 attenuation process
时序基因 temporal gene
激活 activation
顺式作用 cis-acting
真核基因转录调控区 eukaryotic gene transcription control regions
启动子上游调控元件 promoter upstream regulator element
转录调控区 transcription regulation domain
反式调控因子 trans regulation factor
激活结构域 activation domain
糖基化 glycosylation
磷酸化 phosphorylation
转录活化结构域 transcriptional activation domains
转录后水平调控 post transcriptional regulation
翻译水平调控 translation regulation
断裂基因 split gene
异源双链分子 heteroduplex molecule
高度保守序列 highly conserved sequence
营养细胞 vegetative cell
反式作用因子 trans acting factor
传感基因 sensing gene
受体基因 acceptor gene
终止区 terminator
上游启动子元件 upstream promoter element
上游激活序列 upstream activating sequence
沉默子 silencer
小型核内 RNA small nuclear RNA
锌指区 zinc finger region
应答元件 response element

热激应答元件　heat shock response element
金属应答元件　metal response element
热休克蛋白　heat shock protein
热激因子　heat shock factor
金属硫蛋白　metallothionein
胞质受体　cytosol receptor
磷脂酰丝氨酸　phosphatidylserine
胰高血糖素　glucagon
胰泌素　secretin
外周蛋白(表蛋白)　peripherin
核糖核蛋白颗粒　ribonucleoprotein particle
帽子结合蛋白　cap binding protein

1 Heat shock response element (HSE) is a sequence in a promoter or enhancer that is used to activate a gene by an activator induced by heat shock.
热激应答元件:是启动子或增强子的一段序列,通过热激所诱导的激化剂来激活基因。

2 Response element is a sequence in a eukaryotic promoter or enhancer that is recognized by a specific transcription factor.
应答元件:被特异的转录因子所识别的真核生物启动子序列。

3 Intron is a segment of DNA that is transcribed, but later removed from within the transcript by splicing together the sequences (exons) on either side of it.
内含子:是一段 DNA 序列,它转录但将两端的序列(外显子)剪切在一起而被去除。

4 Micro RNAs are very short RNAs that may regulate gene expression.
小 RNA:可调节基因表达的小 RNA。

第九节　分子生物学实验

Section 9　Experiment of Molecular Biology

质粒　plasmid
提取　extract
琼脂糖　agarose
凝胶　gel
电泳　electrophoresis
限制性内切酶　restriction enzyme
缓冲液　buffer solution
溶菌酶　lysozyme
分光光度计　spectrophotometer
溴化乙啶　ethidium bromide
基因片段　gene fragments
克隆　clone
载体　vector
宿主细胞　host cell
转化　transform
感受态细胞　competent cell
聚合酶链式反应　polymerase chain reaction
简并引物　degenerate primers
杂交　hybrid
印迹法　blotting
同步 PCR　synchronous PCR
顺式作用位点　cis-acting site

1 Hybridization describes the pairing of complementary RNA and DNA strands to give an RNA-DNA hybrid.
杂交:使互补 DNA 或 RNA 配对形成 RNA-DNA 杂合体。

2 Cis-acting site affects the activity only of sequences on its own molecule of DNA (or RNA), this property usually implies that the site does not code for protein.
顺式作用位点只影响处于同一 DNA (or RNA) 上的序列,此性质经常暗示该位点不编码蛋白质。

3 Plasmid is a circular, extrachromosomal DNA.
质粒是一个环状的染色体外的 DNA。

第二十二章　基因工程
Chapter 22　Gene Engineering

第一节　绪论
Section 1　Introduction

可移动遗传因子　mobile genetic element
基因　gene
基因作用　gene action
基因激活　gene activation
基因活性　gene activity
基因簇　gene cluster
基因座　gene locus
基因操作　gene manipulation
基因标签　gene tag
基因寻靶　gene targeting
遗传图　genetic map
基因拷贝　gene copy
基因克隆　gene cloning
基因含量　gene content
基因表达　gene expression
基因表达系统　gene expression system
基因融合　gene fusion
基因定位　gene localization
基因突变　gene mutation
结瘤基因　nodulation gene
分枝糊精　branched dextrin
活性　activity
生物碱　alkaloid
抗体工程　antibody engineering
生物信息学　Bio-Informatics
生物信息　biological information
数据库　database
遗传标记　genetic marker
基因组　genome
基因组学　genomics
跳跃基因　jumping gene
开放阅读框 open reading frame

1 Gene：A gene is a unit of heredity in a living organism. It normally resides on a stretch of DNA that codes for a type of protein or for an RNA chain that has a function in the organism. All living things depend on genes，as they specify all proteins and functional RNA chains.
基因：基因是生命有机体中遗传物质的基本单位。正常情况下它存在于 DNA 片段中，DNA 对蛋白质类型或在生物体中发挥功能的 RNA 链进行编码。由于基因指定了蛋白质和功能性 RNA 链，因此所有生物都离不开基因。

2 Gene expression：Gene expression is the process by which information from a gene is used in the synthesis of a functional gene product. These products are often proteins，but in non-protein coding genes such as ribosomal RNA（rRNA），transfer RNA（tRNA）or small nuclear RNA（snRNA）genes，the product is a functional RNA.
基因表达：在基因表达过程中，基因信息用来合成功能基因产品。这些产品通常是蛋白质，但在非蛋白编码基因中，如核糖体 RNA(rRNA) 基因、转移 RNA(tRNA) 基因或核小 RNA(snRNA) 基因，产品是功能 RNA。

3 Genetic marker：A genetic marker is a gene or DNA sequence with a known location on a chromosome that can be used to identify individuals or species. It can be described as a variation（which may arise due to mutation or alteration in the genomic loci）that can be observed.
遗传标记：遗传标记是染色体上位点已知的一组基因或 DNA 序列，可用来辨认个体或物种。可以将遗传标记描述为能观察到的变化(会随着基因组位点的突变或变化而出现)。

4 Genome：Genome contains an organism's entire hereditary information，which is encoded in DNA or RNA depending on the type of organism.
基因组：基因组包含生物体全部的遗传信息，由于生物体种类的不同，而分别编码在 DNA 或 RNA 中。

5 Open reading frame: In molecular genetics, an open reading frame (ORF) is the part of a gene that actually encodes a protein. The transcription termination pause site is located after the ORF, beyond the translation stop codon, because if transcription were to cease before the ribosome reaches the translation stop codon, an incomplete protein would be made.
开放阅读框:在分子遗传学中,开放阅读框(ORF)是基因的组成部分之一,开放阅读框真正编码蛋白质。除翻译终止密码子外,转录终止中断位点位于开放阅读框后,因为如果在核糖体到达翻译终止密码子前就停止转录,那么就会生产出不完整的蛋白质。

第二节 DNA的提取与纯化

Section 2 Extraction and Purification of DNA

提取 extraction
提取率 extraction rate
纯化 purification
共价闭合环状 DNA covalently closed circular DNA
碱基置换 base substitution
结合 conjugation
循环测序法 cycle sequencing
示差杂交 differential hybridization
DNA 指纹图 DNA finger-printing
表达图 expression map
碱基插入 base in sertion
碱基堆积 base tacking
假基因 pseudogene
基体 basal body
必需基因 essential gene
畸形 malformation
描图 mapping
标记 marker
黑色素瘤 melanoma
孟德尔遗传 Mendelian inheritance
单体性 monosomy
小鼠模型 mouse model
突变 mutation
功能性克隆表达 functional cloning expression
富 GC 序列 GC-rich sequence
多数 tRNA major tRNA
主平板 master plate
小卫星 DNA mini-satellite DNA
核酸酶 nuclease
蛋白-蛋白交互作用 protein-protein interaction
基因缺失 gene deletion
基因剔除 gene knock-out
通用转录因子 general transcription factor
基因组印记 genomic imprinting
非组蛋白 nonhistone protein
非组蛋白骨架 nonhistone protein scaffold
随机扩增多态 DNA random amplified polymorphic DNA
密码子 codon
RNA 编辑 RNA editing
亲和层析 affinity chromatography
琼脂 agar
琼脂糖 agarose
反向转运 antiport
反义基因 anti sense gene
反义核酸 anti sense nucleic acid
肌动蛋白纤维 actin filament
主动运输 active transport
腺病毒 adenovirus
粘着带 adhesion belt
协助装配 aided-assembly
选择性剪接 alternate splicing
后期促进因子复合物 anaphase-promoting complex
锚定连接 anchoring junction

1 Mutation: In molecular biology and genetics, mutations are changes in a genomic sequence, such as the DNA sequence of a cell's genome or the DNA or RNA sequence of a virus.
突变:在分子生物学和遗传学中,突变是发生在基因组序列中的变化,比如发生在细胞基因组的DNA序列中或病毒的DNA或RNA序列中的变化。

2 Protein-protein interaction: Protein-protein interactions occur when two or more proteins bind together, often to carry out their biological function. Many of the most important molecular processes in the cell such as DNA replication are carried out by large molecular machines that are built from a large number of protein components organized by their protein — protein interactions.

蛋白-蛋白质交互作用：蛋白质交互作用在两种或以上蛋白质结合时发生，通常旨在执行其生化功能。在细胞中，大量蛋白质元件组成分子机器，通过蛋白质交互作用执行细胞内许多重要分子过程，如DNA复制。

3 Genomic imprinting: Genomic imprinting is a genetic phenomenon by which certain genes are expressed in a parent-of-origin-specific manner. It is an inheritance process independent of the classical Mendelian inheritance.

基因组印记：基因组印记是一种遗传现象，通过基因组印记，可以以亲源效应的方式表达某些基因。它是独立于经典的孟德尔遗传外的遗传过程。

4 Codon: The code defines how sequences of three nucleotides, called codons, specify which amino acid will be added next during protein synthesis.

密码子：编码规定了三个核苷酸序列为一个密码子，指定哪些氨基酸参加下一步蛋白质的合成。

5 RNA editing: The term RNA editing describes those molecular processes in which the information content in an RNA molecule is altered through a chemical change in the base makeup. To date, such changes have been observed in tRNA, rRNA, mRNA and microRNA molecules of eukaryotes but not prokaryotes.

RNA编辑：RNA分子由子碱基组成发生了化学变化，而使其信息内容发生改变的分子过程，叫作RNA编辑。目前在真核生物的tRNA、rRNA、mRNA及小RNA分子中都发现了这种变化，但在原核生物中还没有发现。

第三节 基因文库
Section 3 Gene Library

mRNA差异显示法 mRNA differential display
基因组文库 genomic library
基因文库 gene library
核糖核酸 ribonucleic acid
腺苷脱氨酶缺乏症 adenosine deaminase deficiency
腺病毒 adenovirus
Alagille综合征 Alagille syndrome
等位基因 allele gene
动物模型 animal model
抗体 antibody
常染色体 autosome
常染色体显性 autosomal dominant
Alu家族 Alu family
自主性因子 autonomous element
细菌人工染色体 bacterial artificial chromosome
cDNA微矩阵分析 cDNA micro-array analysis
顺反子 cistron
共整合体 cointegrate
组合库 combinatorial library
感受态细胞 competent cell
互补DNA complementary DNA
插入序列 insertion sequence
文库 library
座位 locus
选择性 selectivity
测序引物 sequencing primer
基因表达的系列分析 serial analysis of gene expression
序列标记位点 sequence-tagged site
性染色体 sex-chromosome
伴性的 sex-linked
体细胞 somatic-cell
光谱核型 spectral karyotype
替代 substitution
自杀基因 suicide gene
综合征 syndrome
先天缺陷 birth-defect
骨髓移植 bone-marrow transplantation
后选基因 candidate gene
cDNA文库 cDNA library
染色体 chromosome
克隆 cloning
密码 codon
天生的 congenital
囊性纤维化 cystic fibrosis
细胞遗传图 cytogenetic map

1 Genomic library: A genomic library, also clone bank or gene bank, is a collection of DNA from a single organism, ideally though not necessarily containing its entire genomic DNA sequence.

基因组文库：基因组文库，也称克隆库或基因库，是单一生物体 DNA 的总和，但理想状态下不一定包含基完整的基因组序列。

2 Bacterial artificial chromosome: A bacterial artificial chromosome (BAC) is a DNA construct, based on a functional fertility plasmid (or F-plasmid), used for transforming and cloning in bacteria. F-plasmids play a crucial role because they contain partition genes that promote the even distribution of plasmids after bacterial cell division. The bacterial artificial chromosome's usual insert size is 150-350 kbp.

细菌人工染色体：简称BAC，是一类以F质粒为基础构建的基因克隆载体，用来在大肠杆菌中克隆分子量较大的外源DNA片段。F-质粒由于含有分割基因，在细菌细胞分裂之后促进质粒传播，因此，它具有重要作用。

3 Complementary DNA: In genetics, complementary DNA (cDNA) is DNA synthesized from a messenger RNA (mRNA) template in a reaction catalyzed by the enzyme reverse transcriptase and the enzyme DNA polymerase.

互补 DNA：在遗传学中，互补 DNA(cDNA) 是在逆转录酶催化的反应中，由信使 DNA 模板合成 DNA。

4 Sequence-tagged site: A sequence-tagged site (or STS) is a short (200 to 500 base pair) DNA sequence that has a single occurrence in the genome and whose location and base sequence are known.

序列标记位点：序列标记位点(或 STS) 是一个短 DNA 序列(200 ~ 500 个碱基对)，在基因组中只出现一次，位置和碱基序列已知。

第四节 基因扩增
Section 4 Gene Amplification

基因扩增 gene amplification
聚合酶链式反应 polymerase chain reaction
保卫基因 protective gene
缺失 deletion
脱氧核糖核酸 deoxyribonucleic acid
DNA 复制 DNA replication
DNA 测序 DNA sequencing
显性的 dominant
双螺旋 double helix
基因治疗 gene therapy
基因转移 gene transfer
遗传密码 genetic code
遗传咨询 genetic counseling
簇 cluster
反转录 PCR reverse transcription PCR
电泳 electrophoresis
重复序列 repetitive sequence
卫星 DNA satellite DNA
转座子 transposon
随机扩增多态 DNA (RAPD) random amplified polymorphic DNA
扩增片段长度多态性(AFLP) amplified fragments length polymorphism
C 值 C-value
分子伴侣 chaperone
多重 PCR multiple PCR
多重 PCR 扩增系统 multiple PCR application system
前体 mRNA pre-mRNA
技术转让 technology transfer
易位 translocation
三体型 trisomy
肿瘤抑制基因 tumor suppressor gene

1 Polymerase chain reaction: The polymerase chain reaction (PCR) is a biochemical technology in molecular biology to amplify a single or a few copies of a piece of DNA across several orders of magnitude, generating thousands to millions of copies of a particular DNA sequence.

聚合酶链式反应：聚合酶链式反应(PCR) 是分子生物学中的一项生物化学技术，通过多次大量增殖能使单一或少数几个拷贝的 DNA 扩增出几千到几百万的拷贝。

2 Genetic code: The genetic code is the set of rules by which information encoded in genetic material

(DNA or mRNA sequences) is translated into proteins (amino acid sequences) by living cells.

遗传密码：遗传密码遵循如下规则，通过活细胞能将遗传物质(DNA 或 mRNA 序列)中编码的信息翻译成蛋白质(氨基酸序列)。

3 Random amplified polymorphic DNA: RAPD stands for random amplification of polymorphic DNA. It is a type of PCR reaction, but the segments of DNA that are amplified are random. The scientist performing RAPD creates several arbitrary, short primers (8-12 nucleotides), then proceeds with the PCR using a large template of genomic DNA, hoping that fragments will amplify. By resolving the resulting patterns, a semi-unique profile can be gleaned from a RAPD reaction.

随机扩增多态性 DNA：随机扩增多态性 DNA 的随机扩增。它是一种聚合酶链式反应，但 DNA 的部分是随意扩增的，为了扩增片段，科学家运用随机扩增多态性 DNA 创造了许多短小的引物(8-12 个核苷酸)，然后使用基因组 DNA 大型模板对聚合酶链进行处理。解析生成的模式后，可以从 RAPD 反应中获得半独特分布图。

第五节　基因重组

Section 5　Gene Recombination

限制　restriction
修饰　modification
成斑率　efficiency of plating
平齐末段　blunt end
黏性末段　cohesive end
同裂酶　isoschizomer
同尾酶　isocaudarner
杂交位点　hybrid site
完全消化作用　complete digestion
局部酶切消化　partial digestion
高效液相色谱　high performance liquid chromatography
专利权　patent
谱系　pedigree
表型　phenotype
物理图谱　physical map
多态性　polymorphism
定位克隆　positional cloning
引物　primer
原核　pronucleus
载体　vector
质粒　plasmid
F 质粒　F plasmid
接合型质粒　conjugative plasmid
非接合型质粒　non-conjugative plasmid
严紧型质粒　stringent plasmid
松弛型质粒　relaxed plasmid
质粒的不亲和性　plasmid incompatibility
不亲和群　incompatibility group
细菌抗药性　bacterial resistance
烈性噬菌体　virulent phage
温和噬菌体　temperate phage
溶源化　lysogenization
多聚衔接物　poly linker
抗终止因子　anti-terminator
插入型载体　insertion vector
置换型载体　replacement vector
存活噬菌体　viable phage
衔接物　linker
双衔接物　double-linker
碱性磷酸酶　alkaline phosphatase
微卫星 DNA　micro-satellite DNA
原核生物　prokaryote

1 Phenotype: A phenotype is the composite of an organism's observable characteristics or traits: such as its morphology, development, biochemical or physiological properties, phenology, behavior, and products of behavior.

表型：表型是生物体能被观察到的特点或特性的组合，比如它的形态、发展、生物或生理特性、物候、行为以及行为产物。

2 Vector: In molecular biology, a vector is a DNA molecule used as a vehicle to transfer foreign genetic material into another cell. The four major types of vectors are plasmids, viral vectors, cosmids, and artificial chromosomes.

载体：在分子生物学中，载体作为一种 DNA 分子，可以被用作把外源遗传材料转移到其他细胞的工具。载体的四种类型包括质粒、病毒载体、粘粒和人造染色体。

3 Plasmid: In microbiology and genetics, a plasmid is a DNA molecule that can replicate independently of, the chromosomal DNA. They are double-stranded and, in many cases, circular.
质粒:在微生物学和遗传学中,质粒是一种脱离染色体的 DNA 分子,但能够独立复制染色体 DNA。质粒是双链结构,并且通常是圆形的。

4 Lysogenization: The process by which a bacterium acquires a phage that becomes integrated into its genome.
溶源化:在溶源化过程中,细菌得到噬菌体并将噬菌体溶入其基因组。

第六节 重组体检测
Section 6 Examination of Recombinants

微丝 micro filament
微丝结合蛋白 micro filament associated protein
微量注射 microinjection
显微操作技术 micromanipulation technology
微粒体 microsome
显微分光光度测定法 microspectrophotometry
微管 microtubule
微管相关蛋白 microtubule associated protein
微管组织中心 microtubule organizing center
微带 mini band
微小染色体 mini chromosome
小卫星染色体 mini satellite DNA
细胞促分裂因子 mitosis promoting factor
有丝分裂器 mitotic apparatus
转化 transformation
转染 transfection
环丝氨酸富集法 cycloserine enrichment
接收器位点 receptor site
原位杂交 in situ hybridization
放射性抗体测定法 radioactive antibody test
双位点检测法 two sites detection method
翻译相 translational phase
DNA-蛋白质筛选法 DNA-southwestern screening
DNA 结合蛋白 DNA-binding protein
杂交选择转译 hybrid selected translation
杂交抑制的转译 hybrid arrested translation
无细胞翻译系统 cell free translation system
杂交释放转译 hybrid released translation
缺口转移 nick translation
鸟枪法 shot gun
沉默子 silencer

1 Micro filament: Micro filaments are the thinnest filaments of the cytoskeleton, a structure found in the cytoplasm of eukaryotic cells.
微丝:微丝是细胞骨架最细的丝,是一种存在于真核细胞细胞质中的结构。

2 Transfection: Transfection is the process of deliberately introducing nucleic acids into cells. The term is used notably for non-viral methods in eukaryotic cells.
转染:转染是有目的地将核酸导入细胞中的过程。该词尤其用于真核生物细胞中的非病毒方法。

3 In situ hybridization: In situ hybridization (ISH) is a technique that allows for precise localization of a specific segment of nucleic acid within a histologic section. The underlying basis of ISH is that nucleic acids, if preserved adequately within a histologic specimen, can be detected through the application of a complementary strand of nucleic acid to which a reporter molecule is attached.
原位杂交:原位杂交(ISH)是一种可应用于组织切片中对特殊核酸片段进行精确定位的技术。ISH 的根本在于只要组织标本中核酸保存完好,就可通过应用能与报告分子相连接的互补核酸链进行检测。

4 DNA-binding protein: DNA-binding proteins are proteins that are composed of DNA-binding domains and thus have a specific or general affinity for either single or double stranded DNA. Sequence-specific DNA-binding proteins generally interact with the major groove of B-DNA, because it exposes more functional groups that identify a base pair.
DNA 结合蛋白:DNA 组合蛋白是由 DNA 结合域构成的蛋白,因此,对某一单链或双链 DNA 具有专一或普遍的亲和性。序列专一 DNA 组合蛋白通常能 与 B-DNA 的主沟结合,因为它能暴露出更多的可识别碱基对的功能团。

第七节 克隆基因的表达
Section 7 Expression of Cloning Gene

操纵子 operator
启动子 promotor
多核糖体 polyribosome
有意义链 sense strand
反意义链 anti-sense strand
抗终止子 antiterminator
衰减子 attenuator
前导序列 leader sequence
间隔序列 spacer sequence
Xa 因子 Xa factor
一致序列 consensus sequence
副作用成分 negatively acting component
分离的不稳定性 segregative instability
重复性基因外回文序列 repetitive extra genic palindromic sequence
Ti 质粒 tumor inducing plasmid
章鱼碱 octopine
胭脂碱 nopaline
受体质粒 acceptor plasmid
穿梭载体 shuttle vector
中间载体 intermediate vector
共合体 cointegrate
接合后体 exconjugant
受纳细胞 permissive cell
胸苷激酶 thymidine kinase
高速微型子弹射击法 high-velocity micro projectile
共转染现象 co-transfection
显微注射法 microinjection

1 Operator：In genetics，an operator is a segment of DNA to which a transcription factor protein binds. It is classically defined in the lac operon as a segment between the promoter and the genes of the operon.

操纵子：在遗传学中，操纵子是一段转录因子蛋白可结合的 DNA 片段，传统上在乳糖操纵子学说中将其定义为启动子和操纵基因间的一个片段。

2 Promotor：In genetics，a promoter is a region of DNA that initiates transcription of a particular gene. Promoters are located near the genes they transcribe，on the same strand and upstream (towards the 5′ region of the sense strand).

启动子：在遗传学中，启动子是能起动某一特定基因转录的 DNA 区域。启动子紧挨能转录的基因，位于同一链上和上游(位于有意义链的 5′端)。

3 Attenuator：Attenuation (in genetics) is a proposed mechanism of control in some bacterial operons which results in premature termination of transcription and which is based on the fact that，in bacteria，transcription and translation can and do proceed simultaneously.

衰减子：在遗传学中，衰减是某些细菌操纵子中的假设控制机制，能导致转录提前终止，而且基于如下事实，即在细菌中转录和翻译能够而且确实同时进行。

4 Co-transfection：Cotransfection refers to the simultaneous transfection with two separate nucleic acid molecules，such as plasmid DNA and siRNA.

共转染现象：共转染是指具有两个独立的核酸分子的同时转染，如质粒 DNA 和 siRNA。

第八节 酵母的基因工程
Section 8 Gene Engineering of Yeast

自主复制序列 autonomously replicating sequence
整合载体 integrating vector
着丝粒 centromere
端粒 telomere
蛋白质印迹 western blot
酵母人工染色体 yeast artificial chromosome
基因置换 gene transplacement
纤维素 cellulose
显微操作器 micromanipulator
同源性重组 homologous recombination
同源染色体 homologous chromosome
同源基因 isogene
转化基因组 transforming genome

1 Autonomously replicating sequence: An autonomously replicating sequence (ARS) contains the origin of replication in the yeast genome. It contains four regions (A, B_1, B_2, and B_3), named in order of their effect on plasmid stability; when these regions are mutated, replication does not initiate.

自主复制序列:自主复制序列(ARS)包含酵母基因组中的复制起点。它包含四个区域(A,B_1,B_2,B_3),依照其在质粒稳定性中的顺序而命名;当这些区域改变时,并不发起复制。

2 Western blot: The western blot (sometimes called the protein immunoblot) is a widely accepted analytical technique used to detect specific proteins in the given sample of tissue homogenate or extract.

蛋白质印迹:蛋白质印迹(或可称为蛋白质免疫印迹)是一种广泛使用的分析法,用来在特定的组织匀浆或组织提取的样品中发现特殊蛋白质。

3 Yeast artificial chromosome: A yeast artificial chromosome (YAC) is a vector used to clone DNA fragments larger than 100 kb and up to 3000 kb. YACs are useful for the physical mapping of complex genomes and for the cloning of large genes.

酵母人工染色体:酵母人工染色体(YAC)是一种克隆DNA片段(大于100 kb,最大可达3000 kb)的载体。酵母人工染色体可用于复杂基因组的物理图谱或克隆大型基因。

4 Homologous recombination: Homologous recombination is a type of genetic recombination in which nucleotide sequences are exchanged between two similar or identical molecules of DNA.

同源性重组:同源性重组是一种基因重组类型。在同源性重组过程中,核苷酸序列在两个相似或相同的DNA分子中进行交换。

第九节 植物基因工程
Section 9 Gene Engineering of Plant

体细胞克隆变异 somaclonal variation
愈伤组织 callus
保育培养 nurse culture
生长素 auxin
细胞分裂素 cytokinin
二元系统 binary system
葡萄糖苷酸 glucuronide
双生病毒 geminivirus
交叉保护 cross-protection
自发突变体 spontaneous mutant
组成型基因表达 constitutive gene expression
组成型异染色质 constitutive heterochromatin
阻抑作用 repression
棉铃虫 bollworm
免疫治疗 immunotherapy
原位杂交 hybridization in situ
插入 insertion
知识产权 intellectual property right

1 Callus: callus is a mass of unorganized parenchyma cells derived from plant tissue (explants) for use in biological research and biotechnology. In plant biology, callus cells are those cells that cover a plant wound.

愈伤组织:愈伤组织是从植物组织(外植体)中衍生出来的大量离体的薄壁细胞,用于生物研究和生物技术中。在植物生物学中,愈伤组织细胞能够覆盖植物伤口。

2 Immunotherapy: Immunotherapy is a medical term defined as the "treatment of disease by inducing, enhancing, or suppressing an immune response".

免疫治疗:免疫疗法是一个医学术语,定义为"通过引发、加强或抑制免疫反应治疗疾病"。

第十节 哺乳动物基因工程
Section 10 Gene Engineering of Mammal

感染 infection
内吞作用 endocytosis
原病毒 provirus
包装细胞 packaging cell
辅助自由 helper free
半不连续复制 Semi-discontinuous replication
DNA 修复 DNA repair
切除修复 excision repair

生殖系 germ line
基因剔除 gene knockout
牛疣病毒 papilloma virus
空泡病毒 simian virus
超级小鼠 super mouse
转基因动物 transgenic animal
转基因小鼠 transgenic mouse
白血病 leukemia
库 library
连接 linkage
淋巴细胞 lymphocyte cell

1 Endocytosis: Endocytosis is a process by which cells absorb molecules (such as proteins) by engulfing them.
内吞作用:在内吞作用过程中,细胞以吞噬分子的形式吸收分子(如蛋白质)。

2 Provirus: A provirus is a virus genome that is integrated into the DNA of a host cell.
原病毒:原病毒是一个病毒基因组,包含在宿主细胞 DNA 中。

3 Transgenic animal: A transgenic animal is one that carries a foreign gene that has been deliberately inserted into its genome. The foreign gene is constructed using recombinant DNA methodology.
转基因动物:转基因动物携带刻意插入其基因组的外源基因。外源基因通过重组 DNA 的方法构建。

第十一节 医药工业基因工程
Section 11 Gene Engineering of Medicine Industry

瞬时表达 transient expression
可溶性受体 soluble receptor
前激素 pre-hormone
血友病 hemophilia
杂合子 heterozygote
高度保守序列 highly conserved sequence
纯合子 homozygote
人工染色体 human artificial chromosome
人类基因组计划 human genome project
人类免疫缺陷病毒 human immuno deficiency virus
获得性免疫缺陷综合征 acquired immuno deficiency syndrome
弱毒疫苗 attenuated vaccine
亚单位疫苗 subunit vaccine
功能元件 functional element
功能基因组学 functional genomics
协同转运 co-transport
共价催化 covalent catalyse
共价修饰 covalent modification
共价调节酶 covalent regulatory enzyme
豇豆花叶病毒 cowpea mosaic virus
单克隆抗体 monoclonal antibody
骨髓瘤 myeloma
效应区域 effector domain
双特异性抗体 bispecific antibody
肿瘤坏死因子 tumor necrose factor
复制起始点 origin of replication
半保留复制 semiconservative replication
复制叉 replication fork
定点突变 site-directed mutagenesis
报告基因 reporter gene
增强子 enhancer
干细胞 embryonic stem cell
基因治疗 gene therapy

1 Transient expression: Temporary expression of a transformed gene in the desired target cells over a relatively brief time span; it does not necessarily indicate integration of the gene into the host chromosome and is not passed onto the next generation.
瞬时表达:瞬时表达是指在较短的时间跨度内,一个转移基因进入预期靶细胞的暂时表达。它不一定表明基因融入宿主染色体的过程,也不会传给下一代。

2 Acquired immuno deficiency syndrome: Acquired immuno deficiency syndrome (AIDS) is a disease of the human immune system caused by the human immunodeficiency virus (HIV).
获得性免疫缺陷综合征:获得性免疫缺陷综合征(AIDS)是由人体免疫缺损病毒(HIV)所引发的一种人体免疫系统疾病。

3 Subunit vaccine: A subunit vaccine is a vaccine that contains isolated proteins from a virus, but lacks viral nucleic acid. These vaccines expose the body to antigens so it can learn to recognize them

without exposing the body to the risk of viral replication and subsequent infection.

亚单位疫苗：亚单位疫苗包含从病毒中分离出来的蛋白质，但缺少病毒核酸。疫苗使人体暴露于抗原之下，人体因此能识别抗原，同时又没有病毒复制和出现感染的风险。

4 Monoclonal antibody: Monoclonal antibodies (mAb or moAb) are monospecific antibodies that are the same because they are made by identical immune cells that are all clones of a unique parent cell. Monoclonal antibodies have monovalent affinity, in that they bind to the same epitope.

单克隆抗体：单克隆抗体(mAb 或 moAb) 和单特异性抗体相同，因为它们是由相同的免疫细胞制造的，免疫细胞是独一无二的母细胞的所有复制品。单克隆抗体能够抵抗病菌的亲和力，它们与相同的抗原表位相连。

第二十三章　仪器分析

Chapter 23　Instrument Analysis

第一节　气相色谱

Section 1　Gas Chromatography

色谱　chromatogram
流动相　mobile phase
固定相　stationary phase
性能　performance
液相色谱　liquid chromatography
薄层层析色谱　thin-layer chromatography
离子交换色谱　ion exchange chromatography
填充柱　Packed Colum
毛细管　capillary
毛细管柱　capillary column
毛细管气相色谱　capillary gas chromatography
毛细管现象　capillary action
理论塔板　theoretical plates
柱温　colum temperature
柱压　colum pressure
洗脱液　eluate
保留时间　retention time
死时间　dead time
保留指数　retention index
标准偏差　standard deviation
半峰宽　half peak width
容量因子　capacity factor
保留因子　retention factor
标准曲线　calibration curve
范德华方程　van der Waals equation
分离柱　separation column
柱效　column performance
极性　polarity
亲和力　affinity
非极性　non-polarity
硅胶　silica gel
阳离子　cation
阴离子　anion
进样器　injector
进样口　injector port
载气　carrier gas
火焰光度检测器　flame photometric detector
火焰电离检测器　flame ionization detector
热导池检测器　thermal conductivity
光离子化检测器　photo ionization detector
电子俘获检测器　electron capture
峰面积　peak area
峰高　peak height
程序升温　temperature programming
塔板理论　plate theory
理论塔板　theoretical plate
速率理论　rate theory
涡流扩散项　eddy diffusion term
纵向扩散项　longitudinal diffusion term
响应时间　response time
线性范围　linear range
定量校正因子　quantitative calibration factor
归一化法　normalization method
内标法　internal standard method
外标法　external standard method

1 Gas Chromatography (GC) is used to separate volatile components of a mixture. A small amount of the sample to be analyzed is drawn up into a syringe. The syringe needle is placed into a hot injector port of the gas chromatograph, and the sample is injected. The injector is set to a temperature higher than the components' boiling points. So components of the mixture evaporate into the gas phase inside the injector. A carrier gas, such as helium, flows through the injector and pushes the gaseous components of the sample onto the GC column. It is within the column that separation of the components takes place. Molecules partition between the carrier gas (the mobile phase) and the high boiling liquid (the stationary phase) happen within the GC column.

气相色谱用于分离混合物中易挥发组分。将少量要分离的样品加入到类似注射器的进样器中，进样器的针插入到气象色谱上的进样口中，并将样品注入进样口中。进样口环境的温度高于样品的沸点，因此样品在进样口由液相变成气相进入到色谱中。在载气的推动作用下，如氦气，变成气相的样品被推入色谱柱中，样品在色谱柱中被分离。在载气（流动相）和色谱柱中高沸点的液体（固定相）间发生分子分离现象。

第二节　原子吸收光谱
Section 2　Atomic Absorption Spectrometry

原子吸收光谱　atomic absorption spectroscopy
原子发射光谱　atomic emission spectroscopy
分光光度计　spectrophotometer
火焰原子化器　flame atomizer
石墨炉　graphite furnace
无火焰原子化器　non-flame atomizer
冷原子吸收法　cold atom absorption
空心阴极灯　hollow cathode lamp
光电倍增管　photomultiplier
雾化器　nebulizer
原子化　atomization
等离子体　plasma
灰化　ash
诱导耦合等离子原子发射光谱　inductively coupled plasma atomic emission spectroscopy
背景吸收　background absorption
多普勒变宽　doppler broadening
自然宽度　natural width
共振线　resonance line
吸收线　absorption line
锐线光源　line source
化学干扰　chemical interference
光谱干扰　spectral interference
电离干扰　ionization interference
物理干扰　physical interference
背景吸收　background absorption
分子吸收　molecular absorption
灵敏度　sensitivity
检出限　detection limit
自吸　self-absorption

■ When using atomic absorption spectrophotometry (AAS) as an analytical technique the absorption of light of free atoms is measured. Therefore it is one of the branches of atomic spectroscopy, together with flame photometry. However — unlike flame photometry — AAS is based on the "first half" of the excitation process, while atoms absorb light getting their electrons from the ground state to a higher energy level.

使用原子吸收光谱法作为测量自由原子吸收光的一种分析手段，其与火焰光谱法均是原子光谱学的分支。而与火焰光谱法不同的是，原子吸收光谱法是基于激发态的"上半程"，即原子吸收光能得到电子从基态跃迁到高能级。

第三节　可见和紫外吸光光度法
Section 3　Visible and Ultraviolet Absorption Spectroscopy

光学的　optical
光谱学　spectroscopy
光谱　spectrum
吸收光谱　absorption spectrum
定性分析　qualitative analysis
定量分析　quantitative analysis
吸收曲线　absorption curve
精度　precision
准确度　accuracy
系统误差　systematic error
随机误差　random error
光源　light source
单色器　monochromator
吸收池　absorption cell
光密度　optical density
参比池　reference cell
灵敏度　sensitivity
单光束　single beam
双光束　double beam
紫外分光光度法　ultraviolet spectrophotometry
分光光度计　spectrophotometer
电子能级　electron energy level
检测器　detector
参数　parameter
光栅　grating
吸光度　absorbency
摩尔吸光系数　molar absorption coefficients
浓度　concentration

折射 refraction
反射 reflection
衍射 diffraction
散射 scattering
入射光束 incident beam
透镜 lenses
棱镜 prism
石英 quartz
波长 wave length
朗伯比尔定律 lambert-beer's law
跃迁 transition
生色团 chromophore
助色团 auxochrome
红移 red shift
蓝移 blue shift
增色效应 hyperchromic effect
减色效应 hypochromic effect
显色剂 chromogenic reagent
激发态 excited state
基态 ground state
键长 bond length

1 Ultraviolet-visible spectroscopy (UV-Vis) refers to absorption spectroscopy or reflectance spectroscopy in the ultraviolet-visible spectral region. This means it uses light in the visible and adjacent (near-UV and near-infrared (NIR) ranges. The absorption or reflectance in the visible range directly affects the perceived color of the chemicals involved. In this region of the electromagnetic spectrum, molecules undergo electronic transitions. This technique is complementary to fluorescence spectroscopy, in that fluorescence deals with transitions from the excited state to the ground state, while absorption measures transitions from the ground state to the excited state.

紫外-可见光谱是指吸收光谱或反射光谱在紫外-可见光区，这表示使用可见光、紫外光(或近红外、近紫外)。在可见光区的吸收或反射光直接影响吸光物质颜色的变化，在此光区吸光物质的分子进行电子跃迁。该技术可与荧光光谱分析方法互补，荧光光谱分析是利用激发态回落到基态发射的光谱，而吸收光谱是利用基态跃迁到激发态吸收的光谱。

参考文献
References

[1] Aguirrezabal L, Bouchier-Combaud S, Radziejwoski A, Dauzat M, Cookson SJ, Granier C. Plasticity to soil water deficit in *Arabidopsis thaliana*: dissection of leaf development into underlying growth dynamic and cellular variables reveals invisible phenotypes. Plant, Cell and Environment, 2006, 29:2216-2227

[2] Aguirrezabal L, Deleens E, Tardieu F. Root elongation rate is accounted for by intercepted PPFD and source-sink relations in field and laboratory-grown sunflower. Plant, Cell and Environment, 1994, 17:443-450

[3] Bacon MA. Water Use Efficiency in Plant Biology. UK:Blackwell Publishing, 2004

[4] Baena-Gonzalez E, Sheen J. Convergent energy and stress signaling. Trends in Plant Science, 2008,13:474-482

[5] Berg JM, Tymoczko JL, Stryer L. Biochemistry. 5th Edition. London:W. H. Freeman, 2002

[6] Bergmann DC, Sack FD. Stomatal development. Annual Review of Plant Biology, 2007, 58:163-181

[7] Briggs D, Walters S. M. Plant variation and Evolution. 3rd Edition. Cambridge University Press, 1997

[8] Cattering JD, Seheult CA. Microbiology. 13th Edition. McGraw-Hill Medical Publishing, 2010

[9] Chapman JL, Reiss MJ. Ecology principles and applications. Cambridge University Press and Tsinghua University Press, 2001

[10] Clark DP, Pazdernik NJ. Molecular Biology: Understanding the Genetic Revolution. Academic Press, 2007

[11] Colin R. Basic Biotechnology. Science Press, 2002

[12] Jordan WR, Gilpin ME and Aber JD. Restoration Ecology. Cambridge University Press, 1993

[13] Klug WS, Cummings MR, Spencer C. Concepts of Genetics. 8th Edition. Upper Saddle River, N. J. :Pearson/Prentice Hall, 2006.

[14] Lack A, Evans D. Instant Notes in Plant Biology. Science Press, 2002

[15] Lodish H, Berk A, Zipursky SL, Matsudaira P, Baltimore D, and Darnell J. Molecular Cell Biology. 4th Edition. London:W. H. Freeman, 2000

[16] Machenzie A. Ecology. Instant Notes in Ecology. Science Press, 1999

[17] Molles MC. Ecology:Concepts and Application. Science Press, 2001

[18] Nelson DL and Cox MM. Lehninger's Principles of Biochemistry. 4th Edition. London:W. H. Freeman, 2004

[19] Pierce BA. Genetics:A Conceptual Approach. 2nd Edition. New York:W. H. Freeman, 2005

[20] Pollard TD, Earnshaw WC, Lippincott-Schwartz J. Cell biology. Philadelphia:Phiadelphia:W. B. Saunders, 2002

[21] Prescott LM, Harley JH, Klein DA. Microbiology. 6th Edition. USA:Mcgraw-Hill (Tx), 2004

[22] Ricklef S, Robert E. Ecology. 4th Edition. W. H. Freeman & Company, 1999

[23] Robert KM. Harpcr's Biochemistry. Science Press, 2001

[24] Schulze ED. Plant Ecology. Springer-Verlag Berlin and Heidelberg Gmb H & Co. K, 2005

[25] Starr C, Taggart R. Grove P, Cell Biology and Genetics. CA:Brooks/Colc, 2001

[26] Watson JD, Baker TA, Bell SP, Gann A. Molecular Biology of the Gene. London:Benjamin Cummings, 2003

